Thermodynamic Optimization of Industrial Energy Systems

Contents

About the Editors

Daniel Flórez-Orrego

Daniel Florez-Orrego obtained his Mechanical Engineer degree from the Faculty of Mines of the National University of Colombia, Campus Medellin (2011). He earned his Master (2014) and Doctor (cum laude 2018) degrees in Mechanical Engineering with emphasis on Energy and Fluids from the Polytechnic School of the University of São Paulo, Brazil, with a 1-year exchange at the École Polytechnique Fédérale de Lausanne, Switzerland. Currently, he works as a Senior Researcher in the area of the Synthesis and Optimization of Industrial, Chemical and Power Generation Processes of the Industrial Process and Energy Systems Engineering (IPESE) group. he was awarded the ABCM Embraer Award 2019 for the Best PhD Mechanical Engineering Thesis and the Honorable Mention in Outstanding PhD Thesis USP Award in 2020. He is a member of the Colombian Association of Electrical and Mechanical Engineers—ACIEM (Antioquia), the American Institute of Chemical Engineers—AIChE, and a Recognized Researcher of the Colombian Ministry of Science Technology and Innovation—MinCiencias. His fields of expertise include industrial process integration; cogeneration, refrigeration and heat pumping; techno-economic and environmental analyses; refineries, biorefineries, fertilizers complexes, hydrogen and synthetic fuels (SNG, DME, MeOH, FT, H2); power to gas systems and reversible solid oxide cells; and combustion, heat transfer, computational fluid dynamics, and reaction design.

Meire Ellen Ribeiro Domingos

Meire Ellen Ribeiro Domingos is a Chemical Engineer from the University of Brasilia, Brazil, who obtained her PhD degree from the Polytechnic School of the University of São Paulo with a 1-year exchange at the École Polytechnique Fédérale de Lausanne, Switzerland. Currently, she works as a Postdoctoral Researcher at École Polytechnique Fédérale de Lausanne (EPFL), Switzerland, in the Industrial Processes and Energy Systems Engineering (IPESE) group. She was awarded the Honorable Mention in Outstanding PhD Thesis USP Award in 2022. She is member of the American Institute of Chemical Engineers—AIChE. She has extensive experience in the area of the Synthesis and Optimization of Industrial, Chemical and Power Generation Processes, with a focus on biomass energy conversion, fuels' production (SNG, DME, MeOH, FT, H2), renewable energy intermittency management, reversible solid oxide cells (power-to-gas and gas-to-power systems), and energy storage.

Rafael Nogueira Nakashima

Rafael Nogueira Nakashima graduated in Mechanical Engineering from the State University of Maringá (2016) and obtained his Master's (2018) and Doctor's (2022) degrees in Mechanical Engineering from the Polytechnic School of the University of Sao Paulo, Brazil. His research field revolves around energy and biofuels, focusing on applying thermodynamic, economic, and environmental assessment methods to evaluate mechanical and chemical engineering processes such as biogas, residues' capitalization, and renewable fuel production. He has extensive expertise in modeling anaerobic digestion processes for biogas production, fuel cells, biomass conversion, and renewable energy integration. Currently, he is a Postdoctoral Researcher in the Department of Energy Conversion and Storage of the Technical University of Denmark.

Preface

The search for more sustainable industrial energy systems has created a pressing need for innovative solutions that reduce resources consumption and minimize environmental impact. This reprint contributes to this search by featuring the application of thermodynamic principles for analyzing, designing, and optimizing energy systems across diverse industrial sectors.

The scope of this reprint encompasses aspects of energy system modeling, exergy analysis, pinch technology, process integration, and advanced optimization techniques. It aims to bridge the gap between theoretical foundations and industrial practice, emphasizing strategies that can improve energy efficiency, reduce costs, and enhance sustainability.

As industries face increasing pressure to reduce their carbon footprint and comply with stringent energy policies, understanding the principles of thermodynamic optimization becomes essential. The primary audience includes engineers, energy managers, and researchers involved in the design and optimization of industrial energy systems. It also serves as a valuable reference for graduate students and researchers seeking to deepen their understanding of thermodynamic optimization.

We express our heartfelt gratitude to all those who have contributed to this reprint, as well as to the reviewers who provided valuable feedback and insights during its development. It is our hope that this reprint will serve as a useful tool for advancing the field of industrial energy optimization and inspire further research and innovation in this vital area.

Daniel Flórez-Orrego, Meire Ellen Ribeiro Domingos, and Rafael Nogueira Nakashima
Guest Editors

Editorial

Editorial "Thermodynamic Optimization of Industrial Energy Systems"

Daniel Florez-Orrego [1,2,*] , **Meire Ellen Ribeiro Domingos** [1] **and Rafael Nogueira Nakashima** [3]

[1] Industrial Process and Energy Systems Engineering, École Polytechnique Fédérale de Lausanne, 1950 Sion, Valais, Switzerland; meire.ribeirodomingos@epfl.ch
[2] Faculty of Mines, National University of Colombia, Av. 80 #65–223, Medellin 1779, Antioquia, Colombia
[3] Department of Energy Conversion and Storage, Technical University of Denmark DTU, 2800 Kongens Lyngby, Denmark; rafnn@dtu.dk
[*] Correspondence: daniel.florezorrego@epfl.ch

Citation: Florez-Orrego, D.; Ribeiro Domingos, M.E.; Nogueira Nakashima, R. Editorial "Thermodynamic Optimization of Industrial Energy Systems". *Entropy* **2024**, *26*, 1047. https://doi.org/10.3390/e26121047

Received: 22 November 2024
Accepted: 25 November 2024
Published: 3 December 2024

Thermodynamic optimization of industrial energy systems is crucial for finding solutions to reduce energy consumption and mitigate losses, leading to environmental and economic benefits. It involves applying thermodynamic principles to enhance the performance of the industrial, chemical and power generation systems, from individual components to entire plants. Among the widely used techniques, heat integration (i.e., the pinch method) and exergy analysis stand out for their ability to pinpoint sources of inefficiency [1]. However, those systems are part of a larger ecosystem, involving complex exchanges and interactions with other energy systems, and with the society and the environment. For this reason, process integration and thermodynamic optimization methods have been extended from a classical perspective of waste heat recovery to include a holistic design and optimization approach. In fact, a systemic dimension of process integration can cover aspects like (i) heat pumps and electrification; (ii) carbon capture, use, and sequestration; (iii) industrial symbiosis and circularity; (iv) urban energy systems' integration; (v) waste management; (vi) power-to-x-to-power; (vii) seasonal energy storage, and (viii) integration to gas and electricity services.

Evaluating the performance of energy conversion processes requires defining sustainable metrics that ensure objective and reproducible assessments, thus enabling a rational basis for assessing innovative integrated concepts on carbon mitigation, electrification, storage, and renewable energy use. In this way, thermodynamic integration and optimization approaches must ensure that any improvement on an industrial system does not trigger a socio-economic or environmental issue elsewhere (burden-shifting). Thus, although biomass energy utilization for fuels' and chemicals' production can be generally seen as a renewable option [2], its use must be carefully examined to certify that the energy technologies and systems involved are compatible with the sustainable energy transition scenarios. For this reason, process integration and optimization have become more and more relevant to assess the performance of biomass energy-based sectors, such as biorefineries. The implementation of carbon capture and storage has gained recent attention as it could lead to net-negative CO_2 emissions. It explains why innovative designs are being explored to make the best use of biomass resources, while observing their sustainability targets.

Meanwhile, systems like biomass-integrated gasification combined cycles (BIGCCs), featuring organic Rankine cycles and absorption refrigeration cycles, aim to demonstrate the potential of the multi-process and multi-product approaches in which waste heat becomes an asset rather than an industrial burden [3]. In the same way, the advanced exergy-based optimization of polygeneration systems using non-conventional working fluid aims to offset the concerns of conventional water-based Rankine cycles, especially in terms of footprint and water scarcity impacts. A high degree of self-power generation of industrial and chemical processes presents new opportunities for electrification efforts towards an industrial heating decarbonization [4]. Novel high-temperature heat pump technologies are being developed to upgrade the low-temperature waste heat available in industrial sites by consuming only a fraction of the high-grade energy input that would be needed if fossil fuels were used instead [5].

Some emerging technologies, such as machine learning for predicting waste heat availability or the application of artificial neural networks for optimizing equipment performance, demonstrate that the novel paradigms of data-driven thermodynamic process optimization have become prevalent [6]. The massification of these powerful computational tools is expected to unveil breakthroughs in energy systems, like unconventional solar power systems and plasma applications for industrial heating electrification [7]. In brief, by combining technical, economic, and sustainability perspectives, this Special Issue aims to bring together different research applications to highlight the role of recent advancements in shaping future energy systems (Figure 1).

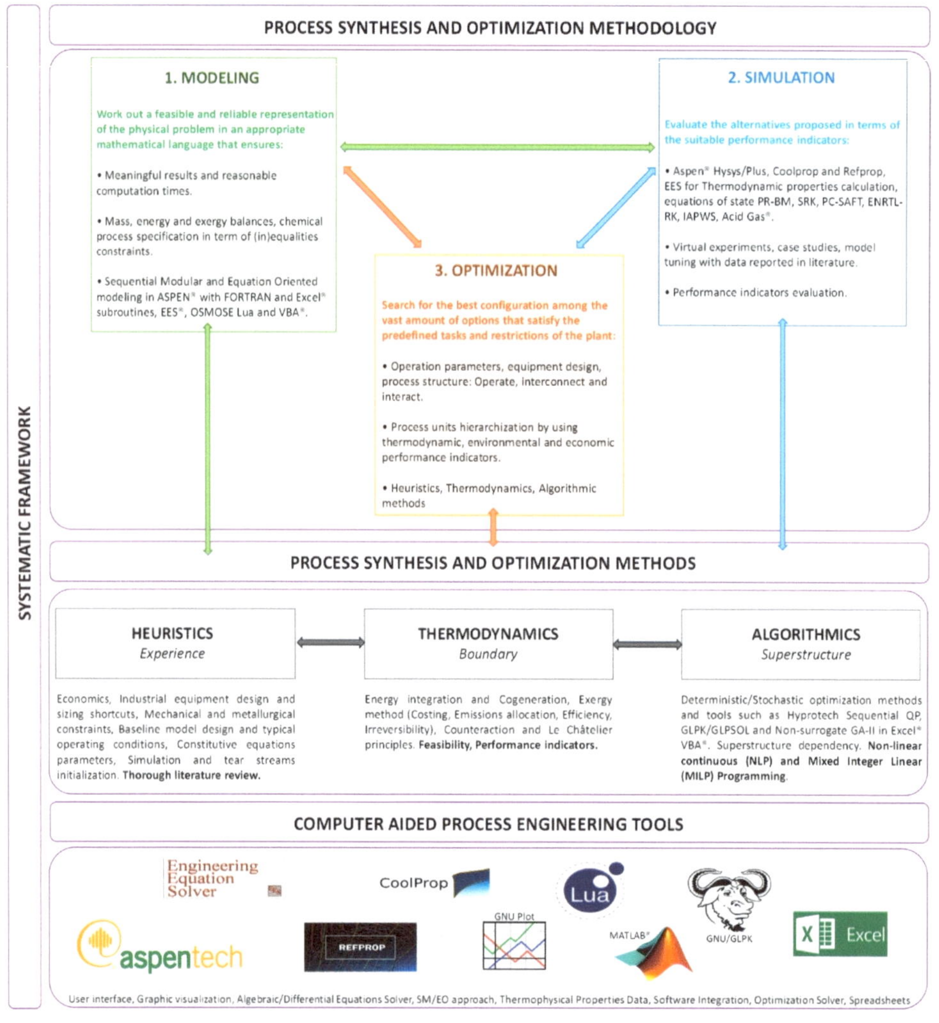

Figure 1. Schematic of thermodynamic optimization and industrial energy systems [8].

This Special Issue, entitled "Thermodynamic Optimization of Industrial Energy Systems", has featured eleven (11) articles on the topics of thermodynamic process design, integration and optimization, leveraging different theoretical and computational methods and tools. Their main contributions are highlighted below:

1. "Advanced Exergy-Based Optimization of a Polygeneration System with CO_2 as Working Fluid" by Jing Luo, Qianxin Zhu and Tatiana Morosuk [9] focuses on optimizing a polygeneration system using CO_2 as the working fluid to produce electricity, refrigeration and heating. Advanced exergy-based methods are implemented, splitting exergy destruction into avoidable and unavoidable parts to identify improvement priorities. Exergoeconomic graphical optimization is conducted at the component level, enhancing system performance. The findings highlight the system potential for energy efficiency and cost-effectiveness, with a 15.4% increase in exergetic efficiency and a 7.1% cost reduction.

2. "Optimization and Tradeoff Analysis for Multiple Configurations of Bio-Energy with Carbon Capture and Storage Systems in Brazilian Sugarcane Ethanol Sector" by Bruno Bunya, César A. R. Sotomonte, Alisson Aparecido Vitoriano Julio, João Luiz Junho Pereira, Túlio Augusto Zucareli de Souza, Matheus Brendon Francisco and Christian J. R. Coronado [10] analyzes bio-energy systems with carbon capture and storage (BECCS) in sugarcane ethanol plants. Some cogeneration setups with chemical absorption were evaluated. A single regenerator system reportedly captures the most CO_2 (51.9%), but reduces plant efficiency by 14.9%. Achieving higher CO_2 capture rates can lead to higher specific emissions (gCO_2/kWh) compared to a base plant. Systems with lower CO_2 capture rates (<51%) ensure overall emission reductions.

3. "Exergoeconomic Analysis and Optimization of a Biomass Integrated Gasification Combined Cycle Based on Externally Fired Gas Turbine, Steam Rankine Cycle, Organic Rankine Cycle, and Absorption Refrigeration Cycle" by Jie Ren, Chen Xu, Zuoqin Qian, Weilong Huang and Baolin Wang [11] explores a novel biomass-based combined cooling and power system. The system integrates an externally fired gas turbine, steam and organic Rankine cycles, and absorption refrigeration cycles. The analysis reveals a thermal efficiency of 70.67%, an exergy efficiency of 39.13%, and a levelized cost of exergy of 11.67 USD/GJ. Adjustments in system parameters can achieve a 5.7% LCOE reduction with minor efficiency trade-offs.

4. "Simultaneous Optimization and Integration of Multiple Process Heat Cascade and Site Utility Selection for the Design of a New Generation of Sugarcane Biorefinery" by Victor Fernandes Garcia and Adriano Viana Ensinas [12] addresses the economic and environmental challenges of sugarcane biorefineries through a novel superstructure model. It integrates heat recovery, utility selection and optimal sizing to design efficient biorefinery configurations. Results show an increase in energy efficiency (from 50.25% to 74.5%) by integrating methanol production to the sugarcane biorefinery via bagasse gasification. Similar results were found for DME production, although the higher power consumption from CO_2 hydrogenation impacts the energy efficiency.

5. "Combining Exergy and Pinch Analysis for the Operating Mode Optimization of a Steam Turbine Cogeneration Plant in Wonji-Shoa, Ethiopia" by Shumet Sendek Sharew, Alessandro Di Pretoro, Abubeker Yimam, Stéphane Negny and Ludovic Montastruc [13] studies the impact of the operating conditions of a steam turbine in an existing cogeneration plant. By combining pinch and exergy analysis, the research highlights the trade-off between heat integration design and exergy losses. The analysis indicates opportunities for heat pump technology integration and energy savings up to 83.44 MW by reducing exergy losses in the cogeneration plant.

6. "Techno–Economic Analysis of the Optimum Configuration for Supercritical Carbon Dioxide Cycles in Concentrating Solar Power Systems" by Rosa P. Merchán, Luis F. González-Portillo and Javier Muñoz-Antón [14] evaluates the techno-economic performance of supercritical carbon dioxide (sCO_2) cycles in concentrating solar power (CSP) systems, focusing on the trade-offs between cost and efficiency. Key factors such

as the turbine inlet temperature, ambient conditions, pressure drops, and turbomachinery efficiency are analyzed, alongside uncertainties in the component and heating costs. The CSP system with partial cooling offers the lowest costs, though under certain conditions, simple cycles or recompression cycles may be more economical.

7. "Exergoeconomic Analysis of a Mechanical Compression Refrigeration Unit Run by an ORC" by Daniel Taban, Valentin Apostol, Lavinia Grosu, Mugur C. Balan, Horatiu Pop, Catalina Dobre and Alexandru Dobrovicescu [15] showcases the exergoeconomic optimization of a vapor compression refrigeration cycle powered by an organic Rankine cycle recovering waste heat from a diesel engine. It focuses on reducing exergy destruction through structural changes, such as preheating the ORC fluid with an internal heat exchanger, improving global exergetic efficiency by 2.03%. Design improvements such as lowering temperature differences in heat exchangers and increasing the compression efficiency can reduce refrigeration unit costs by 59%.

8. "Improved Waste Heat Management and Energy Integration in an Aluminum Annealing Continuous Furnace Using a Machine Learning Approach" by Mohammad Andayesh, Daniel Alexander Flórez-Orrego, Reginald Germanier, Manuele Gatti and François Maréchal [16] focuses on improving energy efficiency in aluminum annealing continuous furnaces to tackle fossil emissions and fuel costs. A heat transfer model based on the machine learning regression of fluid dynamic simulation results predicts the aluminum temperature and heating rates. Two strategies, namely, optimizing furnace temperature profiles and recycling exhaust flue gasses for energy integration, are explored. A maximum reduction in fuel consumption of 20.7% is attained, optimizing energy integration.

9. "Comparative Exergy and Environmental Assessment of the Residual Biomass Gasification Routes for Hydrogen and Ammonia Production" by Gabriel Gomes Vargas, Daniel Alexander Flórez-Orrego and Silvio de Oliveira Junior [17] evaluates the use of biomass for producing hydrogen and ammonia via gasification to reduce fossil fuel consumption. It highlights the environmental and economic potential of converting biomass into valuable fuels with negative carbon emissions. The work reveals exergy inefficiencies from gasification, but also highlights the potential negative emissions for hydrogen (-5.95 kg$_{CO2}$ per kg$_{H2}$) and ammonia (-1.615 kg$_{CO2}$ per kg$_{NH3}$) production.

10. "Modeling and Optimization of Hydraulic and Thermal Performance of a Tesla Valve Using a Numerical Method and Artificial Neural Network" by Kourosh Vaferi, Mohammad Vajdi, Amir Shadian, Hamed Ahadnejad, Farhad Sadegh Moghanlou, Hossein Nami and Haleh Jafarzadeh [18] examines the optimization of a Tesla valve using artificial neural networks. Key geometrical parameters and inlet velocity are used as inputs, whereas the pressure drop ratio and the temperature difference ratio are the outputs. ANN models trained on numerical data achieved high accuracy in predicting responses. The results highlight the potential of the Tesla valve for advanced applications in heat sinks and exchangers.

11. "Design and Performance Evaluation of Integrating the Waste Heat Recovery System (WHRS) for a Silicon Arc Furnace with Plasma Gasification for Medical Waste" by Yuehong Dong, Lai Wei, Sheng Wang, Peiyuan Pan and Heng Chen [19] proposes a waste heat recovery system for a silicon arc furnace with plasma gasification for medical waste treatment (23,040 t/y). The plasma gasifier also disposes of harmful silica particles from polysilicon production. The syngas produced is used to generate power (4.17 MW of power with 33.99% efficiency) and auxiliary heating. An investment of $18.84 million entails a payback period of 3.94 years.

Acknowledgments: The Editors would like to thank the commitment of all the reviewers that contributed their time and thorough suggestions for improving the quality of the published manuscripts in this Special Issue.

Conflicts of Interest: The authors declare that the research was conducted in the absence of any commercial or financial relationships that could be construed as potential conflicts of interest. All claims expressed in this article are solely those of the authors and do not necessarily represent those of their affiliated organizations, or those of the publisher and the reviewers.

References

1. Energie Schweiz. Wärme Konsequent Nutzen: Prozessoptimierung mit Pinch. Swiss Federal Office of Energy. 2024. Available online: https://www.energieschweiz.ch/sich-beraten-lassen/pinch/ (accessed on 10 October 2024).
2. Domingos, M.E.G.; Flórez-Orrego, D.; Teles Dos Santos, M.; Oliveira Junior, S.; Maréchal, F. Process modeling and integration of hydrogen and synthetic natural gas production in a kraft pulp mill via black liquor gasification. *Renew. Energy* **2023**, *219*, 119396. [CrossRef]
3. Moncada, J.; Matallana, L.; Cardona, C. Selection of Process Pathways for Biorefinery Design Using Optimization Tools: A Colombian Case for Conversion of Sugarcane Bagasse to Ethanol, Poly-3-hydroxybutyrate (PHB), and Energy. *Ind. Eng. Chem. Res.* **2013**, *52*, 4132–4145. [CrossRef]
4. Marenco-Porto, C.; Fierro, J.; Nieto-Londoño, C.; Lopera, L.; Escudero-Atehortua, A.; Giraldo, M.; Jouhara, H. Potential savings in the cement industry using waste heat recovery technologies. *Energy* **2023**, *279*, 127810. [CrossRef]
5. Betancur-Arboleda, L. Modeling and Simulating a Heat Recovery System Using Low Temperature Heat Tubes. Master's Thesis, School of Processes and Energy, Faculty de Mines, National University of Colombia Campus Medellin, Medellín, Colombia, 2012. (In Spanish). Supervisor: Farid Chejne.
6. Cortvriendt, L. Data-Driven Systematic Methodology for the Prediction of Optimal Heat Pump Temperature Levels and Refrigerants in Process Integration. Master's Thesis, Katholieke Universiteit Leuven, Leuven, Belgium, 2024. Supervisors: Daniel Florez-Orrego, Dominik Bongartz, François Marechal.
7. Boztas, O. Olefins Production Through Sustainable Pathways: Techno-Economic and Environmental Assessment. Master's Thesis, Engineering School, École Polytechnique Fédérale de Lausanne, Lausanne, Switzerland, 2024. Supervisors: Meire Ellen Ribeiro Domingos, François Marechal.
8. Florez-Orrego, D. Process Synthesis and Optimization of Syngas and Ammonia Production in Nitrogen Fertilizers Complexes: Exergy, Energy Integration and CO_2 Emissions Assessment. Ph.D. Thesis, Polytechnic School, University of Sao Paulo, Sao Paulo, Brazil, 2018. Supervisor: Silvio de Oliveira Junior, François Marechal.
9. Luo, J.; Zhu, Q.; Morosuk, T. Advanced Exergy-Based Optimization of a Polygeneration System with CO_2 as Working Fluid. *Entropy* **2024**, *26*, 886. [CrossRef] [PubMed]
10. Bunya, B.; Sotomonte, C.A.R.; Vitoriano Julio, A.A.; Pereira, J.L.J.; de Souza, T.A.Z.; Francisco, M.B.; Coronado, C.J.R. Optimization and Tradeoff Analysis for Multiple Configurations of Bio-Energy with Carbon Capture and Storage Systems in Brazilian Sugarcane Ethanol Sector. *Entropy* **2024**, *26*, 698. [CrossRef] [PubMed]
11. Ren, J.; Xu, C.; Qian, Z.; Huang, W.; Wang, B. Exergoeconomic Analysis and Optimization of a Biomass Integrated Gasification Combined Cycle Based on Externally Fired Gas Turbine, Steam Rankine Cycle, Organic Rankine Cycle, and Absorption Refrigeration Cycle. *Entropy* **2024**, *26*, 511. [CrossRef] [PubMed]
12. Garcia, V.F.; Ensinas, A.V. Simultaneous Optimization and Integration of Multiple Process Heat Cascade and Site Utility Selection for the Design of a New Generation of Sugarcane Biorefinery. *Entropy* **2024**, *26*, 501. [CrossRef] [PubMed]
13. Sharew, S.S.; Di Pretoro, A.; Yimam, A.; Negny, S.; Montastruc, L. Combining Exergy and Pinch Analysis for the Operating Mode Optimization of a Steam Turbine Cogeneration Plant in Wonji-Shoa, Ethiopia. *Entropy* **2024**, *26*, 453. [CrossRef]
14. Merchán, R.P.; González-Portillo, L.F.; Muñoz-Antón, J. Techno–Economic Analysis of the Optimum Configuration for Supercritical Carbon Dioxide Cycles in Concentrating Solar Power Systems. *Entropy* **2024**, *26*, 124. [CrossRef] [PubMed]
15. Taban, D.; Apostol, V.; Grosu, L.; Balan, M.C.; Pop, H.; Dobre, C.; Dobrovicescu, A. Exergoeconomic Analysis of a Mechanical Compression Refrigeration Unit Run by an ORC. *Entropy* **2023**, *25*, 1531. [CrossRef] [PubMed]
16. Andayesh, M.; Flórez-Orrego, D.A.; Germanier, R.; Gatti, M.; Maréchal, F. Improved Waste Heat Management and Energy Integration in an Aluminum Annealing Continuous Furnace Using a Machine Learning Approach. *Entropy* **2023**, *25*, 1486. [CrossRef] [PubMed]
17. Vargas, G.G.; Flórez-Orrego, D.A.; de Oliveira Junior, S. Comparative Exergy and Environmental Assessment of the Residual Biomass Gasification Routes for Hydrogen and Ammonia Production. *Entropy* **2023**, *25*, 1098. [CrossRef] [PubMed]
18. Vaferi, K.; Vajdi, M.; Shadian, A.; Ahadnejad, H.; Moghanlou, F.S.; Nami, H.; Jafarzadeh, H. Modeling and Optimization of Hydraulic and Thermal Performance of a Tesla Valve Using a Numerical Method and Artificial Neural Network. *Entropy* **2023**, *25*, 967. [CrossRef] [PubMed]
19. Dong, Y.; Wei, L.; Wang, S.; Pan, P.; Chen, H. Design and Performance Evaluation of Integrating the Waste Heat Recovery System (WHRS) for a Silicon Arc Furnace with Plasma Gasification for Medical Waste. *Entropy* **2023**, *25*, 595. [CrossRef] [PubMed]

Article

Exergoeconomic Analysis and Optimization of a Biomass Integrated Gasification Combined Cycle Based on Externally Fired Gas Turbine, Steam Rankine Cycle, Organic Rankine Cycle, and Absorption Refrigeration Cycle

Jie Ren, Chen Xu *, Zuoqin Qian, Weilong Huang and Baolin Wang

School of Naval Architecture, Ocean and Energy Power Engineering, Wuhan University of Technology, Wuhan 430063, China; j.ren@whut.edu.cn (J.R.); qzq@whut.edu.cn (Z.Q.); hwl220@whut.edu.cn (W.H.); wangbaolin1006@163.com (B.W.)
* Correspondence: hesterxc@whut.edu.cn

Abstract: Adopting biomass energy as an alternative to fossil fuels for electricity production presents a viable strategy to address the prevailing energy deficits and environmental concerns, although it faces challenges related to suboptimal energy efficiency levels. This study introduces a novel combined cooling and power (CCP) system, incorporating an externally fired gas turbine (EFGT), steam Rankine cycle (SRC), absorption refrigeration cycle (ARC), and organic Rankine cycle (ORC), aimed at boosting the efficiency of biomass integrated gasification combined cycle systems. Through the development of mathematical models, this research evaluates the system's performance from both thermodynamic and exergoeconomic perspectives. Results show that the system could achieve the thermal efficiency, exergy efficiency, and levelized cost of exergy (LCOE) of 70.67%, 39.13%, and 11.67 USD/GJ, respectively. The analysis identifies the combustion chamber of the EFGT as the component with the highest rate of exergy destruction. Further analysis on parameters indicates that improvements in thermodynamic performance are achievable with increased air compressor pressure ratio and gas turbine inlet temperature, or reduced pinch point temperature difference, while the LCOE can be minimized through adjustments in these parameters. Optimized operation conditions demonstrate a potential 5.7% reduction in LCOE at the expense of a 2.5% decrease in exergy efficiency when compared to the baseline scenario.

Keywords: biomass gasification; combined cooling and power; exergoeconomic analysis; externally fired gas turbine; absorption refrigeration cycle; multi-objective optimization

Citation: Ren, J.; Xu, C.; Qian, Z.; Huang, W.; Wang, B. Exergoeconomic Analysis and Optimization of a Biomass Integrated Gasification Combined Cycle Based on Externally Fired Gas Turbine, Steam Rankine Cycle, Organic Rankine Cycle, and Absorption Refrigeration Cycle. *Entropy* **2024**, *26*, 511. https://doi.org/10.3390/e26060511

Academic Editors: Daniel Flórez-Orrego, Meire Ellen Ribeiro Domingos and Rafael Nogueira Nakashima

Received: 15 April 2024
Revised: 6 June 2024
Accepted: 11 June 2024
Published: 12 June 2024

1. Introduction

Over recent decades, the swift expansion of industrial and economic activities has significantly increased the consumption of fossil fuels, leading to a series of global challenges including environmental degradation, energy scarcity, and the exhaustion of natural resources. Projections suggest that by 2035, the demand for global energy will surge by 37% from its 2013 levels [1]. Fossil fuels, which account for approximately 80% of global energy consumption, are major contributors to the emission of greenhouse gases (GHG) [2]. In response, the global community has enacted various environmental treaties, such as the Montreal Protocol, the Kyoto Protocol, and the Paris Agreement, aiming to reduce GHG emissions and the carbon footprint of nations. Against this backdrop, there is a growing emphasis on the exploration and adoption of renewable energy sources by governments worldwide.

Renewable energy solutions present viable alternatives for mitigating global warming, reducing carbon dioxide emissions, and enhancing the energy independence of countries that rely heavily on imported fossil fuels. Among these renewable resources, biomass stands out for its versatility and sustainability. Biomass derives from a variety of sources including forests, crops, agricultural by-products, and organic waste from industrial, human, and animal activities [3]. It undergoes conversion into more valuable products, including liquid and gaseous fuels, through thermochemical and biochemical processes. Specifically, biomass gasification, a process of thermochemical conversion through partial oxidation, emerges as an optimal strategy for converting biomass into a syngas composed of carbon monoxide, hydrogen, carbon dioxide, gaseous hydrocarbons, and water vapor, along with minor amounts of char and condensable compounds [4]. The biomass integrated gasification combined cycle (BIGCC) has been recognized for its environmental friendliness, operational efficiency, and economic viability in electricity generation, positioning it as a pivotal technology in the shift toward renewable energy sources [5,6].

The adoption of gas turbine (GT) cycles for power generation from bioenergy is on the rise. Nonetheless, employing biogas in traditional internally fired GT cycles presents unique challenges. Gas turbines, being precision machinery, demand the expensive gas cleanup systems for highly purified gas to avoid fuel injector blockage and turbine blade damage [7]. Furthermore, the syngas from biomass gasification, characterized by its low calorific value, necessitates a substantial air intake for combustion to reach desired turbine inlet temperatures that require major modifications of commercially available gas turbines to prevent compressor surge conditions. The above-mentioned problems could be conveniently solved by employing an externally fired gas turbine (EFGT) cycle [8]. The EFGT configuration, where combustion occurs externally at low pressure, enables the use of lower-grade biofuels. In this setup, the turbine is powered by hot compressed air which is heated to the requisite turbine inlet temperature by flue gases from an external combustor via a high-temperature heat exchanger (HTHE) [9]. This configuration, using clean air as the working fluid, not only mitigates maintenance demands but also prolongs the service life of the turbine.

Numerous studies have been undertaken to explore the integration of the EFGT cycle with biomass as an energy source for electricity generation [9–11]. Despite its benefits, the EFGT cycle often faces criticism for its relatively low energy efficiency, primarily attributed to the biomass's inferior calorific value [12]. To enhance the efficiency of biomass-powered plants, additional systems for waste heat recovery, such as the steam Rankine cycle (SRC) and organic Rankine cycle (ORC), have been implemented to recover waste heat and convert it into additional power. Research by Soltani et al. [10] on the thermodynamic performance of a biomass gasification integrated EFGT combined cycle demonstrated potential energy and exergy efficiencies of 46.95% and 39.37%, respectively. Mondal et al. [13] undertook an exergoeconomic analysis of a BIGCC system incorporating an EFGT cycle and a supercritical ORC, achieving energy and exergy efficiencies of 40.77% and 36.30%, respectively. Vera et al. [14] evaluated a small-scale power generation setup comprising a downdraft gasifier, an EFGT cycle, and an ORC, demonstrating that the system could attain a net electrical efficiency of 20.7% when employing isopentane as the working fluid for the ORC. Zhang et al. [15] examined a municipal solid waste (MSW) fueled cogeneration system incorporating an EFGT cycle, an SCO_2 cycle, and a high-temperature organic flash cycle (OFC). They reported the system energy and exergy efficiencies of 75.8% and 41.21% respectively, with a total product cost of 10.2 USD/GJ. Moradi et al. [16] conducted a comparative sensitivity analysis of two micro-scale integrated prime movers based on a GT cycle and an SCO_2 cycle with bottoming ORC units. The study showed that at full load, the average net electric power output of the SCO_2 integrated system is about 25% higher than that of the GT system, although it incurs a 75% higher biomass consumption due to lower net electric efficiency. Sharafi laleh et al. [17] assessed the thermodynamic performance of a biomass gasification-based power plant integrated with an EFGT cycle and an SCO_2 cycle, achieving an energy efficiency of 41.18%.

The combination of biomass energy with multi-generation systems is considered as a strategic approach to boost energy efficiency and satisfy the diverse energy demands of consumers. Advancements in technology have significantly increased the efficiency with which biomass energy is utilized. Biomass-fueled EFGT cycles are supposed to be favorable options for small- to medium-sized multi-generational systems [8,18]. Roy et al. [19] performed techno-economic and environmental analyses of a biomass-based power generation setup integrating a solid oxide fuel cell module (SOFC), an EFGT cycle, and an ORC, with findings indicating potential energy and exergy efficiencies of up to 49.47% and 44.2%, respectively. El-Sattar et al. [20] conducted a thermodynamic study on a combined cooling, heating, and power (CCHP) system including an EFGT cycle, an ORC, and an ARC. They pointed out that toluene is the optimal working fluid for maximizing system thermal efficiency at 43.9%. Roy et al. [21] evaluated a combined power and heating system featuring an EFGT cycle, a biomass gasifier, an SOFC, and a heat recovery steam generator (HRSG), reporting an optimal exergy efficiency and LCOE of 46.58% and 0.0657 USD/kWh, respectively. Zhang et al. [22] analyzed a biomass-fueled cogeneration system, incorporating a gasifier, an EFGT cycle, an SCO_2 cycle, a Stirling engine, and a DWH. They determined the system's optimal exergy efficiency to be 46.48% with a total cost rate of 401.4 USD/h. Xu et al. [23] conducted thermodynamic and exergoeconomic analyses on a biomass-fueled multigeneration system, including a syngas production unit, an SRC, a multi-effect desalination (MED) unit, and a solid oxide electrolyzer cell (SOEC). They reported that the optimal exergy efficiency and unit exergy cost could reach 17.64% and 26 USD/GJ respectively. Du et al. [24] analyzed a biomass-driven multigeneration system comprising a gasification unit, a helium GT cycle, a Kalina cycle, a DWH, a refrigeration unit, and a dual-loop OFC. The results indicated that the system could reach an optimal exergy efficiency of 35.57%, a net present value (NPV) of 15.07 M USD, and a payback period of 3.97 years. Yilmaz et al. [25] proposed a biomass-based multigeneration plant with a GT cycle, an SCO_2 cycle, a multi-stages flash desalination (MSFD) unit, a proton exchange membrane electrolyzer (PEME), and a DWH. They found the energy and exergy efficiencies to be 44.50% and 30.01%, respectively. Zhang et al. [26] proposed a biomass-based multigeneration setup with a GT cycle, an SCO_2 cycle, a double-effect ARC, a DWH, an ORC, and a reverse osmosis (RO) desalination unit. They concluded that the system could attain an optimal exergy efficiency of 38.54%, along with a sum unit cost of product (SUCP) of 30.8 USD/GJ, and an NPV of 75.17 M USD.

The results of previous research have demonstrated that integrating the EFGT cycle with waste heat recovery systems significantly enhances the efficiency of biomass energy utilization. As a general power generation technology, the SRC has been widely adopted to recover the medium- or high-temperature waste heat. Nonetheless, a considerable amount of energy is released into the environment unutilized during the SRC condensation process. Research has suggested the potential of employing low-temperature condensation heat to drive a single-effect ARC [27]. Liang et al. [28,29] developed a CCP system coupling of an SRC and an ARC to capitalize on the waste heat from a marine engine. They discovered that this SRC–ARC configuration markedly elevates the energy utilization efficiency over the basic SRC, with an 84% increase in exergy efficiency under specific conditions of condensation temperature at 323 K and superheat at 100 K. Ahmadi et al. [30] conducted both thermodynamic and exergoenvironmental evaluations of a GT-based trigeneration system integrated with an SRC and a steam-driven ARC, revealing thermal and exergy efficiencies of 75.5% and 47.5%, respectively. Sahoo et al. [31] thermodynamically evaluated a multi-generation system powered by solar and biomass energies, in which an ARC was driven by the residual heat of the SRC, achieving energy and exergy efficiencies of 49.85% and 20.95%, respectively. Nondy et al. [32] compared the thermodynamic performance of four CCP configurations designed for waste heat recovery from a GT cycle, utilizing SRC and ARC as bottoming cycles. They found that the configuration with two ARCs driven, respectively, by steam and exhaust gas is the most appropriate from the energy and exergy viewpoints. Anvari et al. [33] performed an advanced exergetic and exergoeconomic analysis of a CCHP system consisting of a GT cycle, a dual pressure HRSG, and an ARC driven by the low-pressure steam. They identified that nearly 29% of the total exergy destruction and the associated cost rates due to exergy destruction within the system are endogenous-avoidable.

For an enhanced understanding of the current research in the field, several studies related to biomass-based multigeneration system integrated with a GT cycle have been systematically organized in Table 1. The review of the above studies indicates that many researchers have proposed various biomass-based multi-generation systems with the aim of increasing energy utilization efficiency and reducing environmental impact. It is also suggested that the SRC–ARC combined cycles help to further utilize the waste heat and improve the thermodynamic performance. According to the literature review, devising a high-efficient combined cooling and power system based on biomass gasification combined with EFGT cycle has not been extensively investigated up to now. In addition, the coupling of SRC and ARC integrated with the biomass gasification has seldom been considered in the literature. Considering these motivations, this study aims to provide a comprehensive evaluation of a novel CCP system including an EFGT, an SRC, an ARC, and an ORC based on cascade utilization of high-temperature waste heat from syngas combustion. It can be expected that the proposed scheme has great potential to achieve a noticeable energy efficiency compared to the available literature due to better integration of bottoming sub-cycles. The main objectives and contributions of this work can be summarized as follows:

(1) Introduction of a novel biomass gasification-based CCP system to enhance the energy utilization efficiency, alongside the development of comprehensive mathematical models to assess system performance from thermodynamic and exergoeconomic perspectives.
(2) Examination of the influence of critical operational parameters on the performance criteria.
(3) Optimization of the system to determine the optimal operational conditions that maximize exergy efficiency while minimizing the LCOE.

Table 1. Overview of recent research on the configurations and evaluations of multi-generation systems based on biomass gasification.

Researcher	Year	Biomass Fuel	Configuration	Analysis	Result
Zhang et al. [15]	2023	municipal solid waste	EFGT, SCO_2 cycle, OFC	energy, exergy, economic, environmental	energy efficiency of 75.8%, exergy efficiency of 41.21%, net profit of 10.7 M USD, levelized CO_2 emission of 0.518 t/kWh
Moradi et al. [16]	2023	hazelnut shell	GT cycle, SCO_2 cycle, ORC	energy	25% higher electric power output of the SCO_2 integrated system
Sharafi laleh et al. [17]	2024	wood	EFGT, SCO_2 cycle	energy	energy efficiency of 41.8%
Roy et al. [19]	2019	wood, rice husk, paper	EFGT, SOFC, ORC	energy, exergy, economic, environmental	energy efficiency of 49.47%, exergy efficiency of 44.2%
El-Sattar et al. [20]	2020	bagasse	EFGT, ORC, ARC	energy	thermal efficiency of 43.9%
Roy et al. [21]	2020	sawdust	EFGT, SOFC, HRSG	exergy, economic	exergy efficiency of 46.58%, levelized cost of exergy of 0.0657 USD/kWh
Zhang et al. [22]	2022	paddy husk, paper, wood, municipal solid waste	EFGT, SCO_2 cycle, Stirling engine, DWH	energy, exergy, exergoeconomic, environmental	exergy efficiency of 46.48%, total cost rate of 401.4 USD/h

Table 1. *Cont.*

Researcher	Year	Biomass Fuel	Configuration	Analysis	Result
Xu et al. [23]	2022	paddy husk, paper, wood, municipal solid waste	SRC, MED unit, SOEC	energy, exergy, exergoeconomic	exergy efficiency of 17.64%, unit exergy cost of 26 USD/GJ
Du et al. [24]	2024	wood	helium GT cycle, Kalina cycle, DWH, refrigeration unit, dual-loop OFC	energy, exergy, economic	exergy efficiency of 35.57%, NPV of 15.07 M USD, payback period of 3.97 years
Yilmaz et al. [25]	2024	pine sawdust	GT cycle, SCO$_2$ cycle, MSFD unit, PEME, DWH	energy, exergy, environmental	energy efficiency of 44.50%, exergy efficiency of 30.01%
Zhang et al. [26]	2024	carbohydrate	GT cycle, SCO$_2$ cycle, dual-effect ARC, DWH, ORC, RO desalination	energy, exergy, economic	exergy efficiency of 38.54%, SUCP of 30.8 USD/GJ, NPV of 75.17 M USD

2. System Description

Figure 1 illustrates the configuration of the proposed CCP system, which is fed by biomass and encompasses a biomass gasifier, an EFGT, an SRC, an ARC, and an ORC. Within the EFGT cycle, key components include an air compressor (AC), an air preheater (AP), a combustion chamber (CC), and a gas turbine (GT). Ambient air (state 1) undergoes compression in the AC, and this compressed air (state 2) is then heated by the flue gases (state 8) in the AP. The high-temperature air (state 3) expands through the GT, driving the generator to produce electricity. Subsequently, the exhaust air (state 4) flows into the CC, where it reacts with syngas (state 5) from the biomass gasifier (Ga). After rejecting heat to the compressed air in the AP, the flue gas (state 9) is directed through a heat recovery steam generator (HRSG) and a vapor generator (VG), successively activating the bottoming SRC and ORC.

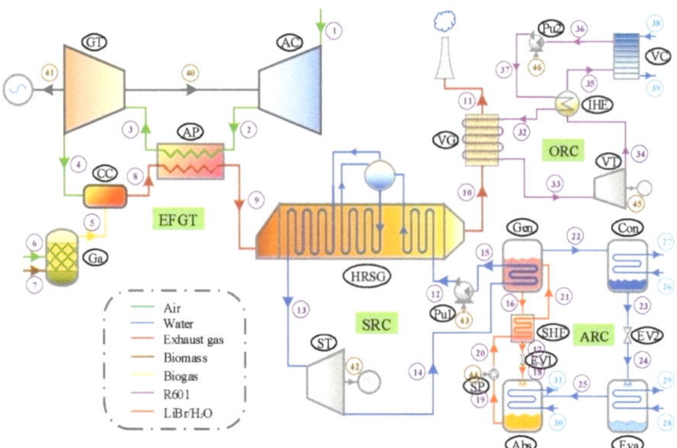

Figure 1. Schematic diagram of the proposed CCP system.

Within the SRC, the pressurized water (state 12) absorbs heat to be converted into superheated vapor (state 13), which is then expanded in the steam turbine (ST) to generate electricity. The resulting exhaust (state 14) serves as the thermal source for a single-effect LiBr-H$_2$O ARC. In the ARC generator (Gen), the dilute solution (state 21) is heated, separating into a concentrated solution (state 16) and refrigerant vapor (state 22). This concentrated solution is then routed through a solution heat exchanger (SHE), warming the returning dilute solution (state 20) back to the generator. Concurrently, the refrigerant vapor condenses in the condenser (Con), and the resulting saturated liquid (state 23) moves to the evaporator (Eva) via an expansion valve (EV2). After absorbing heat in the evaporator, the vaporized refrigerant (state 25) is absorbed by the concentrated solution (state 18) and cooled by the water in the absorber (abs), producing a dilute solution (state 19) that is cycled back through the SHE to the generator.

The exhaust gas is introduced to the bottoming ORC to further exploit its residual thermal energy. The ORC mainly includes the following components: vapor generator (VG), vapor turbine (VT), internal heat exchanger (IHE), condenser (VC), and pump (Pu2). High-pressure vapor (state 33) generated in the VG drives the VT to produce power. The IHE facilitates preheating of the organic liquid (state 37) by the low-pressure vapor (state 34) exiting the VT. This vapor (state 35) condenses into a saturated liquid (state 36) in the VC, releasing heat to the cooling water, before being recirculated by the pump (Pu2) back to the VG via the IHE.

3. Mathematical Modeling

3.1. Assumptions

The system under consideration is conceptualized and analyzed under a set of foundational assumptions as follows [34–36]:

1. Operation of the system is assumed to be in a steady state;
2. Changes in kinetic and potential energy within the system are considered negligible;
3. The system assumes no heat losses across its various components;
4. Pressure variations across piping systems are overlooked;
5. The composition of ambient air is taken as 21% oxygen and 79% nitrogen by volume;
6. Gas mixtures within the system are treated as ideal gases for the purpose of simulation;
7. Within the ARC, fluid streams exit both the evaporator and condenser in a saturated state, and the output solutions from the generator and absorber reach equilibrium at their specific temperatures and concentrations;
8. For the ORC, the working fluid departs the vapor generator as saturated vapor and exits the condenser as saturated liquid;
9. The performance of compressors, pumps, and turbines is modeled with constant isentropic efficiencies.

3.2. Energy Analysis

3.2.1. Biomass Gasifier

This study focuses on an atmospheric downdraft gasifier, utilizing wood chips as fuel and air as the gasifying agent. The composition and higher heating value (HHV) of wood is presented in Table 2. The gasification occurs at high temperatures and involves several key stages: drying, pyrolysis, oxidation, and reduction [37]. To estimate the syngas composition, an equilibrium model is employed, demonstrating a reliable approximation of the gasification process within the downdraft gasifier [38,39]. This model assumes that all chemical reactions are in thermodynamic equilibrium and the pyrolysis products reach equilibrium in the reduction zone prior to exiting the gasifier [40,41]. The comprehensive reaction governing biomass fuel gasification can be succinctly represented as follows [17]:

$$CH_{1.44}O_{0.66} + wH_2O + n_{air,1}(O_2 + 3.76N_2) \rightarrow n_1H_2 + n_2CO + n_3CO_2 + n_4H_2O + n_5CH_4 + n_6N_2 \tag{1}$$

where $CH_aO_bN_c$ represents the chemical composition of the biomass, w indicates the moisture content of the biomass, and $n_{air,1}$ signifies the kilomoles of oxygen from air involved in the reaction. The coefficients n_1 to n_6 denote the kilomoles of the product constituents.

Table 2. Ultimate analysis and higher heating value of wood [17,22].

Biomass	Mass Percentage on Dry Basis (%)						HHV (kJ/kmol)
	C	H	N	S	O	Ash	
wood	50	6	0	0	44	0	449,568

The moisture content in the biomass is typically quantified by its mass-based moisture content (*MC*), calculated using the formula below [15]:

$$w = \frac{MC \times MW_{\text{biomass}}}{MW_{\text{H}_2\text{O}} \times (1 - MC)} \tag{2}$$

where MW_{biomass} and $MW_{\text{H}_2\text{O}}$ refer to the molecular weights of the biomass and water, respectively.

Key reactions occurring during gasification include methane formation and water-gas shift reactions, with the equilibrium constants for these reactions provided as follows [15,23,24]:

$$C + CO_2 \rightleftharpoons 2CO \tag{3}$$

$$CO + H_2O \rightleftharpoons CO_2 + H_2 \tag{4}$$

The equilibrium constants for these reactions are denoted as follows [15,23,24]:

$$K_1 = \frac{n_5}{n_1^{\,2}} \left(\frac{P_{\text{Ga}}/P_{\text{ref}}}{n_{\text{tot}}} \right)^{-1} \tag{5}$$

$$K_2 = \frac{n_1 n_3}{n_2 n_4} \left(\frac{P_{\text{Ga}}/P_{\text{ref}}}{n_{\text{tot}}} \right)^{0} \tag{6}$$

where P_{Ga} is the pressure during gasification. K_1 and K_2 are derived from the Gibbs free energy changes associated with each reaction, calculated by [15,23,24]:

$$\text{In} K_1 = -\frac{\Delta G_1^0}{\overline{R} T_{\text{Ga}}} \tag{7}$$

$$\text{In} K_2 = -\frac{\Delta G_2^0}{\overline{R} T_{\text{Ga}}} \tag{8}$$

where $\overline{R}$ is the universal gas constant, T_{Ga} is the temperature within the gasifier. ΔG_1^0 and ΔG_2^0 are computed using the following equations [15,23,24]:

$$\Delta G_1^0 = \left(\overline{h}_{\text{CH}_4} - T_{\text{Ga}} \overline{s}_{\text{CH}_4}^0 \right) - 2\left(\overline{h}_{\text{H}_2} - T_{\text{Ga}} \overline{s}_{\text{H}_2}^0 \right) \tag{9}$$

$$\Delta G_2^0 = \left(\overline{h}_{\text{CO}_2} - T_{\text{Ga}} \overline{s}_{\text{CO}_2}^0 \right) + \left(\overline{h}_{\text{H}_2} - T_{\text{Ga}} \overline{s}_{\text{H}_2}^0 \right) - \left(\overline{h}_{\text{CO}} - T_{\text{Ga}} \overline{s}_{\text{CO}}^0 \right) - \left(\overline{h}_{\text{H}_2\text{O}} - T_{\text{Ga}} \overline{s}_{\text{H}_2\text{O}}^0 \right) \tag{10}$$

Under the assumption of no heat loss in the gasifier, the energy balance equation governing the gasification process is outlined as follows [15,23,24]:

$$\begin{aligned}
\overline{h}_{f-\text{biomass}}^0 + w \overline{h}_{f-\text{H}_2\text{O}}^0 + n_{\text{air},1} \overline{h}_{\text{air},1} =\ & n_1 \left(\overline{h}_{f-\text{H}_2}^0 + \Delta \overline{h}_{\text{H}_2} \right) + n_2 \left(\overline{h}_{f-\text{CO}}^0 + \Delta \overline{h}_{\text{CO}} \right) + n_3 \left(\overline{h}_{f-\text{CO}_2}^0 + \Delta \overline{h}_{\text{CO}_2} \right) \\
& + n_4 \left(\overline{h}_{f-\text{H}_2\text{O}}^0 + \Delta \overline{h}_{\text{H}_2\text{O}} \right) + n_5 \left(\overline{h}_{f-\text{CH}_4}^0 + \Delta \overline{h}_{\text{CH}_4} \right) + n_6 \left(\overline{h}_{f-\text{N}_2}^0 + \Delta \overline{h}_{\text{N}_2} \right)
\end{aligned} \tag{11}$$

where $\overline{h}_{f-j}^0$ corresponds to the formation enthalpy of the *j*th component, $\Delta \overline{h}_j$ represents the variance in specific enthalpy of the *j*th component at the gasification temperature relative to the reference temperature T_0.

3.2.2. Combustion Chamber

Within the combustion chamber, syngas generated from the gasification process undergoes combustion by reacting with the oxygen in the air supplied by the gas turbine. The chemical reaction is shown as follows under the assumption of complete combustion taking place [15]:

$$n_1H_2 + n_2CO + n_3CO_2 + n_4H_2O + n_5CH_4 + n_6N_2 + n_{air,2}(O_2 + 3.76N_2) \rightarrow n_7CO_2 + n_8H_2O + n_9O_2 + n_{10}N_2 \qquad (12)$$

where $n_{air,2}$ denotes the kilomoles of the oxygen entering the combustion chamber.

For an adiabatic combustion scenario, the energy equation governing the combustion chamber is formulated as follows [15]:

$$\sum_j n_j \left(\overline{h}^0_{f-j} + \Delta\overline{h} \right)_{air} + \sum_j n_j \left(\overline{h}^0_{f-j} + \Delta\overline{h} \right)_{syngas} = \sum_j n_j \left(\overline{h}^0_{f-j} + \Delta\overline{h} \right)_{exh} \qquad (13)$$

where n_j indicates the kilomoles of the jth component in the air, syngas, and exhaust gas.

3.2.3. Other System Components

At steady-state operation, the system is governed by equations representing mass balance, energy balance, and concentration balance for each component, expressed as [26,36]:

$$\sum \dot{m}_{in} = \sum \dot{m}_{out} \qquad (14)$$

$$\dot{Q} + \sum \dot{m}_{in}h_{in} = \dot{W} + \sum \dot{m}_{out}h_{out} \qquad (15)$$

$$\sum \dot{m}_{in}X_{in} = \sum \dot{m}_{out}X_{out} \qquad (16)$$

where X symbolizes the mass concentration of LiBr in the solution.

The equations related to mass and energy balances within the system's components are detailed in Table 3.

Table 3. Mass and energy balance equations for the system components.

Component	Mass and Energy Balance Equations
Air compressor	$\dot{W}_{AC} = \dot{m}_1(h_2 - h_1)$ $\dot{m}_1 = \dot{m}_2$
Air preheater	$\dot{m}_2(h_3 - h_2) = \dot{m}_8(h_8 - h_9)$ $\dot{m}_2 = \dot{m}_3, \ \dot{m}_8 = \dot{m}_9$
Gas turbine	$\dot{W}_{GT} = \dot{m}_3(h_3 - h_4)$ $\dot{m}_3 = \dot{m}_4$
HRSG	$\dot{Q}_{HRSG} = \dot{m}_9(h_9 - h_{10}) = \dot{m}_{12}(h_{13} - h_{12})$ $\dot{m}_9 = \dot{m}_{10}, \ \dot{m}_{12} = \dot{m}_{13}$
Steam turbine	$\dot{W}_{ST} = \dot{m}_{13}(h_{13} - h_{14})$ $\dot{m}_{13} = \dot{m}_{14}$
Pump 1	$\dot{W}_{pu1} = \dot{m}_{12}(h_{12} - h_{15})$ $\dot{m}_{12} = \dot{m}_{15}$
Generator	$\dot{Q}_{gen} = \dot{m}_{14}(h_{14} - h_{15}) = \dot{m}_{16}h_{16} + \dot{m}_{22}h_{22} - \dot{m}_{21}h_{21}$ $\dot{m}_{14} = \dot{m}_{15}, \ \dot{m}_{21} = \dot{m}_{16} + \dot{m}_{22}$
SHE	$\dot{Q}_{SHE} = \dot{m}_{16}(h_{16} - h_{17}) = \dot{m}_{20}(h_{21} - h_{20})$ $\dot{m}_{16} = \dot{m}_{17}, \ \dot{m}_{20} = \dot{m}_{21}$
Absorber	$\dot{Q}_{abs} = \dot{m}_{30}(h_{31} - h_{30}) = \dot{m}_{18}h_{18} + \dot{m}_{25}h_{25} - \dot{m}_{19}h_{19}$ $\dot{m}_{30} = \dot{m}_{31}, \ \dot{m}_{19} = \dot{m}_{18} + \dot{m}_{25}$
Solution pump	$\dot{W}_{SP} = \dot{m}_{19}(h_{20} - h_{19})$ $\dot{m}_{19} = \dot{m}_{20}$

Table 3. *Cont.*

Component	Mass and Energy Balance Equations
Condenser	$\dot{Q}_{\text{con}} = \dot{m}_{22}(h_{22} - h_{23}) = \dot{m}_{26}(h_{27} - h_{26})$ $\dot{m}_{22} = \dot{m}_{23},\ \dot{m}_{26} = \dot{m}_{27}$
Evaporator	$\dot{Q}_{\text{eva}} = \dot{m}_{24}(h_{25} - h_{24}) = \dot{m}_{28}(h_{28} - h_{29})$ $\dot{m}_{24} = \dot{m}_{25},\ \dot{m}_{28} = \dot{m}_{29}$
Vapor generator	$\dot{Q}_{\text{VG}} = \dot{m}_{10}(h_{10} - h_{11}) = \dot{m}_{32}(h_{33} - h_{32})$ $\dot{m}_{10} = \dot{m}_{11},\ \dot{m}_{32} = \dot{m}_{33}$
Vapor turbine	$\dot{W}_{\text{VT}} = \dot{m}_{33}(h_{33} - h_{34})$ $\dot{m}_{33} = \dot{m}_{34}$
IHE	$\dot{Q}_{\text{IHE}} = \dot{m}_{37}(h_{32} - h_{37}) = \dot{m}_{34}(h_{34} - h_{35})$ $\dot{m}_{32} = \dot{m}_{37},\ \dot{m}_{34} = \dot{m}_{35}$
Vapor condenser	$\dot{Q}_{\text{VC}} = \dot{m}_{35}(h_{35} - h_{36}) = \dot{m}_{38}(h_{39} - h_{38})$ $\dot{m}_{35} = \dot{m}_{36},\ \dot{m}_{38} = \dot{m}_{39}$
Pump 2	$\dot{W}_{\text{pu2}} = \dot{m}_{36}(h_{37} - h_{36})$ $\dot{m}_{36} = \dot{m}_{37}$

3.3. Exergy Analysis

The exergy rate balance equation is formulated as follows [42]:

$$\dot{Ex}_F = \dot{Ex}_P + \dot{Ex}_D + \dot{Ex}_L \tag{17}$$

where $\dot{Ex}_F$ and $\dot{Ex}_P$ reflect the input fuel rate and output product rate, respectively. $\dot{Ex}_D$ and $\dot{Ex}_L$ correspond to the exergy destruction rate and exergy loss rate, respectively. The detailed exergy balance equations for the system's components are provided in Table 4.

Disregarding kinetic and potential exergies allows for categorizing the specific exergy of a flow into its physical and chemical components [42]:

$$ex_i = ex_i^{ph} + ex_i^{ch} \tag{18}$$

The physical exergy is defined as [42]:

$$ex_i^{ph} = h_i - h_0 - T_0(s_i - s_0) \tag{19}$$

For an ideal gas mixture, chemical exergy is expressed as [42]:

$$ex_i^{ch} = \sum_i x_i ex_{0,i}^{ch} + \overline{R}T_0 \sum_i x_i \ln x_i \tag{20}$$

where x_i denotes the molar fraction of the ith component, $ex_{0,i}^{ch}$ represents the standard chemical exergy.

The specific chemical exergy of biomass is determined based on its lower heating value (LHV) [10,23,26]:

$$e_{\text{biomass}}^{ch} = \psi \text{LHV}_{\text{biomass}} \tag{21}$$

The coefficient ψ is calculated considering the mass fractions of oxygen (M_O), carbon (M_C), and hydrogen (M_H) within the biomass [10,23,26]:

$$\psi = \frac{1.044 + 0.016\frac{M_H}{M_C} - 0.34493\frac{M_O}{M_C}\left(1 + 0.0531\frac{M_H}{M_C}\right)}{1 - 0.4124\frac{M_O}{M_C}} \tag{22}$$

Table 4. Exergy balance equations for the system components.

Component	Exergy *of* Fuel ($\dot{E}x_F$)	Exergy *of* Product ($\dot{E}x_P$)	Exergy Destruction ($\dot{E}x_D$)
Air compressor	$\dot{E}x_{40}$	$\dot{E}x_2 - \dot{E}x_1$	$\dot{E}x_{F,AC} - \dot{E}x_{P,AC}$
Air preheater	$\dot{E}x_8 - \dot{E}x_9$	$\dot{E}x_3 - \dot{E}x_2$	$\dot{E}x_{F,AP} - \dot{E}x_{P,AP}$
Gas turbine	$\dot{E}x_3 - \dot{E}x_4$	$\dot{E}x_{40} + \dot{E}x_{41}$	$\dot{E}x_{F,GT} - \dot{E}x_{P,GT}$
Combustion chamber	$\dot{E}x_4 + \dot{E}x_5$	$\dot{E}x_8$	$\dot{E}x_{F,CC} - \dot{E}x_{P,CC}$
Biomass gasifier	$\dot{E}x_6 + \dot{E}x_7$	$\dot{E}x_5$	$\dot{E}x_{F,Ga} - \dot{E}x_{P,Ga}$
HRSG	$\dot{E}x_9 - \dot{E}x_{10}$	$\dot{E}x_{13} - \dot{E}x_{12}$	$\dot{E}x_{F,HRSG} - \dot{E}x_{P,HRSG}$
Steam turbine	$\dot{E}x_{13} - \dot{E}x_{14}$	$\dot{E}x_{42}$	$\dot{E}x_{F,ST} - \dot{E}x_{P,ST}$
Pump 1	$\dot{E}x_{43}$	$\dot{E}x_{12} - \dot{E}x_{15}$	$\dot{E}x_{F,Pu1} - \dot{E}x_{P,Pu1}$
Generator	$\dot{E}x_{14} - \dot{E}x_{15}$	$\dot{E}x_{16} + \dot{E}x_{22} - \dot{E}x_{21}$	$\dot{E}x_{F,gen} - \dot{E}x_{P,gen}$
SHE	$\dot{E}x_{16} - \dot{E}x_{17}$	$\dot{E}x_{21} - \dot{E}x_{20}$	$\dot{E}x_{F,SHE} - \dot{E}x_{P,SHE}$
Absorber	$\dot{E}x_{18} + \dot{E}x_{25} - \dot{E}x_{19}$	$\dot{E}x_{31} - \dot{E}x_{30}$	$\dot{E}x_{F,abs} - \dot{E}x_{P,abs}$
Solution pump	$\dot{E}x_{44}$	$\dot{E}x_{20} - \dot{E}x_{19}$	$\dot{E}x_{F,SP} - \dot{E}x_{P,SP}$
Condenser	$\dot{E}x_{22} - \dot{E}x_{23}$	$\dot{E}x_{27} - \dot{E}x_{26}$	$\dot{E}x_{F,con} - \dot{E}x_{P,con}$
Evaporator	$\dot{E}x_{24} - \dot{E}x_{25}$	$\dot{E}x_{29} - \dot{E}x_{28}$	$\dot{E}x_{F,eva} - \dot{E}x_{P,eva}$
Vapor generator	$\dot{E}x_{10} - \dot{E}x_{11}$	$\dot{E}x_{33} - \dot{E}x_{32}$	$\dot{E}x_{F,VG} - \dot{E}x_{P,VG}$
Vapor turbine	$\dot{E}x_{33} - \dot{E}x_{34}$	$\dot{E}x_{45}$	$\dot{E}x_{F,VT} - \dot{E}x_{P,VT}$
IHE	$\dot{E}x_{34} - \dot{E}x_{35}$	$\dot{E}x_{32} - \dot{E}x_{37}$	$\dot{E}x_{F,IHE} - \dot{E}x_{P,IHE}$
Vapor condenser	$\dot{E}x_{35} - \dot{E}x_{36}$	$\dot{E}x_{39} - \dot{E}x_{38}$	$\dot{E}x_{F,VC} - \dot{E}x_{P,VC}$
Pump 2	$\dot{E}x_{46}$	$\dot{E}x_{37} - \dot{E}x_{36}$	$\dot{E}x_{F,pu2} - \dot{E}x_{P,pu2}$

The efficiency in terms of exergy for kth component is defined as [42]:

$$\eta_{ex,k} = \frac{\dot{E}x_{P,k}}{\dot{E}x_{F,k}} \tag{23}$$

The exergy destruction ratio of the kth component is conceptualized as the proportion of that component's exergy destruction relative to the overall exergy destruction within the system [42]:

$$y_{D,k} = \frac{\dot{E}x_{D,k}}{\dot{E}x_{D,tot}} \tag{24}$$

3.4. Exergoeconomic Analysis

Exergoeconomic analysis integrates exergy assessment with economic theories to elucidate the cost generation mechanism and calculate the cost associated with each unit of exergy of the product. The cost balance for kth component is formulated as below [26,42]:

$$\sum \dot{C}_{in,k} + \sum \dot{C}_{q,k} + \dot{Z}_k = \sum \dot{C}_{out,k} + \dot{C}_{w,k} \tag{25}$$

where $\dot{C}_j$ denotes the cost rate (USD/h), $\dot{Z}_k$ signifies the total cost rate encompassing both capital investment and operational and maintenance expenses for the kth component. Table 5 outlines the cost balance and supplementary equations for the system's components.

Table 5. Cost balance and auxiliary equations for the system components.

Component	Cost Balance Equation	Auxiliary Equation
Air compressor	$\dot{C}_1 + \dot{C}_{40} + \dot{Z}_{AC} = \dot{C}_2$	$c_1 = 0$
Air preheater	$\dot{C}_2 + \dot{C}_8 + \dot{Z}_{AP} = \dot{C}_3 + \dot{C}_9$	$\dfrac{\dot{C}_8}{\dot{Ex}_8} = \dfrac{\dot{C}_9}{\dot{Ex}_9}$
Gas turbine	$\dot{C}_3 + \dot{Z}_{GT} = \dot{C}_4 + \dot{C}_{40} + \dot{C}_{41}$	$\dfrac{\dot{C}_3}{\dot{Ex}_3} = \dfrac{\dot{C}_4}{\dot{Ex}_4}, \dfrac{\dot{C}_{40}}{\dot{Ex}_{40}} = \dfrac{\dot{C}_{41}}{\dot{Ex}_{41}}$
Combustion chamber	$\dot{C}_4 + \dot{C}_5 + \dot{Z}_{CC} = \dot{C}_8$	
Biomass gasifier	$\dot{C}_6 + \dot{C}_7 + \dot{Z}_{Ga} = \dot{C}_5$	$c_6 = 0$
HRSG	$\dot{C}_9 + \dot{C}_{12} + \dot{Z}_{HRSG} = \dot{C}_{10} + \dot{C}_{13}$	$\dfrac{\dot{C}_9}{\dot{Ex}_9} = \dfrac{\dot{C}_{10}}{\dot{Ex}_{10}}$
Steam turbine	$\dot{C}_{13} + \dot{Z}_{ST} = \dot{C}_{14} + \dot{C}_{42}$	$\dfrac{\dot{C}_{13}}{\dot{Ex}_{13}} = \dfrac{\dot{C}_{14}}{\dot{Ex}_{14}}$
Pump 1	$\dot{C}_{15} + \dot{C}_{43} + \dot{Z}_{HP} = \dot{C}_{12}$	$\dfrac{\dot{C}_{42}}{\dot{Ex}_{42}} = \dfrac{\dot{C}_{43}}{\dot{Ex}_{43}}$
Generator	$\dot{C}_{14} + \dot{C}_{21} + \dot{Z}_{gen} = \dot{C}_{15} + \dot{C}_{16} + \dot{C}_{22}$	$\dfrac{\dot{C}_{14}}{\dot{Ex}_{14}} = \dfrac{\dot{C}_{15}}{\dot{Ex}_{15}}$ $\dfrac{\dot{C}_{16} - \dot{C}_{21}}{\dot{Ex}_{16} - \dot{Ex}_{21}} = \dfrac{\dot{C}_{22} - \dot{C}_{21}}{\dot{Ex}_{22} - \dot{Ex}_{21}}$
SHE	$\dot{C}_{16} + \dot{C}_{20} + \dot{Z}_{SHE} = \dot{C}_{17} + \dot{C}_{21}$	$\dfrac{\dot{C}_{16}}{\dot{Ex}_{16}} = \dfrac{\dot{C}_{17}}{\dot{Ex}_{17}}$
Absorber	$\dot{C}_{18} + \dot{C}_{25} + \dot{C}_{30} + \dot{Z}_{abs} = \dot{C}_{19} + \dot{C}_{31}$	$\dfrac{\dot{C}_{19}}{\dot{Ex}_{19}} = \dfrac{\dot{C}_{18} + \dot{C}_{25}}{\dot{Ex}_{18} + \dot{Ex}_{25}}, c_{30} = 0$
Solution pump	$\dot{C}_{19} + \dot{C}_{44} + \dot{Z}_{SP} = \dot{C}_{20}$	$\dfrac{\dot{C}_{42}}{\dot{Ex}_{42}} = \dfrac{\dot{C}_{44}}{\dot{Ex}_{44}}$
Condenser	$\dot{C}_{22} + \dot{C}_{26} + \dot{Z}_{con} = \dot{C}_{23} + \dot{C}_{27}$	$\dfrac{\dot{C}_{22}}{\dot{Ex}_{22}} = \dfrac{\dot{C}_{23}}{\dot{Ex}_{23}}, c_{26} = 0$
Evaporator	$\dot{C}_{24} + \dot{C}_{28} + \dot{Z}_{eva} = \dot{C}_{25} + \dot{C}_{29}$	$\dfrac{\dot{C}_{24}}{\dot{Ex}_{24}} = \dfrac{\dot{C}_{25}}{\dot{Ex}_{25}}, c_{28} = 0$
Vapor generator	$\dot{C}_{10} + \dot{C}_{32} + \dot{Z}_{VG} = \dot{C}_{11} + \dot{C}_{33}$	$\dfrac{\dot{C}_{10}}{\dot{Ex}_{10}} = \dfrac{\dot{C}_{11}}{\dot{Ex}_{11}}$
Vapor turbine	$\dot{C}_{33} + \dot{Z}_{VT} = \dot{C}_{34} + \dot{C}_{45}$	$\dfrac{\dot{C}_{33}}{\dot{Ex}_{33}} = \dfrac{\dot{C}_{34}}{\dot{Ex}_{34}}$
IHE	$\dot{C}_{34} + \dot{C}_{37} + \dot{Z}_{IHE} = \dot{C}_{32} + \dot{C}_{35}$	$\dfrac{\dot{C}_{34}}{\dot{Ex}_{34}} = \dfrac{\dot{C}_{35}}{\dot{Ex}_{35}}$
Vapor condenser	$\dot{C}_{35} + \dot{C}_{38} + \dot{Z}_{VC} = \dot{C}_{36} + \dot{C}_{39}$	$\dfrac{\dot{C}_{35}}{\dot{Ex}_{35}} = \dfrac{\dot{C}_{36}}{\dot{Ex}_{36}}, c_{38} = 0$
Pump 2	$\dot{C}_{36} + \dot{C}_{46} + \dot{Z}_{pu2} = \dot{C}_{37}$	$\dfrac{\dot{C}_{45}}{\dot{Ex}_{45}} = \dfrac{\dot{C}_{46}}{\dot{Ex}_{46}}$

The cost rate can be written as [26,42]:

$$\dot{C}_j = c_j \dot{Ex}_j \tag{26}$$

where c stands for the cost per unit of exergy (USD/GJ).

To translate capital investment of the kth component into a cost rate, the equation below is utilized [15]:

$$\dot{Z}_k = \frac{CRF \times \phi_r \times Z_k}{N} \tag{27}$$

where ϕ_r represents maintenance factor (1.06), N refers to the number of operating hours annually (7000), Z_k indicates the capital cost of the kth component. The capital recovery factor (CRF) is determined through the formula presented below [15,26]:

$$CRF = \frac{i_r (1 + i_r)^{n_t}}{(1 + i_r)^{n_t} - 1} \tag{28}$$

where i_r denotes the annual interest rate (15%), n_t is the lifetime of the system (20 years).

The average unit cost of the fuel ($c_{F,k}$), unit cost of the product ($c_{P,k}$), and cost of exergy destruction ($\dot{C}_{D,k}$) for the kth component are defined, respectively, in subsequent equations [42]:

$$c_{F,k} = \frac{\dot{C}_{F,k}}{\dot{Ex}_{F,k}} \tag{29}$$

$$c_{P,k} = \frac{\dot{C}_{P,k}}{\dot{Ex}_{P,k}} \tag{30}$$

$$\dot{C}_{D,k} = c_{F,k}\dot{Ex}_{D,k} \tag{31}$$

The relative cost difference (r_k) and exergoeconomic factor (f_k) for the kth component are characterized by the following definitions [42]:

$$r_k = \frac{c_{P,k} - c_{F,k}}{c_{F,k}} \tag{32}$$

$$f_k = \frac{\dot{Z}_k}{\dot{Z}_k + \dot{C}_{D,k}} \tag{33}$$

Capital costs for system components are preliminarily calculated using cost functions, which are tabulated in Table 6. These costs, based on reference year values (Z_{ref}), require adjustment to current values (Z_{PY}) employing the chemical engineering plant cost index (CEPCI) [15]:

$$Z_{\mathrm{PY}} = Z_{\mathrm{ref}} \times \frac{CEPCI_{\mathrm{PY}}}{CEPCI_{\mathrm{ref}}} \tag{34}$$

Table 6. Cost balance and auxiliary equations for the system components [42–44].

Component	Cost Balance Equation
Air compressor	$Z_{\mathrm{AC}} = 71.1\dot{m}_1/(0.9 - \eta_{\mathrm{is,AC}})\ln(P_2/P_1)$
Air preheater	$Z_{\mathrm{AP}} = 4122\left(\dot{m}_8(h_8 - h_9)/U_{\mathrm{AP}}/\Delta T_{lm,\mathrm{AP}}\right)^{0.6}$
Gas turbine	$Z_{\mathrm{GT}} = 479.34\dot{m}_3/(0.92 - \eta_{\mathrm{is,GT}})\ln(P_3/P_4)[1 + \exp(0.036\,T_3 - 54.4)]$
Combustion chamber	$Z_{\mathrm{CC}} = 46.08\dot{m}_4/(0.995 - P_8/P_4)[1 + \exp(0.018\,T_8 - 26.4)]$
Biomass gasifier	$Z_{\mathrm{Ga}} = 1600\left(\dot{m}_{\mathrm{dry\ biomass}}\,[\mathrm{kg}/\mathrm{h}]\right)^{0.67}$
HRSG	$Z_{\mathrm{HRSG}} = 6570\sum_i\left(\dot{Q}_i/\Delta T_{lm,i}\right)^{0.8} + 21{,}276\dot{m}_{12} + 1184.4\dot{m}_9^{1.2}$
Steam turbine	$Z_{\mathrm{ST}} = 6000\left(\dot{W}_{\mathrm{ST}}\right)^{0.7}$
Pump 1	$Z_{\mathrm{pu1}} = 3540\left(\dot{W}_{\mathrm{pu1}}\right)^{0.71}$
Generator	$Z_{\mathrm{gen}} = 17{,}500(A_{\mathrm{gen}}/100)^{0.6}$
SHE	$Z_{\mathrm{SHE}} = 12{,}000(A_{\mathrm{SHE}}/100)^{0.6}$
Absorber	$Z_{\mathrm{abs}} = 16{,}500(A_{\mathrm{abs}}/100)^{0.6}$
Condenser	$Z_{\mathrm{con}} = 8000(A_{\mathrm{con}}/100)^{0.6}$
Evaporator	$Z_{\mathrm{eva}} = 16{,}000(A_{\mathrm{eva}}/100)^{0.6}$
Solution pump	$Z_{\mathrm{SP}} = 2100\left(\dot{W}_{\mathrm{SP}}/10\right)^{0.26}((1 - \eta_{\mathrm{is,SP}})/\eta_{\mathrm{is,SP}})^{0.5}$
Vapor generator	$Z_{\mathrm{VG}} = 130(A_{\mathrm{VG}}/0.093)^{0.78}$
Vapor turbine	$Z_{\mathrm{VT}} = 6000\left(\dot{W}_{\mathrm{VT}}\right)^{0.7}$
IHE	$Z_{\mathrm{IHE}} = 1.3(190 + 310A_{\mathrm{IHE}})$
Vapor condenser	$Z_{\mathrm{VC}} = 1773\dot{m}_{35}$
Pump 2	$Z_{\mathrm{pu2}} = 3540\left(\dot{W}_{\mathrm{pu2}}\right)^{0.71}$

Capital cost estimations for heat exchangers necessitate initial determination of heat transfer areas (A_k), calculated by the following equation [15,26]:

$$A_k = \frac{\dot{Q}_k}{U_k\Delta T_{lm,k}} \tag{35}$$

where $\Delta T_{lm,k}$ denotes the logarithmic mean temperature difference, U_k represents the heat transfer coefficient. The determination of heat transfer coefficients for heat exchangers is elaborated in Appendix B.

3.5. Overall Performance Assessment

The thermal efficiency of the proposed system is calculated as:

$$\eta_{\text{th}} = \frac{\dot{W}_{\text{GT}} + \dot{W}_{\text{ST}} + \dot{W}_{\text{VT}} - \dot{W}_{\text{AC}} - \dot{W}_{\text{pu1}} - \dot{W}_{\text{pu2}} + \dot{Q}_{\text{eva}}}{\dot{m}_{\text{biomass}} \text{LHV}_{\text{biomass}}} \tag{36}$$

The exergy efficiency of the proposed system is computed by:

$$\eta_{\text{ex}} = \frac{\dot{W}_{\text{GT}} + \dot{W}_{\text{ST}} + \dot{W}_{\text{VT}} - \dot{W}_{\text{AC}} - \dot{W}_{\text{pu1}} - \dot{W}_{\text{pu2}} + \left(\dot{Ex}_{29} - \dot{Ex}_{28} \right)}{\dot{Ex}_1 + \dot{Ex}_6 + \dot{Ex}_7} \tag{37}$$

The levelized cost of exergy (LCOE) is adopted as the criterion for evaluating the exergoeconomic performance of the system, formally defined as:

$$LCOE_{\text{sys}} = \frac{\dot{C}_{41} + \dot{C}_{42} + \dot{C}_{45} - \dot{C}_{43} - \dot{C}_{44} - \dot{C}_{46} + \dot{C}_{29}}{\dot{W}_{\text{GT}} + \dot{W}_{\text{ST}} + \dot{W}_{\text{VT}} - \dot{W}_{\text{AC}} - \dot{W}_{\text{pu1}} - \dot{W}_{\text{pu2}} + \dot{Ex}_{29}} \tag{38}$$

3.6. Multi-Objective Optimization

Optimization plays a critical role in enhancing the performance of energy system designs, particularly in thermal systems where design objectives often present conflicting requirements, making it challenging to achieve an optimal solution that meets all criteria simultaneously. To navigate these complexities, multi-objective optimization techniques are frequently employed. This approach involves defining objective functions, decision variables, and their respective boundaries, which can be described as follows [45]:

$$\min F(X) = [f_1(X), f_2(X), \dots, f_k(X)]^T \tag{39}$$

subject to

$$g_i(X) \leq 0, \, i = 1, \dots, m \tag{40}$$

$$h_j(X) = 0, \, j = 1, \dots, n \tag{41}$$

$$X_{k,min} \leq X_k \leq X_{k,max} \tag{42}$$

where X, $F(X)$, and $f(X)$ indicate the vectors of decision variables, multi-objective function, and single-objective function, respectively; $g_i(X)$ and $h_i(X)$ represent the inequality and equality constraints, respectively; $X_{k,min}$ and $X_{k,max}$ stand for the bottom and top bounds of the kth decision variables, respectively.

In the current research, the genetic algorithm (GA) is utilized to address the multi-objective optimization issue. This technique begins by creating an initial population of solution candidates, which undergoes evolution through random selection from the existing population. This population is then evolved using a series of operations including selection, mutation, crossover, and inheritance. Over successive generations, the most favorable solutions emerge and are compiled into a Pareto frontier, with each point on this frontier representing a viable optimal solution [46]. The ultimate solution is identified using TOPSIS (Technique for Order of Preference by Similarity to Ideal Solution) decision making [47]. The TOPSIS method introduces two hypothetical solutions: the "ideal point", which signifies the optimal values for each objective, and the "non-ideal point", representing the worst values. The solution that lies nearest to the ideal point and furthest from the non-ideal point is adjudged the ultimate optimal solution. The methodology for constructing the decision matrix and computing the distance of each solution to the ideal and non-ideal points is detailed as follows [23,26]:

$$F_{ij} = \frac{x_{ij}}{\sqrt{\sum_{i=1}^{m} x_{ij}^2}} \tag{43}$$

$$D_{i+} = \sqrt{\sum_{j=1}^{n} \left(F_{ij} - F_{ij}^{\text{ideal}} \right)^2} \tag{44}$$

$$D_{i-} = \sqrt{\sum_{j=1}^{n} \left(F_{ij} - F_{ij}^{\text{non-ideal}} \right)^2} \tag{45}$$

The relative closeness is defined as:

$$Cl_i = \frac{D_{i-}}{D_{i+} + D_{i-}} \tag{46}$$

Finally, the solution with maximum Cl_i is considered as the desired final solution.

4. Results and Discussion

4.1. Model Validation

To validate the mathematical models applied to the proposed system, this study compares simulation outcomes for various components, including the EFGT, biomass gasifier, ORC, and ARC, against findings reported in prior research. Computational models are constructed utilizing MATLAB R2018b software for simulation purposes, and the thermophysical properties of the working fluids are sourced from the REFPROP 9.0 database. The comparison, detailed in Tables 7–10, reveals a satisfactory concordance between the results of this study and those documented in existing literature.

Table 7. Results comparison between the present work and Ref. [48] for the EFGT cycle.

State	Substance	P (kPa)		T (K)		$\dot{m}$ (kg/s)	
		Ref. [48]	Present Work	Ref. [48]	Present Work	Ref. [48]	Present Work
1	Air	101.3	101.3	298.15	298.15	9.45	9.84
2	Air	911.7	911.7	589.9	583.84	9.45	9.84
3	Air	884.35	884.35	1400	1400	9.45	9.84
4	Air	103.83	103.88	877.6	886.18	9.45	9.84
5	Syngas	101.3	101.3	1073.15	1073.15	2.789	2.792
8	Comb. gas	102.82	102.84	1562	1578.6	12.24	12.63
9	Comb. gas	101.3	101.3	1000	1000	12.24	12.63

Table 8. Comparison of the component percentages of the syngas calculated by the present work with those reported in the literature (wood: $CH_{1.44}O_{0.66}$, MC = 20%, T_{Ga} = 1073.15 K).

Constituent	Roy et al. [19]	Cao et al. [49]	Present Work
H_2 (%)	21.63	21.66	21.50
CO (%)	20.25	20.25	20.21
CH_4 (%)	0.98	1.011	0.95
CO_2 (%)	12.48	12.36	12.50
N_2 (%)	44.94	44.72	44.84

Table 9. Results comparison between the present work and Ref. [50] for the ORC cycle with IHE using R601 as working fluid.

Parameter	T_{eva} (K)	T_{con} (K)	P_{eva} (kPa)	P_{con} (kPa)	$\dot{m}$ (kg/s)	η_{th} (%)
Ref. [50]	373.15	303.15	5.963	0.828	16.331	13.84
This work	373.15	303.15	5.927	0.820	16.382	13.84

Table 10. Results comparison between the present work and Ref. [32] for the single-effect LiBr-H_2O ARC at same operating conditions ($\dot{Q}_{eva}$ = 3.51 kW, T_{gen} = 363.15 K, T_{eva} = 280.15 K, T_{abs} = 313.15 K, T_{con} = 313.15 K, ε_{SHE} = 0.8).

Parameter	Ref. [32]	This Work
Heat capacity of generator (kW)	4.5999	4.6000
Heat capacity of condenser (kW)	3.7432	3.7420
Heat capacity of absorber (kW)	4.368	4.368
Evaporator pressure (kPa)	1.0021	1.0021
Condenser pressure (kPa)	7.3844	7.3849
Weak solution concentration (%)	62.33	62.15
Strong solution concentration (%)	56.72	56.66
Refrigerant mass flow rate (kg/s)	0.0015	0.0015
Weak solution mass flow rate (kg/s)	0.0151	0.0154
Strong solution mass flow rate (kg/s)	0.0166	0.0169
Coefficient of performance	0.763	0.763

4.2. Base Case Results

Table A1 outlines the base case input parameters for these subsystems, which enable the derivation of simulation outcomes by solving the equations previously described. The characteristics of each fluid stream, encompassing both thermodynamic and economic aspects, are summarized in Table A2. Table A3 details the distribution of exergy and exergoeconomic parameters across the system's components. Notably, the steam turbine exhibits the highest capital cost rate, succeeded by the air preheater and air compressor. In terms of exergy efficiency, the gas turbine, air preheater, and air compressor exhibit superior performance, whereas the absorber in ARC displays the lowest efficiency.

Figures 2 and 3 depict the Sankey diagrams for the exergy and cost rate flows of the proposed system, respectively. As shown in Figure 2, the input exergy rate from wood biomass fuel surpasses other sources, amounting to 34,044.75 kW. The combustion chamber is identified as the main contributor to the system's total exergy destruction. In Figure 3, the air outlet of the air preheater exhibits the highest cost rate within the system at 778.75 USD/h, succeeded by the flue gas outlet of the combustion chamber at 594.83 USD/h.

Thermodynamic and exergoeconomic analyses of the system, as summarized in Table 11, demonstrate the system's capability to generate a net power output of 12,950.2 kW and a cooling capacity of 7738.4 kW. Additionally, the system achieves total energy and exergy efficiencies of 70.67% and 39.13%, respectively. The analysis also shows a disparity in the unit cost of power production, with the ORC turbine at 31.50 USD/GJ, significantly higher than both the SRC turbine at 15.60 USD/GJ and the gas turbine at 8.60 USD/GJ. Given that the gas turbine and SRC turbine contribute significantly more power than the ORC turbine and that the unit cost of cooling production is lower than that of power generation, the LCOE of the overall system is determined to be 11.67 USD/GJ, reflecting a balanced cost-efficiency ratio.

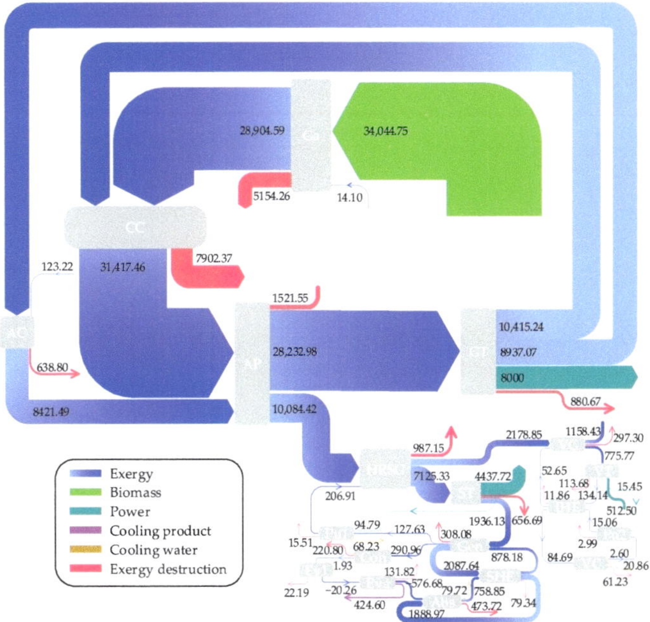

Figure 2. Exergy flow diagram of the system.

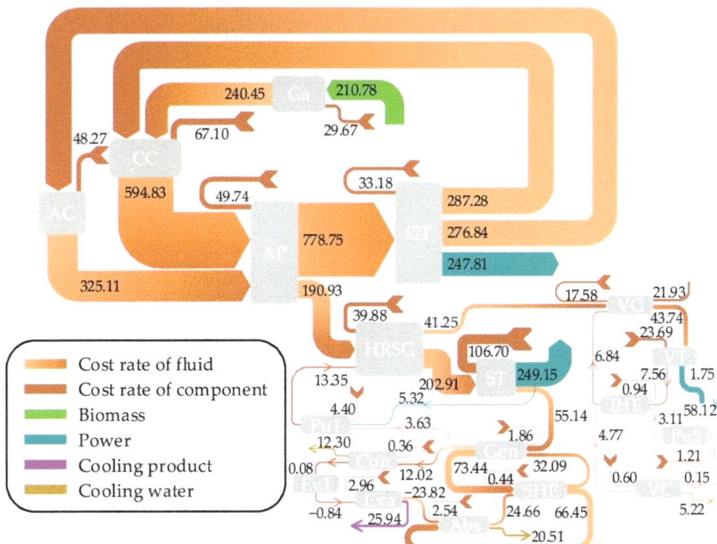

Figure 3. Cost rate flow diagram of the system.

Table 11. Thermodynamic and exergoeconomic evaluation results of the base case.

Performance Parameters	Unit	Value
SRC turbine work $\left(\dot{W}_{ST}\right)$	kW	4532.51
SRC pump consumed power $\left(\dot{W}_{Pu1}\right)$	kW	94.79
ORC turbine work $\left(\dot{W}_{VT}\right)$	kW	527.95
ORC pump consumed power $\left(\dot{W}_{VP}\right)$	kW	15.45
Net power output $\left(\dot{W}_{net}\right)$	kW	12,950.2
Cooling output $\left(\dot{Q}_{eva}\right)$	kW	7738.4
Thermal efficiency (η_{th})	%	70.67
Exergy efficiency (η_{ex})	%	39.13
Unit cost of the GT − produced power (c_{GT})	USD/GJ	8.60
Unit cost of the SRC − produced power (c_{SRC})	USD/GJ	15.60
Unit cost of the ORC − produced power (c_{ORC})	USD/GJ	31.50
Unit cost of exergy production for cooling (c_{eva})	USD/GJ	8.23
LCOE of the system $(LCOE_{sys})$	USD/GJ	11.67

4.3. Parametric Study

4.3.1. Effect of Air Compressor Pressure Ratio on the System Performance

Figure 4 illustrates the influence of air compressor pressure ratio (PR_{AC}) on the system performance. According to the figure, the thermal efficiency rises considerably as the PR_{AC} augments, while the exergy efficiency increases gently and reaches a peak value of 39.1%. The LCOE of the system attains its lowest at 11.64 USD/GJ for a PR_{AC} value around 9, beyond which it begins to ascend. Additionally, both the net power and cooling capacity present upward trends as the PR_{AC} rises. This trend is attributed to the augmented thermal energy available to the subsequent cycles, driven by the elevation in flue gas temperature at the air preheater exit under a constant CETD. Despite a slight rise in biomass consumption, the total energy output's augmentation surpasses the increase in biomass input, thus elevating thermal efficiency.

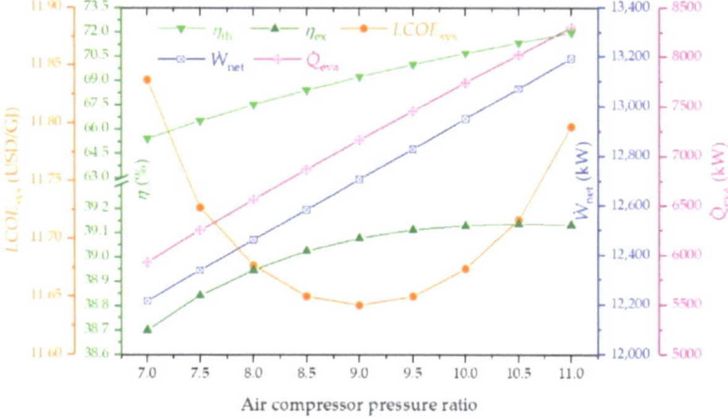

Figure 4. Effect of air compressor pressure ratio on the thermal efficiency, exergy efficiency, LCOE, net power output, and cooling output of the system.

4.3.2. Effect of Gas Turbine Inlet Temperature on the System Performance

Figure 5 examines the impact of gas turbine inlet temperature (GTIT) on system performance. This figure reveals that both thermal and exergy efficiencies improve with an ascending GTIT, whereas net power and cooling capacities experience a marked decrease. The LCOE of the system attains its lowest point at a GTIT of approximately 1400 K. The rationale behind these observations lies in the augmented enthalpy difference across the gas turbine as GTIT increases, which in turn significantly reduces the mass flow rates of air and flue gas to maintain a constant power output of EFGT. Consequently, the supply of thermal heat to the subsequent cycles diminishes, leading to the decline of net power and cooling outputs. However, the decrease in biomass consumption, coupled with unchanged efficiencies of the bottoming cycles, contributes to overall increases in both thermal and exergy efficiencies. Nonetheless, the increment in thermal and exergy efficiencies of the EFGT cycle, due to reduced biomass fuel consumption, contributes to an overall increase in the system's efficiencies, as the efficiencies of bottoming cycles remain unchanged.

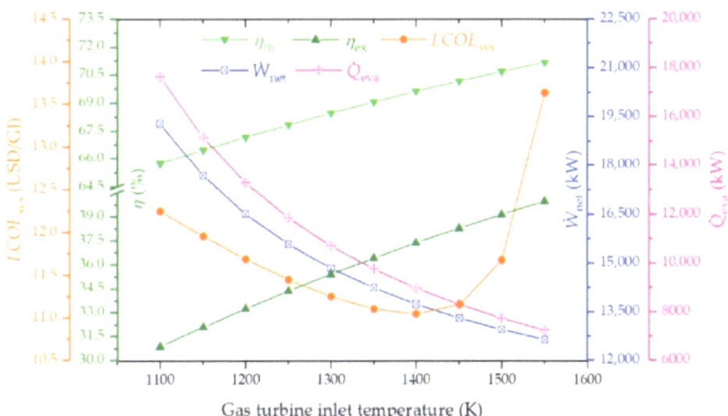

Figure 5. Effect of gas turbine inlet temperature on the thermal efficiency, exergy efficiency, LCOE, net power output, and cooling output of the proposed system.

4.3.3. Effect of Pinch Point Temperature Difference in HRSG on the System Performance

Figure 6 displays the impact of pinch point temperature difference in HRSG on the system performance. It is observed that elevating the pinch point temperature difference leads to declines in both thermal and exergy efficiencies, alongside reductions in net power and cooling capacity. Conversely, the LCOE initially decreases, reaching a minimum, before it starts to ascend. This trend can be attributed to the widened temperature gap between the high-temperature flue gases and the working fluid in the HRSG, which amplifies exergy destruction and diminishes the thermal energy supplied to the SRC. Consequently, this reduction in heat transfer causes net power and cooling outputs to decline, adversely affecting both thermal and exergy efficiencies.

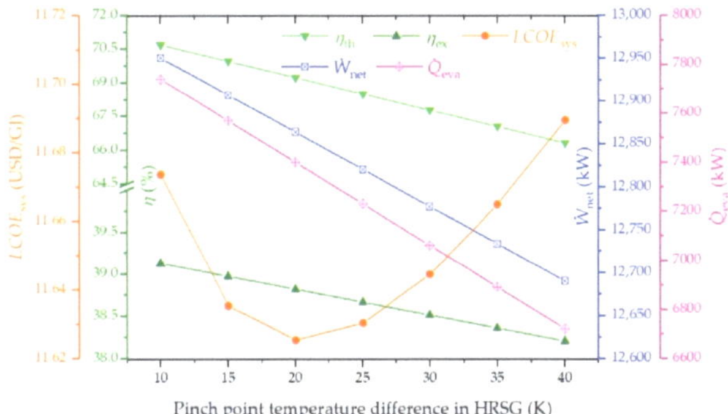

Figure 6. Effect of pinch point temperature difference in HRSG on the thermal efficiency, exergy efficiency, LCOE, net power output, and cooling output of the proposed system.

4.3.4. Effect of Steam Turbine Inlet Pressure on the System Performance

The impact of the steam turbine inlet pressure (STIP) on system performance is illustrated in Figure 7. It is observed that exergy efficiency enhances with a rise in STIP, whereas thermal efficiency exhibits a gradual decline. The LCOE demonstrates a decrease to a minimum point, subsequently increasing. Additionally, an increase in STIP leads to a boost in net power output due to the enhanced efficiency of the SRC. Conversely, cooling capacity experiences a downturn, attributed to a decreased availability of condensation heat for the ARC. This results in an improvement in exergy efficiency, as the generation of electricity, which is of a higher quality, outweighs the cooling production. The overall thermal efficiency of the system is thus a function of the combined outputs of power and cooling capacity.

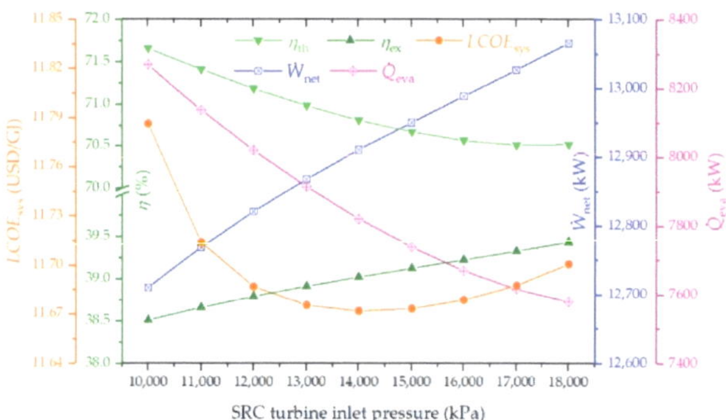

Figure 7. Effect of steam turbine inlet pressure on the thermal efficiency, exergy efficiency, LCOE, net power output, and cooling output of the proposed system.

4.3.5. Effect of SRC Condenser Temperature on the System Performance

Figure 8 depicts the relationship between the SRC condenser temperature and its impact on system metrics. An inverse relationship is noted between the SRC condenser temperature and both the LCOE and exergy efficiency, while thermal efficiency initially rises before showing a decline. This behavior is attributable to several factors. An elevation in the SRC condenser temperature leads to a reduction in net power due to a diminished SRC efficiency. Concurrently, cooling capacity experiences a boost owing to the enhanced COP and increased thermal energy supply to the ARC. Therefore, the variation in thermal efficiency is influenced by the cumulative effect on output power and cooling capacity, whereas the exergy efficiency witnesses a downturn primarily because the reduction in net power has a more pronounced impact than the increase in cooling capacity.

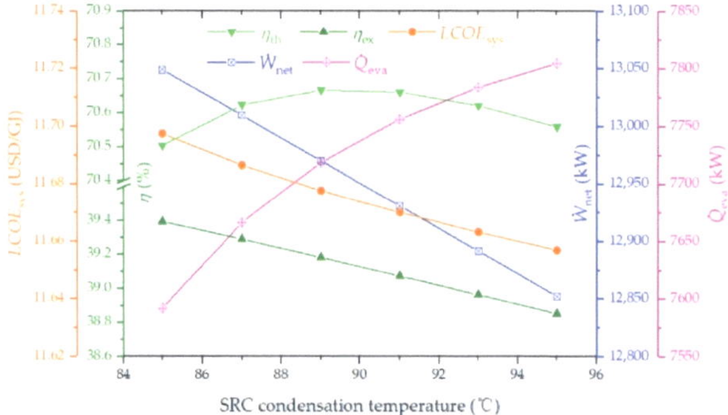

Figure 8. Effect of SRC condenser temperature on the thermal efficiency, exergy efficiency, LCOE, net power output, and cooling output of the proposed system.

4.3.6. Effect of ORC Turbine Inlet Pressure on the System Performance

The influence of ORC turbine inlet pressure on system performance is sketched in Figure 9. This figure reveals that thermal and exergy efficiencies, LCOE, and output power all experience marginal improvements with the elevation of ORC turbine inlet pressure. Notably, changes in ORC turbine inlet pressure do not affect the performance of the topping cycles. The ORC efficiency improves with higher turbine inlet pressure, facilitating an increase in generated power. Nevertheless, given that the contribution of power from the ORC to the overall system is relatively modest, its impact on the overall system performance is minimal.

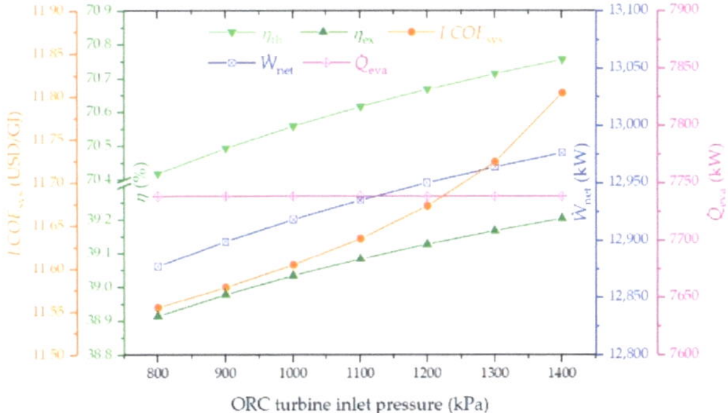

Figure 9. Effect of ORC turbine inlet pressure on the thermal efficiency, exergy efficiency, LCOE, net power output, and cooling output of the proposed system.

4.4. Optimization Results

In examining the system performance from both thermodynamic and economic perspectives, the study adopts exergy efficiency and LCOE as its main performance indicators. Through detailed parametric scrutiny, essential operational parameters are delineated as decision-making variables, with their respective ranges provided in Table 12. Utilizing MATLAB R2018b software, a specialized algorithm is created to implement the GA method aimed at optimizing the two objectives. The resulting optimal solutions are depicted as a scattered set across the Pareto frontier in Figure 10, where each marker denotes a potentially optimal configuration, revealing the inherent trade-off between the objectives. Optimal thermodynamic efficiency is achieved at point A, characterized by an exergy efficiency peak of 39.40%, whereas the most favorable economic outcome is observed at point B, showcasing the lowest LCOE at 10.59 USD/GJ. Given this context, the TOPSIS method is employed to determine the ultimate optimal point on the Pareto front, which is identified at point C, balancing an exergy efficiency of 38.15% with an LCOE of 11.01 USD/GJ. The objective function values and the decision variables for points A, B, and C on the Pareto frontier are detailed in Table 13.

Table 12. Selected decision variables of the proposed system and their limits.

Parameter	Unit	Range
PR_{AC}	-	$6 \leq PR_{AC} \leq 16$
T_3	K	$1100 \leq T_3 \leq 1500$
CETD	K	$200 \leq CETD \leq 300$
$\Delta T_{PP,\,HRSG}$	K	$10 \leq \Delta T_{PP,\,HRSG} \leq 50$
P_{13}	kPa	$10{,}000 \leq P_{13} \leq 18{,}000$
T_{15}	K	$358.15 \leq T_{15} \leq 368.15$
P_{33}	kPa	$400 \leq P_{33} \leq 2000$

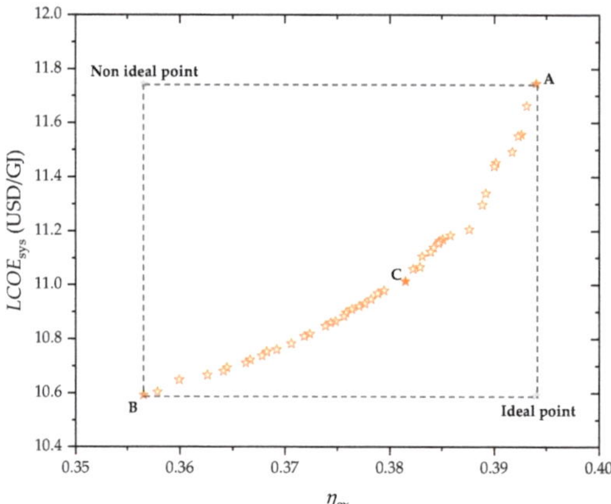

Figure 10. Distribution of the optimal points on the Pareto frontier.

Table 13. The values of decision variables and objective functions at points A, B, and C.

Parameter	A	B	C
PR_{AC}	11.03	7.86	10.62
T_3 (K)	1479.2	1374.1	1450.2
CETD (K)	217.7	279.5	256.2
$\Delta T_{PP,\,HRSG}$ (K)	19.97	11.71	14.44
P_{13} (kPa)	16,509.9	10,257.5	16,642.3
T_{15} (K)	362.2	361.0	359.7
P_{33} (kPa)	1811.8	459.4	567.4
$\dot{W}_{net}$ (kW)	12,821.4	13,582.6	13,660.5
$\dot{Q}_{eva}$ (kW)	6863.4	9807.7	8771.8
η_{ex} (%)	39.40	35.66	38.15
$LCOE_{sys}$ (USD/GJ)	11.74	10.59	11.01

4.5. Comprative Study

As a final step in presenting the results, a comparison with previously published data is conducted. Under identical operating conditions of the EFGT cycle ($\eta_{is,AC}= 0.87$, $\eta_{is,GT}= 0.89$), the energy and exergy efficiencies as well as cost of products are compared in Table 14. According to Table 14, the designed plant exhibits moderate thermodynamic performance and slightly inferior economic characteristics when assessed against various other systems. When contrasted with the findings from Ref. [22], the current system demonstrates superior energy and exergy efficiencies under the base case conditions. However, in comparison to Ref. [15], it records slightly lower efficiencies under the same simulation conditions. Additionally, the cost of products is marginally higher than those in Refs. [15,22], which is primarily due to the structural design of the system and the methodology used for cost calculation. The systems referenced in Refs. [15,22] are configured to generate both power and heating load, whereas the present study incorporates an ARC to provide a cooling load. Moreover, in Refs. [15,22], municipal solid waste is employed as biomass fuel, in contrast to the wood used in this study, leading to different conversion efficiencies and syngas compositions during the biomass gasification process. Relative to Ref. [17], the system in this study exhibits significantly higher thermal efficiency. This improvement is attributed to the use of additional bottoming cycles for recovering heat from combustion gases generated by biomass-based fuels.

Table 14. Comparison of thermodynamic efficiency and economic performance in current and previous investigations.

Parameter	Ref. [22]	This Work	Ref. [15]	This Work	Ref. [17]	This Work
PR_{AC}	10		10		7	
GTIT (K)	1573.15		1300		1300	
Energy efficiency (%)	67.26%	71.72%	75.8%	68.93%	41.18%	64.62%
Exergy efficiency (%)	41.08%	41.31%	41.21%	36.58%	-	37.97%
Cost of products (USD/GJ)	17.17	19.32	10.2	11.74	-	-

5. Conclusions

This research introduces an innovative combined cooling and power system, integrating an EFGT, an SRC, an ORC, and an ARC to enhance biomass energy utilization, and its performance is evaluated from the thermodynamic and exergoeconomic perspectives. A thorough parametric study is performed to ascertain the impact of various design parameters on system performance, while multi-objective optimization focuses on maximizing exergy efficiency and minimizing the LCOE. The main conclusions can be drawn as follows:

- For the baseline scenario, the system exhibits a thermal efficiency of 70.67%, an exergy efficiency of 39.13%, and an LCOE of 11.67 USD/GJ, alongside generating a net power of 12,950.2 kW and a cooling output of 7738.4 kW.
- Exergy analysis revealed that the highest rate of exergy destruction occurs in the combustion chamber, followed closely by the biomass gasifier. The gas turbine and the absorber demonstrated the best and poorest performances from exergy viewpoint among the system components, respectively.
- The inlet temperature of the gas turbine emerged as a critical factor affecting the system performance. Elevating GTIT significantly boosts both thermal and exergy efficiencies, despite a notable reduction in net power and cooling outputs.
- Superior thermodynamic performance is achieved at a higher air compressor pressure ratio and a gas turbine inlet temperature, or at a lower pinch point temperature difference in the HRSG. Optimizing these parameters also leads to minimized LCOE.
- Under optimal conditions, the CCP system demonstrates a 5.7% reduction in LCOE and a 2.5% decrease in exergy efficiency compared to the baseline scenario, highlighting a trade-off between different optimization criteria. This balance suggests that the optimal solution varies depending on specific engineering applications' requirements.

Future research could concentrate on the enhancement of integrated energy systems by incorporating additional energy sources or subsystems to expand product diversity and improve system functionality. Efforts should be directed toward minimizing exergy destruction rates and maximizing energy utilization to enhance system efficiency, alongside environmental assessment to evaluate operational sustainability. Subsequent studies should also include comparative analyses of diverse biomass feedstocks in gasifiers, exploration of alternative ORC working fluids, and investigation of advanced power generation technologies such as transcritical ORCs and supercritical CO_2 Brayton cycles to enhance system performance. Practical feasibility assessments and experimental validations with real-world devices are also imperative for advancing the application of developed systems.

Author Contributions: Conceptualization, J.R. and C.X.; methodology, J.R. and C.X.; software, J.R. and C.X.; validation, J.R.; formal analysis, J.R., B.W. and W.H.; investigation, J.R., B.W. and W.H.; writing—original draft preparation, J.R. and W.H.; writing—review and editing, C.X. and Z.Q.; project administration, Z.Q.; funding acquisition, Z.Q. All authors have read and agreed to the published version of the manuscript.

Funding: This research was funded by the Hubei Province Technological Innovation Special Fund, grant number 2019BKJ002.

Institutional Review Board Statement: Not applicable.

Data Availability Statement: Data are contained within the article.

Conflicts of Interest: The authors declare no conflicts of interest.

Nomenclature

A	area (m^2)		COP	coefficient of performance
c	cost per exergy unit (USD·GJ^{-1})		CRF	capital recovery factor
$\dot{C}$	cost rate (USD·h^{-1})		EFGT	externally fired gas turbine
ex	exergy per unit mass (kW·kg^{-1})		EV	expansion valve
$\dot{Ex}$	exergy rate (kW)		eva	evaporator
f	exergoeconomic factor		GA	genetic algorithm
h	specific enthalpy (kJ·kg^{-1})		Ga	gasifier
i_r	annual interest rate (%)		gen	generator
K	equilibrium constant		GT	gas turbine
$\dot{m}$	mass flow rate (kg·s^{-1})		GTIT	gas turbine inlet temperature
n	kilomoles of component (kmol)		HRSG	heat recovery steam generator
N	annual operating hours (h)		is	isentropic
n_t	lifetime of the system		IHE	internal heat exchanger
P	pressure (kPa)		LCOE	levelized cost of exergy
$\dot{Q}$	heat transfer rate (kW)		LHV	lower heating value
r	relative cost difference		MW	molecular weight
s	specific entropy (kJ·kg^{-1}·K^{-1})		ORC	organic Rankine cycle
T	temperature (K)		PR	pressure ratio
U	heat transfer coefficient (W·m^{-2}·K^{-1})		pu	pump
$\dot{W}$	power (kW)		SHE	solution heat exchanger
y_D	exergy destruction ratio (%)		SP	solution pump
Z	investment cost (USD)		SRC	steam Rankine cycle
$\dot{Z}$	investment cost rate (USD/h)		ST	steam turbine
			STIP	steam turbine inlet pressure
Subscript and abbreviations			VC	vapor condenser
0	dead state		VG	vapor generator
1,2,...	state points		VT	vapor turbine
abs	absorber			
AC	air compressor		**Greek Symbols**	
AP	air preheater		Δ	difference
ARC	absorption refrigeration cycle		η	efficiency
CC	combustion chamber		ε	heat exchanger effectiveness
CETD	cold end temperature difference		ϕ_r	maintenance factor
con	condenser		ψ	chemical exergy coefficient

Appendix A

For the proposed system, input parameters for the subsystems under basic operating conditions are provided in Table A1. Table A2 details the thermodynamic and exergoeconomic properties of key state points, including temperature, pressure, mass flow rate, specific enthalpy, specific entropy, exergy, cost rate, and cost per exergy unit. Utilizing these parameters, the exergy and exergoeconomic indicators for the system components are calculated and presented in Table A3.

Table A1. Input parameters for the proposed system.

Parameter	Value	Unit
Reference temperature (T_0)	298.15	K
Reference pressure (P_0)	101.3	kPa
EFGT [12,51]		
Air compressor isentropic efficiency ($\eta_{is,AC}$)	86	%
Air compressor pressure ratio (PR_{AC})	10	-
Gas turbine isentropic efficiency ($\eta_{is,GT}$)	86	%
Gas turbine inlet temperature (T_3)	1500	K
Cold end temperature difference (CETD)	245	K
Pressure drop of the cold side in the AP	5	%
Pressure drop of the hot side in the AP	3	%
Pressure drop of the flue gas in the CC	1	%
Pressure drop of the flue gas in the HRSG	5	%
Pressure drop of the flue gas in the VG	5	%
Net power output of the EFGT ($\dot{W}_{EFGT}$)	8000	kW
SRC [52,53]		
Turbine inlet pressure (P_{13})	15,000	kPa
Pinch point temperature difference of HRSG ($\Delta T_{PP,HRSG}$)	30	K
Condenser temperature (T_{15})	363.15	K
Steam quality at outlet of the ST	0.9	-
Vapor turbine isentropic efficiency ($\eta_{is,ST}$)	85	%
Pump isentropic efficiency ($\eta_{is,pu1}$)	80	%
ARC [32]		
Generator temperature (T_{16})	358.15	K
Absorber temperature (T_{19})	308.15	K
Condenser temperature (T_{23})	308.15	K
Evaporator temperature (T_{25})	278.15	K
Effectiveness of solution heat exchanger	70	%
Cooling water inlet/outlet temperature in condenser (T_{26}/T_{27})	298.15/303.15	K
Cooling water inlet/outlet temperature in evaporator (T_{28}/T_{29})	285.15/280.15	K
Cooling water inlet/outlet temperature in absorber (T_{30}/T_{31})	298.15/303.15	K
ORC [45]		
Turbine inlet pressure (P_{33})	1200	kPa
Condenser temperature (T_{36})	308.15	K
Effectiveness of IHE (ε_{SHE})	90	%
Turbine isentropic efficiency ($\eta_{is,VT}$)	80	%
Pump isentropic efficiency ($\eta_{is,pu2}$)	80	%
Cooling water inlet/outlet temperature in condenser (T_{38}/T_{39})	298.15/303.15	K

Table A2. Thermodynamic properties and costs of the streams for the proposed system.

State	Fluid	T (K)	P (kPa)	$\dot{m}$ (kg/s)	h (kJ/kg)	s (kJ·kg^{-1}·K^{-1})	$\dot{Ex}$ (kW)	$\dot{C}$ (USD/h)	c (USD/GJ)
1	Air	298.15	101.3	27.67	0	6.888	123.22	0	0
2	Air	605.05	1013.0	27.67	323.05	6.966	8421.49	325.11	10.72
3	Air	1500	962.35	27.67	1352.89	8.018	28,232.98	778.75	7.66
4	Air	978.07	116.88	27.67	740.68	8.125	10,415.24	287.28	7.66
5	Syngas	1073.15	101.3	5.12	−2710.35	10.10	28,904.59	240.45	2.31
6	Air	298.15	101.3	3.17	0	6.888	14.10	0	0
7	Biomass	298.15	101.3	1.95	−7104.72	-	34,044.75	210.78	1.72
8	Comb. gas	1557.97	115.72	32.78	201.88	8.840	31,417.46	594.83	5.26
9	Comb. gas	850.05	112.24	32.78	−667.18	8.108	10,084.42	190.93	5.26
10	Comb. gas	463.28	106.63	32.78	−1110.75	7.429	2178.85	41.25	5.26
11	Comb. gas	378.15	101.3	32.78	−1203.13	7.224	1158.43	21.93	5.26
12	Water	364.99	15,000	4.92	396.31	1.203	206.91	13.35	17.93
13	Water	787.84	15,000	4.92	3352.79	6.402	7125.33	202.91	7.91
14	Water	363.15	70.18	4.92	2431.28	6.850	1936.13	55.14	7.91
15	Water	363.15	70.18	4.92	377.04	1.193	127.63	3.63	7.91
16	LiBr/H$_2$O	358.15	5.63	24.93	217.14	0.463	2087.64	73.44	9.77

Table A2. *Cont.*

State	Fluid	T (K)	P (kPa)	$\dot{m}$ (kg/s)	h (kJ/kg)	s (kJ·kg^{-1}·K^{-1})	$\dot{E}x$ (kW)	$\dot{C}$ (USD/h)	c (USD/GJ)
17	LiBr/H$_2$O	323.15	5.63	24.93	152.58	0.273	1888.97	66.45	9.77
18	LiBr/H$_2$O	323.15	0.87	24.93	152.58	0.273	1888.97	66.45	9.77
19	LiBr/H$_2$O	308.15	0.87	28.21	85.37	0.211	758.85	24.65	9.02
20	LiBr/H$_2$O	308.15	5.63	28.21	85.37	0.211	758.85	24.66	9.03
21	LiBr/H$_2$O	336.17	5.63	28.21	142.44	0.389	878.18	32.09	10.15
22	Water	358.15	5.63	3.27	2659.54	8.637	290.96	12.02	11.48
23	Water	308.15	5.63	3.27	146.63	0.505	1.93	0.08	11.48
24	Water	278.15	0.87	3.27	146.63	0.528	−20.26	−0.84	11.48
25	Water	278.15	0.87	3.27	2510.06	9.025	−576.68	−23.82	11.48
26	Water	298.15	101.3	393.63	104.92	0.367	0	0	0
27	Water	303.15	101.3	393.63	125.82	0.437	68.23	12.30	50.09
28	Water	285.15	101.3	368.84	50.51	0.181	450.89	0	0
29	Water	280.15	101.3	368.84	29.53	0.106	875.49	25.94	8.23
30	Water	298.15	101.3	459.98	104.92	0.367	0	0	0
31	Water	303.15	101.3	459.98	125.82	0.437	79.72	20.51	71.49
32	R601	337.80	1200	6.86	70.48	0.212	52.65	6.84	36.07
33	R601	407.57	1200	6.86	512.19	1.340	775.77	43.74	15.66
34	R601	351.24	97.70	6.86	435.19	1.396	134.14	7.56	15.66
35	R601	312.98	97.70	6.86	364.45	1.183	84.69	4.77	15.66
36	R601	308.15	97.70	6.86	−2.52	−0.008	2.60	0.15	15.66
37	R601	308.73	1200	6.86	−0.26	−0.007	15.06	3.11	57.37
38	Water	298.15	101.3	120.37	104.92	0.367	0	0	0
39	Water	303.15	101.3	120.37	125.82	0.437	20.86	5.22	69.56

Table A3. Exergy and exergoeconomic parameters of the system.

Component	$\dot{E}x_{F,k}$ (kW)	$\dot{E}x_{P,k}$ (kW)	$\dot{E}x_{D,k}$ (kW)	$\eta_{ex,k}$ (%)	$\dot{Z}_k$ (USD/h)	$\dot{C}_{D,k}$ (USD/h)	f_k	r_k
Air compressor	8937.07	8298.28	638.80	92.85	48.27	19.79	70.93	0.265
Air preheater	21,333.04	19,811.49	1521.55	92.87	49.74	28.81	63.32	0.209
Gas turbine	17,817.75	16,937.07	880.67	95.06	33.18	24.29	57.73	0.123
Combustion chamber	39,319.83	31,417.46	7902.37	79.90	67.10	106.06	38.75	0.411
Biomass gasifier	34,058.85	28,904.59	5154.26	84.87	29.67	31.90	48.19	0.344
HRSG	7905.57	6918.42	987.15	87.51	39.88	18.69	68.09	0.447
Steam turbine	5189.20	4532.51	656.69	87.35	106.70	18.70	85.09	0.972
Pump 1	94.79	79.29	15.51	83.64	4.40	0.87	83.47	1.183
Generator	1808.51	1500.43	308.08	82.97	1.86	8.77	17.54	0.249
SHE	198.67	119.33	79.34	60.06	0.44	2.79	13.70	0.770
Absorber	553.45	79.72	473.73	14.41	2.54	15.39	14.17	6.923
Condenser	289.03	68.23	220.80	23.61	0.36	9.12	3.83	3.365
Evaporator	556.42	424.60	131.82	76.31	2.96	5.45	35.21	0.479
Vapor generator	1020.42	723.11	297.30	70.86	17.58	5.63	75.75	1.695
Vapor turbine	641.63	527.95	113.68	82.28	23.69	6.41	78.71	1.011
IHE	49.45	37.59	11.86	76.01	0.94	0.67	58.36	0.758
Vapor condenser	82.09	20.86	61.23	25.41	0.60	3.45	14.73	3.442
Pump 2	15.45	12.47	2.99	80.68	1.21	0.34	78.18	1.097

Appendix B

The vapor generator stands as a critical component within the ORC, exerting a significant impact on the system efficiency. Within the vapor generator, the working fluid is heated by absorbing energy from the high-temperature exhaust gases. This process is characterized by a substantial disparity in the heat transfer coefficients between the exhaust (hot) side and the working fluid (cold) side. Given these conditions, a fin-and-tube heat exchanger (FTHE) is chosen for its superior heat transfer capabilities and enhanced stability during the recovery of waste heat from exhaust gases [46]. The geometric dimensions of the FTHE are detailed in Table A4.

Table A4. Geometric dimensions of the fin-and-tube heat exchanger.

Item	Value	Unit
Tube inner diameter, d_i	20	mm
Tube outer diameter, d_o	25	mm
Tube pitch, S_{Tu}	60	mm
Fin height, H_F	12.5	mm
Fin thichness, δ_F	1	mm
Fin pitch, Y_F	4	mm
Fouling factor [54,55]		
Exhaust gas, r_{exh}	1.7×10^{-4}	$m^2 \cdot K^{-1} \cdot W$
Refrigerant (liquid), r_{liq}	1.761×10^{-4}	$m^2 \cdot K^{-1} \cdot W$
Refrigerant (vapor), r_{vap}	3.522×10^{-4}	$m^2 \cdot K^{-1} \cdot W$
Refrigerant (two-phase), r_{tp}	6.7×10^{-4}	$m^2 \cdot K^{-1} \cdot W$
Tube row alignment	Staggered type	
Tube and fin material	Stainless steel 316L	

For the generation of saturated vapor, the vapor generator is mainly divided into preheating section and evaporating section The overall heat transfer coefficients for each section can be calculated by [56]:

$$\frac{1}{U_{FTHE}} = \frac{\gamma}{\alpha_i} + r_i\gamma + \frac{\delta_{Tu}\gamma}{\lambda_{Tu}} + \frac{r_o}{\eta} + \frac{1}{\alpha_o\eta_o} \tag{A1}$$

where α_i and α_o stand for heat transfer coefficient inside and outside the tube, respectively; γ represents the rib effect coefficient; λ_{Tu} denotes the thermal conductivity of the tube material; r_i and r_o refer to fouling resistances inside and outside the tube, respectively; η_o indicates the outside overall surface efficiency.

The Young correlation is employed to calculate the heat transfer coefficient of exhaust gas [57]:

$$Nu_{exh} = 0.1378(\frac{d_b G_{max}}{\mu_{exh}})^{0.718} Pr_{exh}^{1/3} \left(\frac{Y_F}{H_F}\right)^{0.296} \tag{A2}$$

For the single-phase flow in the tube side, the Gnielinski correlation is used [58]:

$$Nu_{wf} = \frac{(f/8)(Re_{wf}-1000)Pr_{wf}}{1 + 12.7\sqrt{f/8}\left(Pr_{wf}^{2/3}-1\right)} \left[1 + \left(\frac{d_i}{L_{Tu}}\right)^{2/3}\right] c_t \tag{A3}$$

$$f = (1.82 \lg Re_{wf} - 1.64)^{-2} \tag{A4}$$

For liquid state:

$$c_t = (\frac{Pr_{wf}}{Pr_{wall}})^{0.01}, \quad \frac{Pr_{wf}}{Pr_{wall}} = 0.05 \sim 20 \tag{A5}$$

For vapor state:

$$c_t = (\frac{T_{wf}}{T_{wall}})^{0.45}, \quad \frac{T_{wf}}{T_{wall}} = 0.5 \sim 1.5 \tag{A6}$$

The thermodynamic properties of ORC working fluid vary with the vapor quality for the two-phase flow on the tube side. In order to estimate the heat transfer area, the two-phase section is discretized and divided into 50 small parts so that the thermodynamic properties of the working fluid in each small part are considered to be constant. The convective heat transfer coefficient of the two-phase flow can be calculated by the Liu and Winterton correlation [59]:

$$h_{wf} = \sqrt{\left(F_{tp}h_{liq}\right)^2 + \left(S_{tp}h_{pool}\right)^2} \tag{A7}$$

For the single:

$$F_{tp} = \left[1 + x_{wf}Pr_{liq}\left(\frac{\rho_{liq}}{\rho_{vap}} - 1\right)\right]^{0.35} \tag{A8}$$

$$S_{tp} = \left(1 + 0.055F_{tp}^{0.1}Re_{liq}^{0.16}\right)^{-1} \tag{A9}$$

$$h_{liq} = 0.023(\lambda_{liq}/d_o)Re_{liq}^{0.8}Pr_{liq}^{0.4} \tag{A10}$$

$$h_{pool} = 55Pr_{wf}^{0.12}q_{wf}^{2/3}(-\lg pr)^{-0.55}MW_{wf}^{-0.5} \tag{A11}$$

As for the ARC, heat transfer coefficients of the main components are determined by adopting the values from the literature [35,60], as summarized in Table A5.

Table A5. Heat transfer coefficients of the components in the ARC.

Component	Heat Transfer Coefficient (W·m^{-2}·K^{-1})
Generator	1500
Condenser	2500
Evaporator	1500
Absorber	700
SHE	1000

References

1. Yaghoubi, E.; Xiong, Q.; Doranehgard, M.H.; Yeganeh, M.M.; Shahriari, G.; Bidabadi, M. The effect of different operational parameters on hydrogen rich syngas production from biomass gasification in a dual fluidized bed gasifier. *Chem. Eng. Process. Process Intensif.* **2018**, *126*, 210–221. [CrossRef]
2. Medeiros, D.L.; Sales, E.A.; Kiperstok, A. Energy production from microalgae biomass: Carbon footprint and energy balance. *J. Clean. Prod.* **2015**, *96*, 493–500. [CrossRef]
3. Saidur, R.; Abdelaziz, E.A.; Demirbas, A.; Hossain, M.S.; Mekhilef, S. A review on biomass as a fuel for boilers. *Renew. Sustain. Energy Rev.* **2011**, *15*, 2262–2289. [CrossRef]
4. Puig-Arnavat, M.; Bruno, J.C.; Coronas, A. Review and analysis of biomass gasification models. *Renew. Sustain. Energy Rev.* **2010**, *14*, 2841–2851. [CrossRef]
5. Siddiqui, O.; Dincer, I.; Yilbas, B.S. Development of a novel renewable energy system integrated with biomass gasification combined cycle for cleaner production purposes. *J. Clean. Prod.* **2019**, *241*, 118345. [CrossRef]
6. Zang, G.; Tejasvi, S.; Ratner, A.; Lora, E.S. A comparative study of biomass integrated gasification combined cycle power systems: Performance analysis. *Bioresour. Technol.* **2018**, *255*, 246–256. [CrossRef] [PubMed]
7. Soltani, S.; Mahmoudi, S.M.S.; Yari, M.; Rosen, M.A. Thermodynamic analyses of a biomass integrated fired combined cycle. *Appl. Therm. Eng.* **2013**, *59*, 60–68. [CrossRef]
8. Datta, A.; Ganguly, R.; Sarkar, L. Energy and exergy analyses of an externally fired gas turbine (EFGT) cycle integrated with biomass gasifier for distributed power generation. *Energy* **2010**, *35*, 341–350. [CrossRef]
9. Al-Attab, K.A.; Zainal, Z.A. Externally fired gas turbine technology: A review. *Appl. Energy* **2015**, *138*, 474–487. [CrossRef]
10. Soltani, S.; Mahmoudi, S.M.S.; Yari, M.; Rosen, M.A. Thermodynamic analyses of an externally fired gas turbine combined cycle integrated with a biomass gasification plant. *Energy Convers. Manag.* **2013**, *70*, 107–115. [CrossRef]
11. Soltani, S.; Yari, M.; Mahmoudi, S.M.S.; Morosuk, T.; Rosen, M.A. Advanced exergy analysis applied to an externally-fired combined-cycle power plant integrated with a biomass gasification unit. *Energy* **2013**, *59*, 775–780. [CrossRef]
12. Soltani, S.; Mahmoudi, S.M.S.; Yari, M.; Morosuk, T.; Rosen, M.A.; Zare, V. A comparative exergoeconomic analysis of two biomass and co-firing combined power plants. *Energy Convers. Manag.* **2013**, *76*, 83–91. [CrossRef]
13. Mondal, P.; Ghosh, S. Exergo-economic analysis of a 1-MW biomass-based combined cycle plant with externally fired gas turbine cycle and supercritical organic Rankine cycle. *Clean Technol. Environ. Policy* **2017**, *19*, 1475–1486. [CrossRef]
14. Vera, D.; Jurado, F.; Carpio, J.; Kamel, S. Biomass gasification coupled to an EFGT-ORC combined system to maximize the electrical energy generation: A case applied to the olive oil industry. *Energy* **2018**, *144*, 41–53. [CrossRef]
15. Zhang, W.; Chen, F.; Shen, H.; Cai, J.; Liu, Y.; Zhang, J.; Wang, X.; Heydarian, D. Design and analysis of an innovative biomass-powered cogeneration system based on organic flash and supercritical carbon dioxide cycles. *Alex. Eng. J.* **2023**, *80*, 623–647. [CrossRef]
16. Moradi, R.; Cioccolanti, L.; Del Zotto, L.; Renzi, M. Comparative sensitivity analysis of micro-scale gas turbine and supercritical CO_2 systems with bottoming organic Rankine cycles fed by the biomass gasification for decentralized trigeneration. *Energy* **2023**, *266*, 126491. [CrossRef]

17. Sharafi Laleh, S.; Fatemi Alavi, S.H.; Soltani, S.; Mahmoudi, S.M.S.; Rosen, M.A. A novel supercritical carbon dioxide combined cycle fueled by biomass: Thermodynamic assessment. *Renew. Energy* **2024**, *222*, 119874. [CrossRef]
18. Kautz, M.; Hansen, U. The externally-fired gas-turbine (EFGT-Cycle) for decentralized use of biomass. *Appl. Energy* **2007**, *84*, 795–805. [CrossRef]
19. Roy, D.; Samanta, S.; Ghosh, S. Techno-economic and environmental analyses of a biomass based system employing solid oxide fuel cell, externally fired gas turbine and organic Rankine cycle. *J. Clean. Prod.* **2019**, *225*, 36–57. [CrossRef]
20. El-Sattar, H.A.; Kamel, S.; Vera, D.; Jurado, F. Tri-generation biomass system based on externally fired gas turbine, organic rankine cycle and absorption chiller. *J. Clean. Prod.* **2020**, *260*, 121068. [CrossRef]
21. Roy, D.; Samanta, S.; Ghosh, S. Performance optimization through response surface methodology of an integrated biomass gasification based combined heat and power plant employing solid oxide fuel cell and externally fired gas turbine. *Energy Convers. Manag.* **2020**, *222*, 113182. [CrossRef]
22. Zhang, G.; Li, H.; Xiao, C.; Sobhani, B. Multi-aspect analysis and multi-objective optimization of a novel biomass-driven heat and power cogeneration system; utilization of grey wolf optimizer. *J. Clean. Prod.* **2022**, *355*, 131442. [CrossRef]
23. Xu, Y.-P.; Lin, Z.-H.; Ma, T.-X.; She, C.; Xing, S.-M.; Qi, L.-Y.; Farkoush, S.G.; Pan, J. Optimization of a biomass-driven Rankine cycle integrated with multi-effect desalination, and solid oxide electrolyzer for power, hydrogen, and freshwater production. *Desalination* **2022**, *525*, 115486. [CrossRef]
24. Du, J.; Zou, Y.; Dahlak, A. Process development and multi-criteria optimization of a biomass gasification unit combined with a novel CCHP model using helium gas turbine, kalina cycles, and dual-loop organic flash cycle. *Energy* **2024**, *291*, 130319. [CrossRef]
25. Yilmaz, F.; Ozturk, M.; Selbas, R. Design and performance evaluation of a biomass-based multigeneration plant with supercritical CO_2 brayton cycle for sustainable communities. *Int. J. Hydrogen Energy* **2024**, *59*, 1540–1554. [CrossRef]
26. Zhang, T.; Sobhani, B. Multi-criteria assessment and optimization of a biomass-based cascade heat integration toward a novel multigeneration process: Application of a MOPSO method. *Appl. Therm. Eng.* **2024**, *240*, 122254. [CrossRef]
27. Shu, G.; Liang, Y.; Wei, H.; Tian, H.; Zhao, J.; Liu, L. A review of waste heat recovery on two-stroke IC engine aboard ships. *Renew. Sustain. Energy Rev.* **2013**, *19*, 385–401. [CrossRef]
28. Liang, Y.; Shu, G.; Tian, H.; Liang, X.; Wei, H.; Liu, L. Analysis of an electricity–cooling cogeneration system based on RC–ARS combined cycle aboard ship. *Energy Convers. Manag.* **2013**, *76*, 1053–1060. [CrossRef]
29. Liang, Y.; Shu, G.; Tian, H.; Wei, H.; Liang, X.; Liu, L.; Wang, X. Theoretical analysis of a novel electricity–cooling cogeneration system (ECCS) based on cascade use of waste heat of marine engine. *Energy Convers. Manag.* **2014**, *85*, 888–894. [CrossRef]
30. Ahmadi, P.; Rosen, M.A.; Dincer, I. Greenhouse gas emission and exergo-environmental analyses of a trigeneration energy system. *Int. J. Greenh. Gas Control* **2011**, *5*, 1540–1549. [CrossRef]
31. Sahoo, U.; Kumar, R.; Singh, S.K.; Tripathi, A.K. Energy, exergy, economic analysis and optimization of polygeneration hybrid solar-biomass system. *Appl. Therm. Eng.* **2018**, *145*, 685–692. [CrossRef]
32. Nondy, J.; Gogoi, T.K. Comparative performance analysis of four different combined power and cooling systems integrated with a topping gas turbine plant. *Energy Convers. Manag.* **2020**, *223*, 113242. [CrossRef]
33. Anvari, S.; Khoshbakhti Saray, R.; Bahlouli, K. Conventional and advanced exergetic and exergoeconomic analyses applied to a tri-generation cycle for heat, cold and power production. *Energy* **2015**, *91*, 925–939. [CrossRef]
34. Zoghi, M.; Habibi, H.; Yousefi Choubari, A.; Ehyaei, M.A. Exergoeconomic and environmental analyses of a novel multi-generation system including five subsystems for efficient waste heat recovery of a regenerative gas turbine cycle with hybridization of solar power tower and biomass gasifier. *Energy Convers. Manag.* **2021**, *228*, 113702. [CrossRef]
35. Mussati, S.F.; Gernaey, K.V.; Morosuk, T.; Mussati, M.C. NLP modeling for the optimization of LiBr-H_2O absorption refrigeration systems with exergy loss rate, heat transfer area, and cost as single objective functions. *Energy Convers. Manag.* **2016**, *127*, 526–544. [CrossRef]
36. Balafkandeh, S.; Zare, V.; Gholamian, E. Multi-objective optimization of a tri-generation system based on biomass gasification/digestion combined with S-CO_2 cycle and absorption chiller. *Energy Convers. Manag.* **2019**, *200*, 112057. [CrossRef]
37. Tezer, Ö.; Karabağ, N.; Öngen, A.; Çolpan, C.Ö.; Ayol, A. Biomass gasification for sustainable energy production: A review. *Int. J. Hydrogen Energy* **2022**, *47*, 15419–15433. [CrossRef]
38. Baruah, D.; Baruah, D.C. Modeling of biomass gasification: A review. *Renew. Sustain. Energy Rev.* **2014**, *39*, 806–815. [CrossRef]
39. Safarian, S.; Unnþórsson, R.; Richter, C. A review of biomass gasification modelling. *Renew. Sustain. Energy Rev.* **2019**, *110*, 378–391. [CrossRef]
40. Silva, I.P.; Lima, R.M.A.; Silva, G.F.; Ruzene, D.S.; Silva, D.P. Thermodynamic equilibrium model based on stoichiometric method for biomass gasification: A review of model modifications. *Renew. Sustain. Energy Rev.* **2019**, *114*, 109305. [CrossRef]
41. Zainal, Z.A.; Ali, R.; Lean, C.H.; Seetharamu, K.N. Prediction of performance of a downdraft gasifier using equilibrium modeling for different biomass materials. *Energy Convers. Manag.* **2001**, *42*, 1499–1515. [CrossRef]
42. Bejan, A.; Tsatsaronis, G.; Moran, M. *Thermal Design and Optimization*; John Wiley & Sons: Hoboken, NJ, USA, 1996.
43. Misra, R.D.; Sahoo, P.K.; Sahoo, S.; Gupta, A. Thermoeconomic optimization of a single effect water/LiBr vapour absorption refrigeration system. *Int. J. Refrig.* **2003**, *26*, 158–169. [CrossRef]
44. Nazari, N.; Heidarnejad, P.; Porkhial, S. Multi-objective optimization of a combined steam-organic Rankine cycle based on exergy and exergo-economic analysis for waste heat recovery application. *Energy Convers. Manag.* **2016**, *127*, 366–379. [CrossRef]

45. Feng, Y.; Zhang, Y.; Li, B.; Yang, J.; Shi, Y. Comparison between regenerative organic Rankine cycle (RORC) and basic organic Rankine cycle (BORC) based on thermoeconomic multi-objective optimization considering exergy efficiency and levelized energy cost (LEC). *Energy Convers. Manag.* **2015**, *96*, 58–71. [CrossRef]
46. Yang, F.; Zhang, H.; Bei, C.; Song, S.; Wang, E. Parametric optimization and performance analysis of ORC (organic Rankine cycle) for diesel engine waste heat recovery with a fin-and-tube evaporator. *Energy* **2015**, *91*, 128–141. [CrossRef]
47. Behzadian, M.; Khanmohammadi Otaghsara, S.; Yazdani, M.; Ignatius, J. A state-of the-art survey of TOPSIS applications. *Expert Syst. Appl.* **2012**, *39*, 13051–13069. [CrossRef]
48. Moharamian, A.; Soltani, S.; Rosen, M.A.; Mahmoudi, S.M.S.; Morosuk, T. A comparative thermoeconomic evaluation of three biomass and biomass-natural gas fired combined cycles using organic Rankine cycles. *J. Clean. Prod.* **2017**, *161*, 524–544. [CrossRef]
49. Cao, Y.; Mihardjo, L.W.W.; Dahari, M.; Tlili, I. Waste heat from a biomass fueled gas turbine for power generation via an ORC or compressor inlet cooling via an absorption refrigeration cycle: A thermoeconomic comparison. *Appl. Therm. Eng.* **2021**, *182*, 116117. [CrossRef]
50. Saleh, B.; Koglbauer, G.; Wendland, M.; Fischer, J. Working fluids for low-temperature organic Rankine cycles. *Energy* **2007**, *32*, 1210–1221. [CrossRef]
51. Hosseini, S.E.; Barzegaravval, H.; Wahid, M.A.; Ganjehkaviri, A.; Sies, M.M. Thermodynamic assessment of integrated biogas-based micro-power generation system. *Energy Convers. Manag.* **2016**, *128*, 104–119. [CrossRef]
52. Al-Sulaiman, F.A. Energy and sizing analyses of parabolic trough solar collector integrated with steam and binary vapor cycles. *Energy* **2013**, *58*, 561–570. [CrossRef]
53. Bet Sarkis, R.; Zare, V. Proposal and analysis of two novel integrated configurations for hybrid solar-biomass power generation systems: Thermodynamic and economic evaluation. *Energy Convers. Manag.* **2018**, *160*, 411–425. [CrossRef]
54. Kazemi, N.; Samadi, F. Thermodynamic, economic and thermo-economic optimization of a new proposed organic Rankine cycle for energy production from geothermal resources. *Energy Convers. Manag.* **2016**, *121*, 391–401. [CrossRef]
55. Tian, H.; Chang, L.; Gao, Y.; Shu, G.; Zhao, M.; Yan, N. Thermo-economic analysis of zeotropic mixtures based on siloxanes for engine waste heat recovery using a dual-loop organic Rankine cycle (DORC). *Energy Convers. Manag.* **2017**, *136*, 11–26. [CrossRef]
56. Yang, F.; Zhang, H.; Song, S.; Bei, C.; Wang, H.; Wang, E. Thermoeconomic multi-objective optimization of an organic Rankine cycle for exhaust waste heat recovery of a diesel engine. *Energy* **2015**, *93*, 2208–2228. [CrossRef]
57. Zhang, H.G.; Wang, E.H.; Fan, B.Y. Heat transfer analysis of a finned-tube evaporator for engine exhaust heat recovery. *Energy Convers. Manag.* **2013**, *65*, 438–447. [CrossRef]
58. Gnielinski, V. New equations for heat mass transfer in turbulent pipe and channel flows. *Int. Chem. Eng.* **1976**, *16*, 359–368.
59. Liu, Z.; Winterton, R.H.S. A general correlation for saturated and subcooled flow boiling in tubes and annuli, based on a nucleate pool boiling equation. *Int. J. Heat Mass Transf.* **1991**, *34*, 2759–2766. [CrossRef]
60. Mazzei, M.S.; Mussati, M.C.; Mussati, S.F. NLP model-based optimal design of LiBr–H_2O absorption refrigeration systems. *Int. J. Refrig.* **2014**, *38*, 58–70. [CrossRef]

Article

Techno–Economic Analysis of the Optimum Configuration for Supercritical Carbon Dioxide Cycles in Concentrating Solar Power Systems

Rosa P. Merchán [1], Luis F. González-Portillo [2] and Javier Muñoz-Antón [2,*]

[1] Department of Applied Physics and IUFFYM, University of Salamanca, 37008 Salamanca, Spain; rpmerchan@usal.es
[2] Departamento de Ingeniería Energética, Universidad Politécnica de Madrid, 28006 Madrid, Spain; lf.gonzalez@upm.es
* Correspondence: javier.munoz.anton@upm.es; Tel.: +34-910677196

Abstract: There is a general agreement among researchers that supercritical carbon dioxide (sCO_2) cycles will be part of the next generation of thermal power plants, especially in concentrating solar power (CSP) plants. While certain studies focus on maximizing the efficiency of these cycles in the hope of achieving a reduction in electricity costs, it is important to note that this assumption does not always hold true. This work provides a comprehensive analysis of the differences between minimizing the cost and maximizing the efficiency for the most remarkable sCO_2 cycles. The analysis considers the most important physical uncertainties surrounding CSP and sCO_2 cycles, such as turbine inlet temperature, ambient temperature, pressure drop and turbomachinery efficiency. Moreover, the uncertainties related to cost are also analyzed, being divided into uncertainties of sCO_2 component costs and uncertainties of heating costs. The CSP system with partial cooling (sometimes with reheating and sometimes without it) is the cheapest configuration in the analyzed cases. However, the differences in cost are generally below 5% (and sometimes neglectable), while the differences in efficiency are significantly larger and below 15%. Besides the much lower efficiency of systems with simple cycle, if the heating cost is low enough, their cost could be even lower than the cost of the system with partial cooling. Systems with recompression cycles could also achieve costs below systems with partial cooling if the design's ambient temperature and the pressure drop are low.

Keywords: concentrating solar power; supercritical CO_2; optimum configuration; techno–economic assessment; optimization

Citation: Merchán, R.P.; González-Portillo, L.F.; Muñoz-Antón, J. Techno–Economic Analysis of the Optimum Configuration for Supercritical Carbon Dioxide Cycles in Concentrating Solar Power Systems. *Entropy* **2024**, *26*, 124. https://doi.org/10.3390/e26020124

Academic Editors: Daniel Flórez-Orrego, Meire Ellen Ribeiro Domingos and Rafael Nogueira Nakashima

Received: 23 November 2023
Revised: 26 January 2024
Accepted: 28 January 2024
Published: 31 January 2024

1. Introduction

Supercritical CO_2 (sCO_2) Brayton cycles have a great potential for the cost reduction of thermal power plants due to their high efficiencies [1], but also due to their potential lower cost in comparison to steam cycles [2]. Steam cycles are more "cost effective and more thermally efficient" in current thermal power plants [3]. However, higher turbine inlet temperatures make sCO_2 cycles more efficient and more cost-effective than steam cycles [4–9]. One of the potential uses for sCO_2 cycles is its integration in high-temperature Concentrating Solar Power (CSP) plants [10,11], where the levelized cost of electricity (LCoE) reduction is expected to be between 15.6% and 67.7% in comparison to current CSP systems [4]. This cost reduction is essential in order to achieve further spread of commercial CSP plants [9].

Since there are several possible sCO_2 cycle layouts, most studies have tried to identify the best configuration by analyzing thermal and/or exergetic efficiency [12]. A common outcome of these recent studies is that the recompression cycle presents higher efficiency than the simple one [10,13–16]. More specifically, White et al. concluded that the process efficiency could be moderately improved by including reheating and/or compressor intercooling [13]. Focusing on the integration with CSP plant types, two studies proved that, within a solar central receiver system, higher net specific work output and thermal efficiency are achieved in the recompression cycle, compared to the simple regenerative [14,15], to the precompression and to the split expansion cycle, due to the reduced compressor work [15]. Nevertheless, Neises et al. highlighted that their results do not tip the scales in the partial cooling cycle favor over the recompression one, but they recommended further studies of the partial cooling cycle with the aim of clarifying its advantages [17]. Furthermore, Padilla et al. demonstrated that the simple cycle could compete with the conventional regenerative steam Rankine cycle with respect to solar central tower systems [16].

On the other hand, the study of those sCO_2 Brayton cycles, which are adapted for CSP systems, looking for the best layouts is not so usual from a techno–economic perspective, at least to the authors' knowledge. This fact has been pointed out by Alfani et al., who indeed put the focus of their study on the trade-off between system cost and efficiency and highlighted the recompressed cycle with intercooling as the most promising cycle by means of a multi-objective genetic algorithm [18]. Other methods have been employed too, like the stochastic approach performed by Meybodi et al. [19]. However, other outcomes are found in the literature, like results from Marchionni et al.'s study, which showed that recompression configurations with reheating are linked with higher CAPEX per unit of electric power despite the higher efficiency [20]. Additionally, Cheang et al. concluded that the partial cooling cycle would constitute the "best" layout for CSP, looking at both maximum efficiency and minimum cost, though they are not competitive against current steam cycles [3]. Nonetheless, those analyses were made with a maximum temperature of 580 °C, which is not the ideal temperature for supercritical Brayton cycles. On the contrary, another study highlighted the advantages of sCO_2 Brayton cycles over steam ones when integrating into CSP plants [4]. Moreover, Ho et al. concluded that, given a certain output power, higher efficiency sCO_2 cycles are related to lower solar and power-block component costs [21]. Nevertheless, some of those studies [18,21] did not take into account the inherent uncertainties in cost and performance; only Crespi et al. have considered those cost uncertainties [22,23]. They found that sCO_2 cycles are associated with similar, or even lower, LCoEs than traditional steam CSP plants, which supports sCO_2 Brayton cycles as a viable option to improve CSP competitiveness [22,23].

When analyzing the different sCO_2 layouts for CSP systems, the studies in the literature have focussed on different features. Neises et al. analyzed total recuperator conductance for the different cycles and how it influences cycle thermal efficiency [17]. Atif et al. studied how heat absorbed by a central receiver can induce turbine inlet temperature variability [14]. Turchi et al. assessed thermal efficiency by looking at turbine and compressor inlet temperatures [24]. The ambient temperature effect on cycle performance was evaluated by Neises et al. as a part of an off-design conditions test [25]. In addition, pressure drop has also been studied as a key parameter of those cycles [26].

This work provides a comprehensive analysis of the impact of the aforementioned features on the most remarkable cycles found in the literature. The objective is to identify the optimal configurations of sCO2 cycles that minimize the capital costs of CSP systems operating across various conditions. By exploring different ranges of conditions, the most cost-effective configurations for CSP systems are determined. The most significant physical uncertainties in these systems are addressed, from ambient temperature and turbine inlet temperature to pressure drop and turbomachinery efficiency. The effects of cost uncertainties are also analyzed, which have an enormous influence on the results, and which have not been previously analyzed according to the author's knowledge. Although the main objective of this study is to optimize the CSP systems to minimize the capital cost, the systems are also optimized separately to maximize efficiency with the purpose of better understanding the importance of techno–economic analysis. Conventionally, some studies have maximized efficiency and specific power [27] since they were expected to minimize electricity costs [8]; however, this was not always true [28]. This work provides a comprehensive analysis of the differences between minimizing the cost and maximizing the efficiency for the most remarkable cycles found in the literature. The methodology employed is described in Section 2, where the model and its optimization are shown. Minimum-cost systems and maximum-efficiency systems are compared in Section 3. This section gathers all results with their corresponding discussion, and Section 4 is devoted to the main conclusions.

2. Methodology

This section describes the methodology followed in this study to optimize the CSP systems with sCO_2 cycle. First, the system layouts are outlined, and the model used to simulate them is benchmarked against results from the literature. Then, the analyzed variables and their boundaries are explained. The last part of the section describes the optimization method used to obtain systems with minimum cost on the one hand and maximum efficiency on the other.

2.1. Cycle Configurations

The following six configurations are considered in this comprehensive study, based on the most common cycles found in literature: simple, recompression and partial cooling layouts, all of them with and without reheating [3,4,11,13–17,20,29]. The components diagrams of these cycles are shown in Figure 1, meanwhile, Figure 2 shows examples of their T-s diagrams.

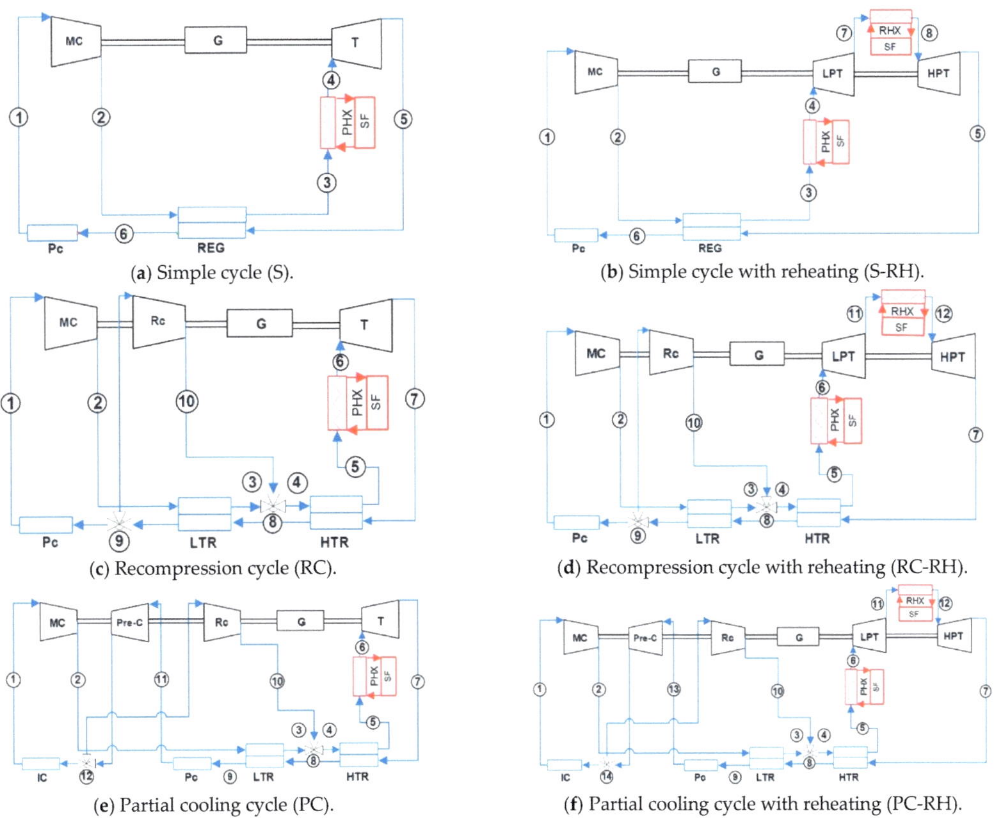

Figure 1. Component diagrams of the analyzed cycles.

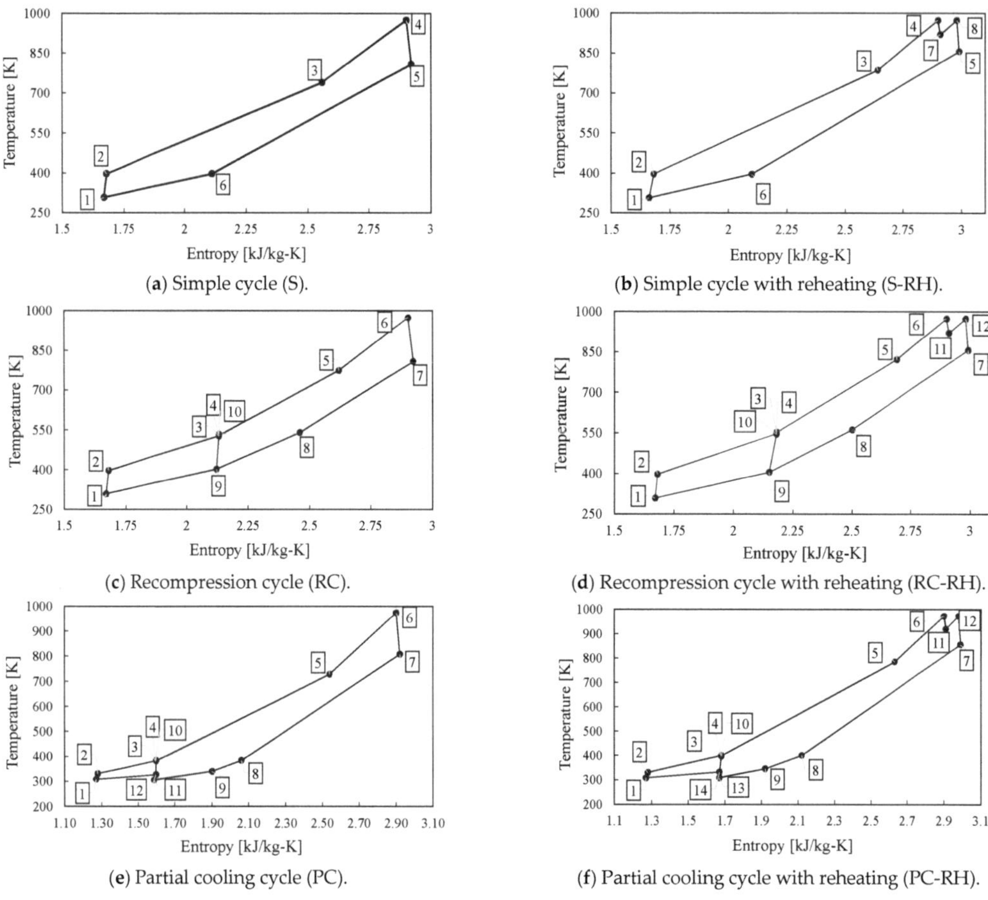

(**a**) Simple cycle (S).

(**b**) Simple cycle with reheating (S-RH).

(**c**) Recompression cycle (RC).

(**d**) Recompression cycle with reheating (RC-RH).

(**e**) Partial cooling cycle (PC).

(**f**) Partial cooling cycle with reheating (PC-RH).

Figure 2. T-s diagrams of the analyzed cycles.

In the simple Brayton cycle (see Figures 1a and 2a), the sCO_2 enters a main compressor (MC), then the excess heat from the turbine outlet is absorbed by the sCO_2 thanks to a regenerator (REG). Next, heat from the solar field (SF) is transferred to the fluid by means of a primary heat exchanger (PHX). After that heat absorption, the sCO_2 is expanded in a turbine (T), and the excess heat is exchanged via the regenerator and via a precooler (Pc) with the ambient.

Regarding the recompression Brayton cycle (see Figures 1c and 2c), a second compressor, the recompressor (Rc), is added to the layout when compared with the simple cycle. Additionally, the regenerator is divided into two different components: the low-temperature regenerator (LTR) and high-temperature regenerator (HTR). In this way, the hot outlet of the LTR is divided into two flows. One of the fractions of sCO_2 (split ratio, SR) is compressed in the main compressor after being precooled, meanwhile, the other fraction is not precooled and is directly compressed in the recompressor. Then, the latter flow merges with the main flow of sCO_2 after the cold outlet of the LTR and the whole flow exchanges heat in the HTR.

For the partial cooling Brayton cycle (see Figures 1e and 2e), the same layout as in a recompression cycle is considered with the inclusion of a third compressor, the precompressor (Pre-C), and an intercooler (IC). In this case, the hot exit of the LTR is precooled and precompressed before being split into two flow fractions. One of them (split ratio, SR) is then cooled in the intercooler and compressed in the main compressor, reaching the cold inlet of the LTR, while the other one is directly recompressed and sent to the cold outlet of the LTR.

Concerning the reheating layouts (see Figures 1b,d,f and 2b,d,f), two expansion processes are considered. Therefore, a low-pressure turbine (LPT) and a high-pressure turbine (HPT) are included in the cycle, while a reheater (RHX) is placed between these two turbines.

2.2. Model and Benchmark

The model has been built into EES software [30], version 10.836, which employs the REFPROP version 10.0 [31] database to obtain the thermodynamic properties of sCO_2. The main part of the system is the sCO_2 cycle. The mass and energy balance equations used to model the six cycle layouts are based on Padilla's work [16]. The conductance (UA) of the heat exchangers was calculated by dividing the heat exchanger into a certain number of parts, see Table 1, "REG steps number". Additionally, this discretization of the heat exchangers was useful for dealing with possible issues related to the temperature pinch point. The following main assumptions were considered for the development of the cycle model:

- the cycle operates in steady-state conditions;
- pressure drops are considered in each component as a percentage of the working pressure;
- compressors and turbines work in adiabatic conditions.

Table 1. The parameters of the energy model divided between fixed, optimized, and analyzed.

Fixed Parameter	Value	Optimized Variable	Value
High pressure	25 MPa	REG/global effectiveness	Optimized
Minimum pinch point temperature difference	5 K	Pressure ratio	Optimized
REG steps number	20	Compressor inlet temperature	Optimized
HTR/LTR/IC steps number	10	Split ratio (RC and PC)	Optimized
Pc steps number	5	Pressure at the outlet of Pre-C (PC)	Optimized
Cycle power	50 MW	Pressure at RHX inlet (RH)	Optimized

Analyzed Variable	Reference Value	Lower Bound	Upper Bound
Ambient temperature	35 °C	20 °C	40 °C
Turbine inlet temperature	700 °C	500 °C	900 °C
Heating cost	1500 USD/kW$_t$	1000 USD/kW$_t$	2000 USD/kW$_t$
Pressure drop	1.5%	0%	4.5%
Turbomachinery efficiency	0.90	0.7	1

The cycle energy model has been benchmarked by comparing the results obtained with our model against results from the literature [16,24]. Figure 3 shows the comparison of the efficiencies obtained for different cycle layouts and different turbine inlet temperatures. The cycle efficiency has been maximized in our model, as in Turchi et al. [24,32]. All six proposed cycles showed a relative deviation lower than 2% [7,15].

In this study, cycle features were divided among parameters that were fixed, variables that were optimized, and variables that were aimed to be analyzed, as is shown in Table 1. High pressure, power output, and minimum pinch point temperature were set to values commonly used in the literature [24]. The range of values analyzed in the analyzed variables is further explained in Section 2.3, and the optimization process for the optimized variables is in Section 2.4.

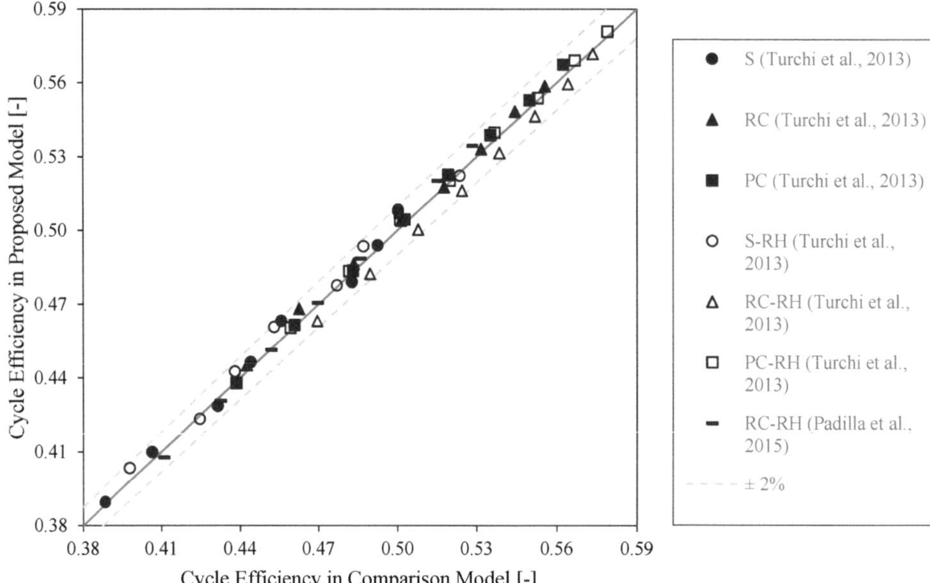

Figure 3. Parity plot between the literature results [16,24] and proposed model.

A cost model has been included in the code along with the energy model for the adequate economic evaluation of the cycles (see Appendix A). The cost of the following components is calculated according to the correlations from Weiland et al. [33]: recuperators, direct air coolers, turbomachinery, gearboxes, generators, and motors. The cost of the sCO_2 piping is estimated by adding an extra 10% to the cycle cost. The maximum temperature of each component, as well as a parameter related to the scale of the corresponding component, are introduced as inputs of these correlations. In this way, the cost of each cycle component is computed by means of the cost correlations and scaled by the cycle power. Although the turbine cost correlation is valid until 730 °C, it is employed as an approximation for the study of higher temperatures.

Solar field and primary heat exchanger costs are introduced in the system model as a unique parameter, "heating cost". "Heating cost" integrates the cost uncertainties of these systems under one parameter. Since the cost of the solar field and the primary heat exchangers have high uncertainties (the type of receiver, type of heat exchanger, solar multiple…), the use of "heating cost" allows us to analyze the best cycle layout under different cost conditions in a simple way, which is one on the main novelties of this work.

2.3. Variables to Analyze

The best configuration of system layouts is calculated by optimizing the optimized variables from Table 1. These optimizations are made for different values of the following parameters: ambient temperature, turbine inlet temperature, heating cost, pressure drop, and turbomachinery efficiency. Table 1 gathers the design point and the range bounds of these parameters.

The ambient temperature mainly depends on the weather conditions. This study sets the conditions for the design point. Although weather conditions change over time, previous studies suggest that cycle performance at any off-design ambient temperature conditions are similar to the respective on-design behavior [25]. This study analyzes the CSP systems for different ambient temperatures between 30 °C and 40 °C, which are the temperatures commonly analyzed in the literature [17,21,24,25,33–36]. Moreover, the range between 20 °C to 30 °C is also analyzed to generalize the results by covering wet-cooling applications too [37]. The compressor inlet temperature will be calculated to reduce the plant-specific cost, setting a minimum of 32 °C [37] in order to keep in the supercritical region.

The compressor inlet temperature, which directly depends on the ambient temperature, is optimized for every system. This optimization found that, for ambient temperatures of 30–40 °C, the optimum compressor inlet temperature obtained to minimize the cost and maximize the efficiency was always 5 K greater than the ambient temperature. This result is because of the considered restrictions of Table 1: the minimum pinch point temperature is set to 5 K. In this work, a minimum pinch point temperature difference of 5 K is enforced in the heat exchangers, following the approach of Turchi et al. [24]. Ambient and compressor inlet temperatures are linked by a minimum pinch point of 5 K. The compressor inlet temperature has been optimized since it can be changed according to the design of the component, however, the ambient temperature cannot, thus, it is considered as an analyzed variable that depends on the weather conditions.

Regarding the turbine inlet temperature, the analysis of the range 500–900 °C is common in nuclear reactors [4], and very similar values are used for the analysis in CSP applications, 500–850 °C [16,24,26,38]. Achieving these turbine inlet temperatures is subject to reaching higher temperatures in the heat transfer fluid. The highest temperatures could be achieved with particle receivers [21] or liquid metals [39], and the lowest temperature with current commercial molten salts [40].

The temperatures achieved in the solar receiver highly affect the cost of the solar field and the heat exchanger. These two costs are analyzed together in this study under the parameter heating cost. This cost not only depends on the temperatures achieved in the system, but also on the solar field size (commonly analyzed by the solar multiple). The heating cost with current CSP costs could vary between 1535 USD/kW$_t$ and 2329 USD/kW$_t$ for solar multiples 2 and 3, respectively [41]. Although these costs could be greatly reduced to 818 USD/kW$_t$ and 1253 USD/kW$_t$, respectively, if the Sunshot goals are achieved [41]. Particle-based systems for the next generation of CSP plants with high solar multiple show heating costs of 1600 USD/kW$_t$ [42]. These costs increase to 1944 USD/kWt without the heat exchanger in the study from Cheang et al. [3]. The system advisor model [43] suggests 1249 USD/kWt in the system with molten salts solar tower without including land and primary heat exchanger cost. After the literature review, the range of values selected for the analysis is 1000–2000 USD/kW$_t$, and the design value 1500 USD/kW$_t$.

The turbomachinery efficiency also has several uncertainties. Very high values around 0.9 are commonly used in the literature [17,44]. However, DOE recently suggested values of 0.8 and 0.87 for compressor and turbine efficiency, respectively [45]. The reality is that there is a lack of experimental tests to corroborate these values. Thus, this study analyzes turbomachinery efficiencies from 0.7 to consider worse scenarios than expected, which could happen for small powers.

The pressure drop is difficult to estimate since it depends on several factors. Vendors estimate pressure drops in sCO$_2$ recuperators from 0.7 to 4 bars, and from 0.5 to 1.5 bars on the sCO$_2$ side of the dry cooler [33]. Seo et al. have modelized a sCO$_2$ heat exchanger whose maximum outlet pressure is 200 bar at operating conditions and with a 0.75% pressure drop [46]. Additionally, both Siddiqui et al. [26] and NETL (National Energy Technology Laboratory) [47] have accounted for the effect of pressure drops in the cycle within the interval 0–4%. Reznicek measured experimentally the pressure drop for sCO$_2$ recompression Brayton cycles, getting a value of 2.5% for LTR hot-side as baseline and bounds of 0.3–6.6% [48]. Moreover, Padilla et al. considered pressure drops of 1%, 2.5%, and 5% in the solar central receiver and fixed a pressure drop in heat exchangers of 82.74 kPa [16]. Finally, Zhang et al. tested the impact of different pressure drop fractions (0–3%), concluding that they have a significant effect on energy performance [49]. Thus, following the mentioned approaches, a design pressure drop of 1.5% is selected, with lower and upper bounds of 0% and 4.5%, respectively.

2.4. Cycle Optimization

The objective of the study is to find the minimum-cost system layout under different conditions. Thus, the systems are optimized to obtain the configuration with minimum electricity specific cost (in USD/kW$_e$). The configurations needed to achieve the maximum cycle efficiency are obtained for comparison. The variables for each optimization are regenerator effectiveness, pressure ratio, compressor inlet temperature, split ratio (for recompression and partial cooling cycles), pressure at the precompressor outlet (if partial cooling cycles), and pressure at the reheater inlet (in the case of reheating cycles).

The optimization method used is the "Genetic method" from EES, which uses "a robust optimization algorithm that is designed to reliably locate a global optimum even in the presence of local optima" [30]. It is essential that the algorithm used for this type of optimization can find local optima. The reason is that small changes close to the critical point can lead to very different tendencies [50] that mislead the optimizer. Other simpler algorithms that were unable to find local optima showed much worse results. Even using the "Genetic method", several iterations were sometimes needed to find the optimum values.

The procedure considers as input variables the analyzed variables in Table 1 and the optimized solution refers to the values of "optimized variables" of Table 1 that reach the optimum result.

3. Results

This section analyzes the impact of the main variables affecting the optimum cost of the CSP system. The first sections analyze the effect of the variables from Table 1, and the last section analyzes the uncertainty of cycle component costs. The objective is to find the cycle configuration with the minimum specific cost in the different cases. The differences between maximizing efficiency and minimizing cost are used to understand the advantages and disadvantages of each configuration in terms of cost. The results shown in this section are cycle efficiencies and system costs. Conductances and pressure ratios linked to those results can be found in Appendix B.

3.1. Turbine Inlet Temperature

Figure 4 shows cycle efficiency and specific cost of the analyzed cycles as a function of turbine inlet temperature when the efficiency is maximized. The efficiency of all cycles increased with the turbine inlet temperature (as did the Carnot efficiency $\eta_C = 1 - T_{cold}/T_{hot}$). The reheating (RH) is useful to slightly increase the efficiency of all configurations. As a result, the partial cooling cycle with reheating (PC-RH) was the most efficient layout.

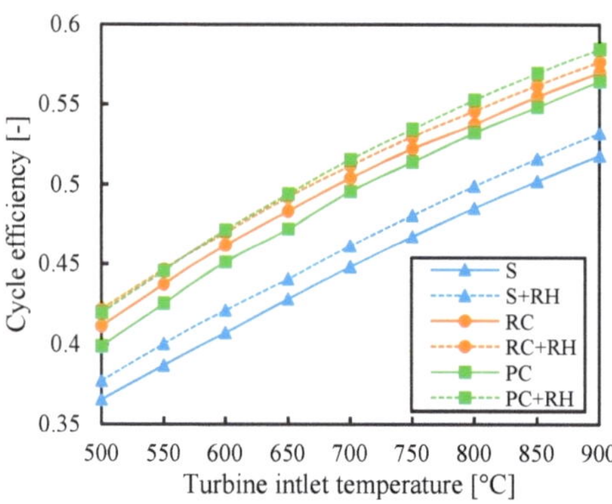

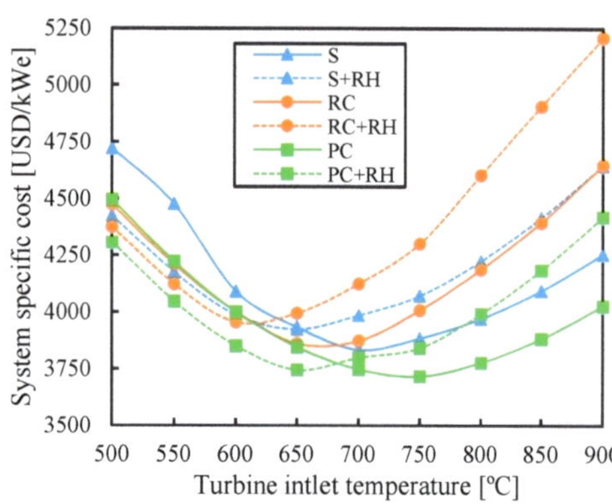

Figure 4. Cycle efficiency and system-specific cost as a function of turbine inlet temperature when the efficiency is maximized. Legend: S (simple), RC (recompression), PC (partial cooling), and RH (reheating).

Previous studies [16,18] have chosen the recompression cycle as the cycle with the greatest potential by assuming that its high efficiency would lead to a lower specific cost. However, Figure 4 suggests that this may not be true. If the system costs of the cycle configurations obtained for maximizing the efficiency are calculated, the recompression cycle with reheating (RC-RH) is shown to be the most expensive for medium-high values of turbine inlet temperature. The large heat exchangers needed to achieve high efficiencies in recompression cycles are the main contributors to this high cost (see Appendix B.1). Since the conductance (used to measure the heat exchanger size) is not limited to maximize efficiency, some systems, such as the one with the recompression cycle, can reach very high costs.

Figure 4 also shows that the reheating (RH) can increase the cycle-specific cost despite increasing the efficiency when the turbine inlet temperature is high. The main reason is the higher cost of the HTR when there is reheating. The HTR works at higher temperatures when there is reheating, which highly increases the cost, especially at high temperatures. So when the turbine inlet temperature increases, the high cost of the regenerator outweighs the benefit of the higher efficiency of reheating. Another reason for the higher cost of reheating is the higher cost of two turbines instead of one. In the cases of low turbine inlet temperature, this greater cost is outweighed by the higher efficiency, and the cycle with reheating is cheaper.

Component costs increase substantially when its maximum temperature is above 550 °C [33]. These cost increments overcome the benefit of the higher efficiencies at high turbine inlet temperatures and lead to higher system-specific costs. On the other hand, systems with low turbine inlet temperatures can use cheaper materials, but the smaller cycle efficiencies involve greater system-specific costs due to the bigger (and so more expensive) solar fields that are needed. As a result, there is a minimum system-specific cost for each layout found at turbine inlet temperatures between 650 °C and 750 °C.

Figure 5 shows the results obtained when the main optimization objective is to minimize the system-specific cost. The system with simple cycle and reheating (S-RH) achieves lower costs than the simple system without reheating (S) when the turbine inlet temperature is less than 700 °C. However, the optimum configuration of the simple cycle with reheating (S-RH) deletes the reheating at higher turbine inlet temperatures, which makes it a simple cycle without reheating (S). The same occurs for the recompression (RC) and partial cooling (PC) cycles. The lines of systems with and without reheating overlap at high temperatures because reheating cannot reduce the system cost.

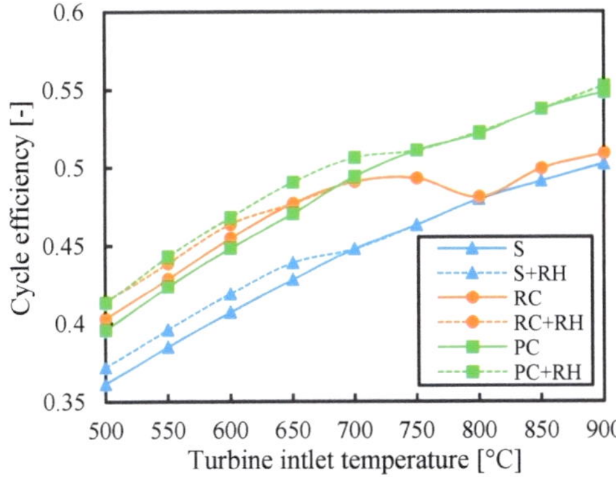

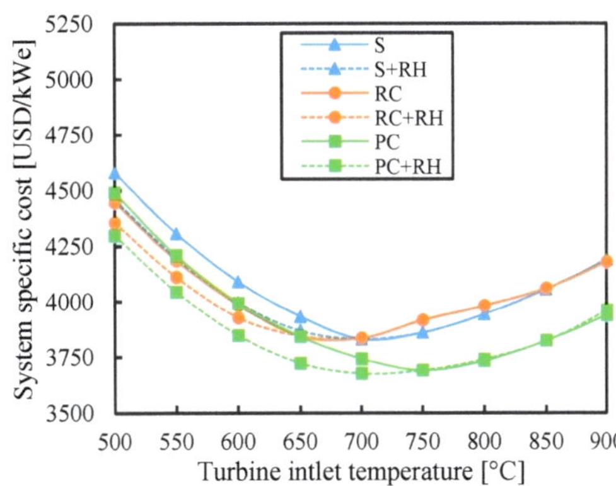

Figure 5. Cycle efficiency and system-specific cost as a function of turbine inlet temperature when the system cost is minimized. Legend: S (simple), RC (recompression), PC (partial cooling), and RH (reheating).

The lines of simple (S) and recompression (RC) systems overlapped at high turbine inlet temperatures. In these cases, the minimum-cost configuration of the system with a recompression cycle did not include a secondary compressor (i.e., all the fluid went through the main compressor), which made the recompression (RC) cycle a simple (S) cycle. This means that the system with recompression (RC) cycle at high turbine inlet temperatures achieved low specific costs at the expense of reducing the cycle efficiency up to the simple (S) cycle values.

The specific costs obtained for the system with recompression (RC) cycle by minimizing the cost (Figure 5) were significantly lower than those obtained when maximizing the efficiency (Figure 4) at high turbine inlet temperatures. The minimum-cost configuration was quite different from the configuration obtained for maximizing efficiency (see Appendix B.1). The costs obtained for the system with partial cooling (PC) and for the simple system (S) did not change as much as in the system with the recompression (RC) cycle from Figure 4 to Figure 5. The cycle configuration in these systems was more similar when the efficiency was maximized and when the cost was minimized.

Figure 5 shows that the minimum cost is achieved by the partial cooling cycle: with reheating for low temperatures and without it for high temperatures. The cost difference between the simple (S) cycle and the partial cooling configurations is in the range of 160–285 USD/kW$_e$, which represents a difference of 4.3–6.6% (where the smallest difference corresponds to the turbine inlet temperature of 700 °C). These differences are much lower than the expected ones when the maximum cycle efficiencies were compared (Figure 4). In this case, the relative difference in efficiency between the maximum efficiency cycle (partial cooling cycle with reheating, PC-RH) and the minimum efficiency cycle (simple cycle, S) was between 11.4% and 13.7% (which corresponds to an absolute difference of 5.4–6.8%).

3.2. Ambient Temperature

A first analysis for evaluating the relationship between ambient temperature and compressor inlet temperature was performed. In this case, ambient temperature was fixed at 35 °C, while compressor inlet temperature was varied in such a way that the difference between them changed in the range [0.1–10] °C. The system-specific cost of a simple cycle without reheating was minimized by optimizing pressure ratio and regenerator effectiveness. Figure 6 shows that a decrease in this temperature difference led to a decrease in the system-specific cost, except when the temperature difference was really small. Hence, the election of the compressor inlet temperature was made based on the minimum pinch point temperature difference (set to 5 °C, see Table 1).

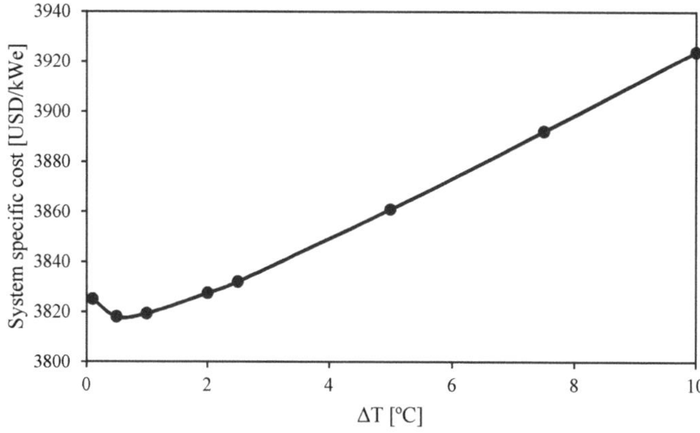

Figure 6. System-specific cost as a function of the temperature difference between compressor inlet temperature and ambient temperature when the system cost is minimized for the simple cycle without reheating.

Figure 7 shows cycle efficiency and system-specific cost of the analyzed systems as a function of ambient temperature when the efficiency is maximized. The lower the ambient temperature, the higher the efficiency (as does the Carnot efficiency $\eta_C = 1 - T_{cold}/T_{hot}$). The recompression (RC) cycle achieved the highest increase in efficiency by lowering the ambient temperature from 40 °C to 30 °C, with a relative increase of 5% (absolute increase of 2.6%). These values are similar to the ones obtained by the partial cooling (PC), 4.2% relative increase, but far away from the 2.6% relative increase obtained with the simple (S) cycle. Lower ambient temperatures involved greater proximity to the critical point, which led to higher irreversibilities inside the regenerator that partial cooling and recompression cycles can better compensate.

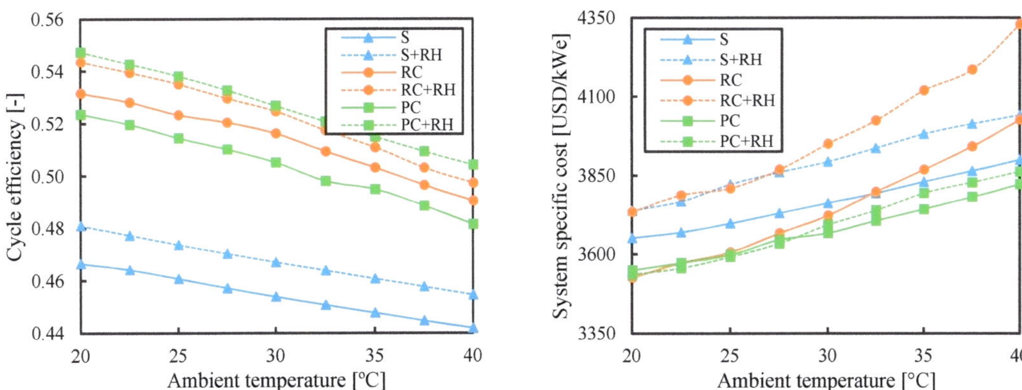

Figure 7. Cycle efficiency and system-specific cost as a function of ambient temperature when the efficiency is maximized. Legend: S (simple), RC (recompression), PC (partial cooling), and RH (reheating).

The partial cooling cycle with reheating (PC-RH) was the most efficient when the efficiency was maximized and also the least expensive for low ambient temperatures. However, for high ambient temperatures, removing the reheating reduces the system costs. The second most efficient layout was the recompression cycle with reheating (RC-RH), but it was also the most expensive. Adding reheating clearly increased the efficiency for all cycles. But this raise was more significant for the partial cooling (PC) layout, which can achieve efficiencies around 4% higher, in relative terms, if the reheating is considered.

If the objective is to minimize cost, Figure 8 shows that the partial cooling with reheating (PC-RH) cycle can be the cheapest both at low ambient temperatures and at high temperatures. The system with the simple (S) cycle was the most expensive for ambient temperatures lower than 35 °C, while the recompression (RC) one surpassed it. The cost of the recompression cycle (RC) grew faster at high ambient temperatures.

The simple cycles with and without reheating (S-RH and S) overlapped in the whole range of ambient temperatures (with the only exception of the lowest ambient temperature: 20 °C) in the same way as the recompression cycle with and without reheating (RC-RH and RC). The partial cooling cycle (PC) was the only configuration that could take the benefit of reheating. This occurred due to the 700 °C used as the turbine inlet temperature for this analysis. While the turbine inlet temperature affected the benefit of reheating (as shown in Figure 5), the ambient temperature did not.

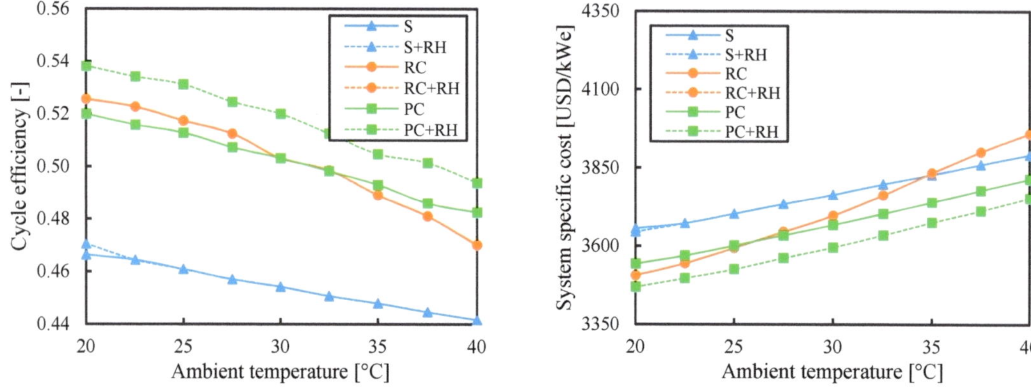

Figure 8. Cycle efficiency and system-specific cost as a function of ambient temperature when the system cost is minimized. Legend: S (simple), RC (recompression), PC (partial cooling), and RH (reheating).

3.3. Heating Cost

Since the heating cost in CSP (composed of the solar field and primary heat exchanger) is the greatest cost of the system [18], the objective of increasing the cycle efficiency has always been to reduce the effect of the heating cost on the system cost. However, previous figures have shown that the highest efficiency cycle configuration may not be the cheapest one. Figure 9 analyzes the influence of the heating cost when the efficiency is maximized. The heating cost had no impact on the cycle efficiency since the former was not considered in the optimization to maximize efficiency. However, it had a big impact on the system cost. For example, systems with a simple (S) cycle showed the lowest costs when the heating cost was 1000 USD/kW_t, but they were more expensive than systems with partial cooling (PC) when the heating cost was 2000 USD/kW_t.

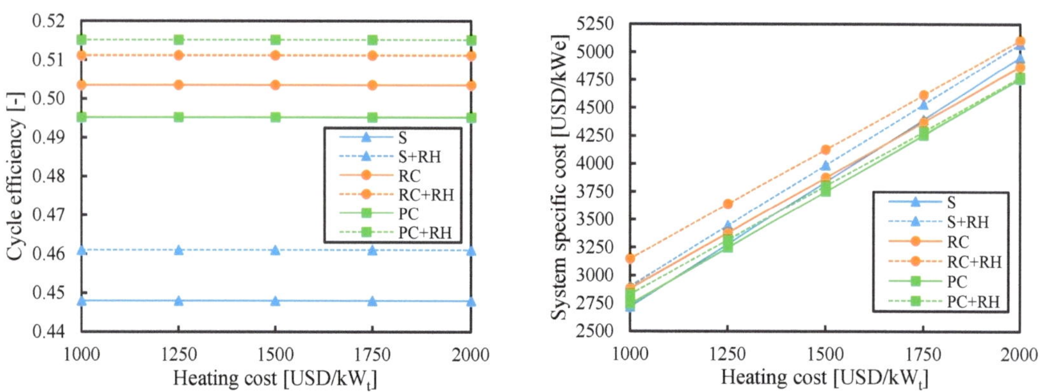

Figure 9. Cycle efficiency and system specific cost as a function of heating cost when the efficiency is maximized. Legend: S (simple), RC (recompression), PC (partial cooling), and RH (reheating).

Figure 10 shows the influence of the heating cost when the system cost was minimized. The efficiency of the simple (S) cycle was the same as in Figure 9 since minimizing cost and maximizing efficiency led to the same cycle configuration. The efficiency of cycles with partial cooling (PC and PC + RH) was slightly reduced from the results obtained when the efficiency was maximized (Figure 9). Moreover, the cycle efficiency of these cycles slightly increased with the heating cost since increasing the efficiency was more relevant when the heating was more expensive. This effect was even more pronounced in the cycle with recompression (RC).

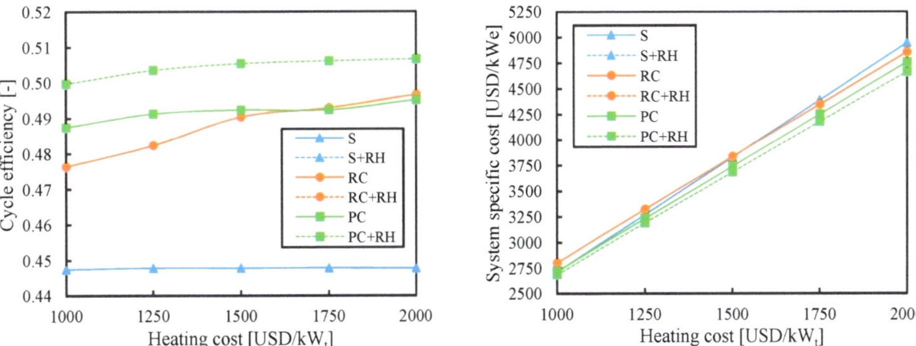

Figure 10. Cycle efficiency and system-specific cost as a function of heating cost when the system cost is minimized. Legend: S (simple), RC (recompression), PC (partial cooling), and RH (reheating).

Although the most economic cycle was the one with partial cooling and reheating, the difference in cost with the simple cycle was almost negligible (28 USD/kW$_e$, i.e., 1%) when the heating cost was small. This difference increased up to 283 USD/kW$_e$ (i.e., 6%) when the heating cost was high. More complex systems take more advantage of their high efficiency when the heating is more expensive.

3.4. Pressure Drop

The cycle efficiency decreased with the pressure drop and the specific cost increased. However, the pressure drop is a value with high variability in the analysis of power cycles. This value depends on the heat exchangers design and the connections between the different components. This study gives a specific relative pressure drop to each heat exchanger. This is a way to penalize the cycle configurations with a greater number of components.

Figures 11 and 12 show the results obtained when the cycle efficiency was maximized and when the cost was minimized, respectively, for different values of pressure drop. The pressure drop had a bigger impact on the recompression cycle. The simple cycle could better deal with pressure drop due to the smaller number of components, and the cycles with partial cooling were less affected by pressure drop because of the higher pressure ratios.

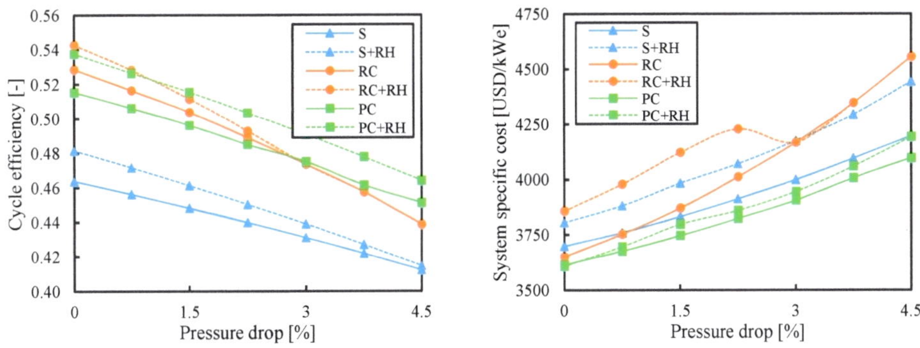

Figure 11. Cycle efficiency and system-specific cost as a function of pressure drop when the efficiency is maximized. Legend: S (simple), RC (recompression), PC (partial cooling), and RH (reheating).

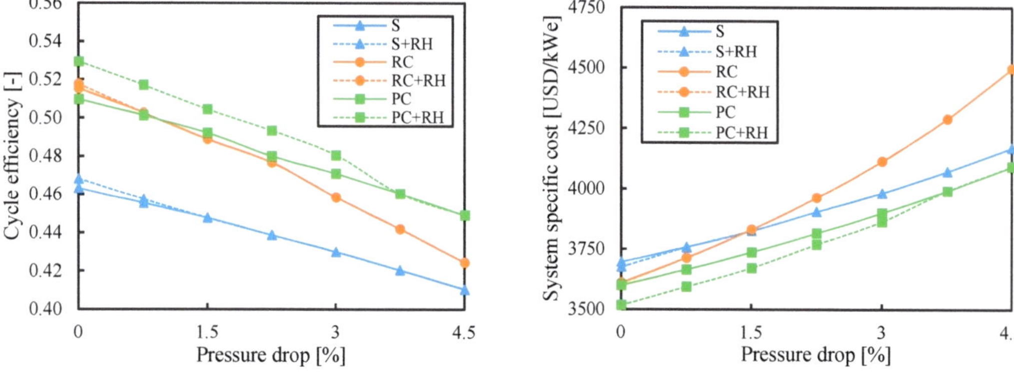

Figure 12. Cycle efficiency and system-specific cost as a function of pressure drop when the system cost is minimized. Legend: S (simple), RC (recompression), PC (partial cooling), and RH (reheating).

Note that these results were obtained for an ambient temperature of 35 °C. However, if the ambient temperature was reduced to 25 °C or 30 °C, the cost of the recompression cycle would be closer to the configuration with partial cooling (see Figure 8) or even below if the pressure drops were kept below 1.5%. In this case, the recompression cycle could be preferred over the cycles with precooling and reheating to reduce the system complexity.

3.5. Turbomachinery Efficiency

Cycle efficiency increased with the turbomachinery efficiency and the system specific cost decreased. However, it is interesting to compare the influence of this variable on the different cycle layouts. Although the most common values for compressor and turbine efficiency used to simulate sCO$_2$ cycles were 0.89 and 0.93, respectively, [44] they could be lower in real systems.

Figures 13 and 14 show the results obtained when the cycle efficiency was maximized and when the cost was minimized, respectively. Figure 13 shows that the more complex systems obtained a higher benefit in efficiency with higher turbomachinery efficiencies. The highest cycle efficiencies were achieved by the partial cooling cycle with reheating for turbomachinery efficiency higher than 0.85. On the other hand, for lower turbomachinery efficiencies, the recompression cycle with reheating reached better cycle efficiencies.

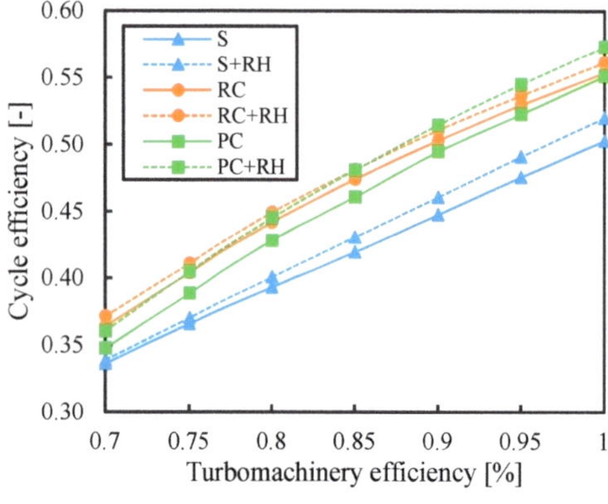
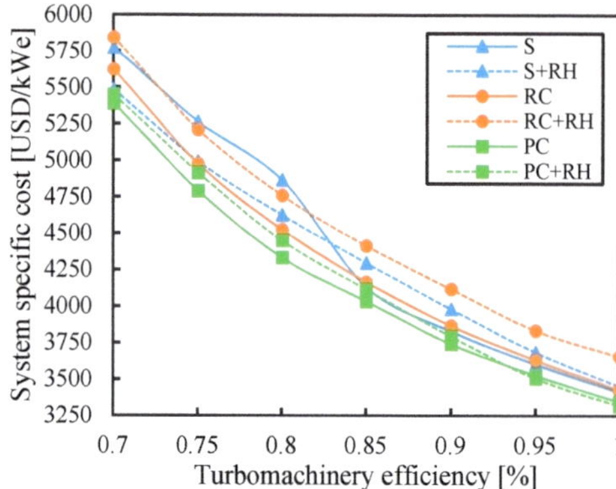

Figure 13. Cycle efficiency and system-specific cost as a function of turbomachinery efficiency when the efficiency is maximized. Legend: S (simple), RC (recompression), PC (partial cooling), and RH (reheating).

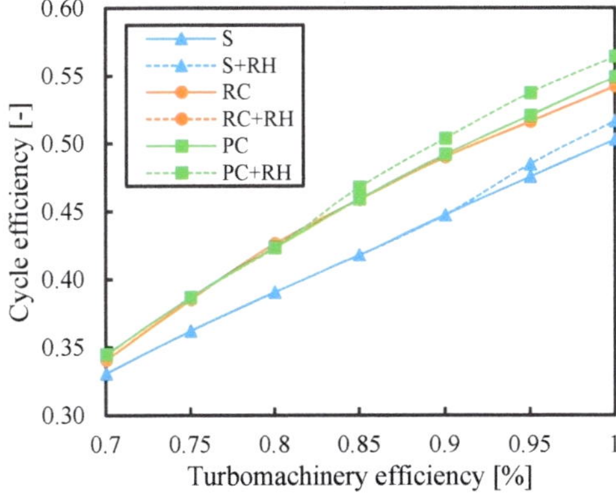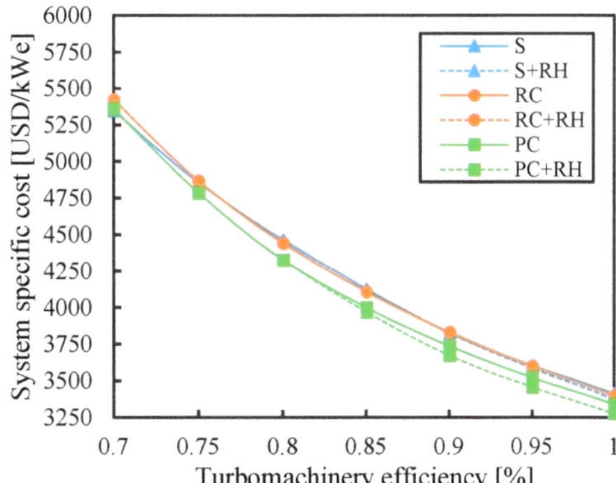

Figure 14. Cycle efficiency and system-specific cost as a function of turbomachinery efficiency when the system cost is minimized. Legend: S (simple), RC (recompression), PC (partial cooling), and RH (reheating).

The cost differences between the systems were small at low turbomachinery efficiencies when the cost was minimized in Figure 14. In the range of efficiencies expected for sCO_2 turbomachines (0.85 to 0.95), the cost difference between the most expensive and cheapest system was 147–157 USD/kW$_e$ (3.8–4.1%). However, if the efficiency was 0.7, this difference would be almost negligible, and the simple cycle could be selected as the most economical choice. For turbomachinery efficiencies lower than 0.8, minimum system costs were found for the partial cooling (PC) cycle. Then, at higher turbomachinery efficiencies, adding the reheating to the partial cooling decreased the system cost by up to 1.9% in relative terms.

3.6. Cost Uncertainties

Any techno–economic analysis is full of uncertainties that are difficult to address. Some of the uncertainties related to sCO_2 power cycles, such as the heat source cost, have been analyzed in this paper. However, there is a highly relevant uncertainty that has not been analyzed yet: the uncertainty of the cost correlations. Table 2 shows this uncertainty for each cycle component [33]. These uncertainties are used to obtain the minimum cost configuration of each layout in the following scenarios at the reference conditions from Table 1:

- reference case: no uncertainty used;
- cheap HX and expensive TM: lower bounds applied to heat exchangers (HX) and upper bounds applied to turbomachinery (TM);
- expensive HX and cheap TM: upper uncertainties applied to heat exchangers (HX) and lower uncertainty applied to turbomachinery (TM);
- expensive HX and expensive TM: upper uncertainties applied to heat exchangers (HX) and turbomachinery (TM);
- cheap HX and cheap TM: lower uncertainties applied to heat exchangers (HX) and turbomachinery (TM).

Table 2. Uncertainties of the cycle component costs [33].

Component	Component Type	Lower Bounds	Upper Bounds
Recuperator	Heat exchanger	−31%	38%
Cooler	Heat exchanger	−25%	28%
Turbine	Turbomachine	−25%	30%
Compressor	Turbomachine	−39%	48%

Figure 15 compares the specific cost of the analyzed systems for the different scenarios. The minimum cost layout is the system with partial cooling and reheating (PC-RH) in all the scenarios: 2.5–6.0% higher costs were found for the recompression (RC) cycle and 2.9–5.5% for the simple (S) cycle, depending on the case scenario. Those differences were big enough to consider the partial cooling cycle with reheating (PC-RH) as the recommended option for the first generation of sCO$_2$ cycles with the reference conditions from Table 1.

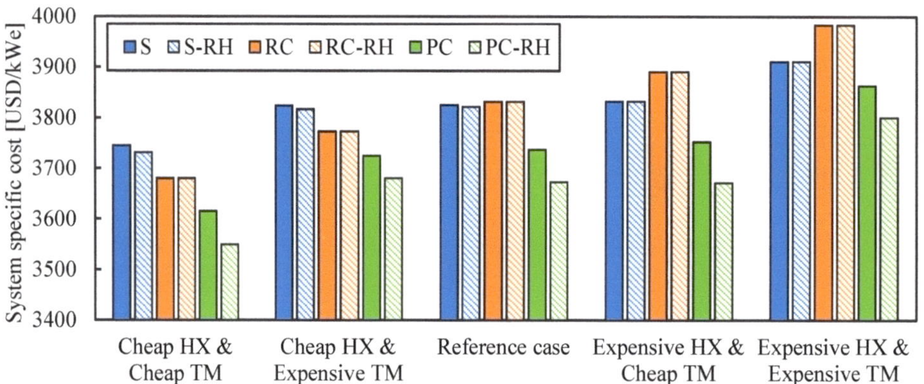

Figure 15. System-specific cost of studied cycles for different scenarios that account for the cost uncertainties. Legend: S (simple), RC (recompression), PC (partial cooling), and RH (reheating).

Although the minimum cost configuration for the reference case was the one with partial cooling, it is interesting to observe the differences between the simple and recompression cycles for the different scenarios. As previously mentioned, these two configurations could achieve lower costs under different conditions. For example, the simple cycle would benefit more from low heating costs and the recompression cycle from low ambient temperatures. For the scenarios with cheap heat exchangers, the recompression (RC) cycle gives lower costs than the simple (S) cycle. However, in the case of scenarios with expensive heat exchangers, the simple (S) cycle performs better than the recompression (RC) regarding system costs.

4. Conclusions

This study shows the great utility of techno–economic analysis in terms of capital costs in CSP systems with sCO$_2$ cycles. Although maximizing the cycle efficiency can show the potential of the cycle, the optimum cycle configuration can substantially change if the objective is to minimize the system cost. Most of the time, this more economical configuration is obtained at the expense of reducing the cycle efficiency, but it gives a better guide to choosing the best system configuration, which depends on the boundary conditions. Hence, this study also provides a comprehensive analysis of the differences between minimizing the cost and maximizing the efficiency for the most remarkable sCO$_2$ cycles.

The CSP system with partial cooling (sometimes with reheating and sometimes without it) is the cheapest configuration in the shown cases. Nevertheless, the differences in cost are generally below 5% (and sometimes neglectable), while the differences in efficiency are significantly larger and below 15%. The optimum turbine inlet temperature is 700 °C, which perfectly fits the next generation of solar receivers that can achieve temperatures above 720 °C. In case the receiver cannot reach these temperatures, the most economical layout would keep being the partial cooling cycle with reheating. Although it would make no sense to increase the turbine inlet temperature above 700 °C because the system cost would grow, it is remarkable that then reheating would not be useful to reduce the system cost.

Although the cheapest configuration resulting from this paper contains cycles with partial cooling, systems with simple cycles could achieve very similar or even lower costs if the heating cost is low or if the turbomachinery efficiency is also low. If the heating costs were as low as 1000 USD/kW$_t$, (which could happen in CSP systems with small solar multiple), the system cost with simple cycles would be very similar to the ones obtained with partial cooling but with a fairly simpler cycle. If the turbomachinery efficiency was as low as 0.7, the costs of both systems would also be very similar. If the two conditions of low heating cost and low turbomachinery efficiency were combined, the cost of systems with simple cycles could be even lower than the cost of the systems with partial cooling. The Solar Power Gen3 Demonstration Roadmap from the National Renewable Energy Laboratory (NREL) [51] states that the operation of high-temperature CSP plants should be performed with the help of systems that are as simple as possible [52]. Therefore, in the search of that simplicity, the sCO$_2$ simple Brayton configuration would be preferred in those cases when really similar costs are found.

Systems with recompression cycles could also achieve very similar or even lower costs than systems with partial cooling if the design ambient temperature and the pressure drop were reduced. The ambient temperature is mainly given by the weather conditions. For low ambient temperatures such as 20 °C, the recompression cycle is only 1% more expensive than the partial cooling cycle with reheating, which could tip the scales in favor of the recompression due to its simpler layout. Moreover, small pressure drops would benefit more from the recompression cycle than the partial cooling cycle. Thus, a combination of design ambient temperatures below 30 °C and low-pressure drop would lead to the recompression cycle to be the most economical one.

In summary, the results shown in this study can help guide the choice of the most economical cycle layout, not only for CSP systems but also for other thermal power plants. A deeper analysis of the pressure drop as a function of heat exchanger size could improve the uncertainty of the results shown in this paper. However, the tendencies will probably be still the same.

Author Contributions: Methodology, R.P.M. and J.M.-A.; Validation, R.P.M. and L.F.G.-P.; Formal analysis, L.F.G.-P.; Investigation, R.P.M.; Data curation, R.P.M., L.F.G.-P. and J.M.-A.; Writing—original draft, R.P.M.; Writing—review & editing, L.F.G.-P. and J.M.-A.; Supervision, L.F.G.-P. and J.M.-A.; Funding acquisition, J.M.-A. All authors have read and agreed to the published version of the manuscript.

Funding: This article has been partially supported by the project "ACES 2030 CM: Energía solar de concentración" (S2018/EMT-4319) granted by Comunidad de Madrid through European Structural Funds. R. P. Merchán acknowledges a postdoctoral contract co-financed by the European NextGenerationEU fund, Spanish "Plan de Recuperación, Transformación y Resilencia" fund, Spanish Ministry of Universities, and Universidad de Salamanca ("Ayudas para la recualificación del sistema universitario español 2021–2022").

Institutional Review Board Statement: Not applicable.

Data Availability Statement: Data are contained within the article.

Nomenclature

CAPEX	capital expenditure	RC	recompression cycle
EES	engineering equation solver	RC-RH	recompression cycle with reheating
HPT	high-pressure turbine	Rc	recompressor
HTR	high-temperature regenerator	REG	regenerator
IC	intercooler	RH	reheating
LCoE	levelized cost of electricity	RHX	reheater
LPT	low-pressure turbine	S	simple cycle
LTR	low-temperature regenerator	S-RH	simple cycle with reheating

MC	main compressor		sCO_2	supercritical CO_2
NETL	National Energy Technology Laboratory		SF	solar field
PC	partial cooling cycle		SR	split ratio
PC-RH	partial cooling cycle with reheating		T	turbine
Pc	precooler		TIT	turbine inlet temperature
PHX	primary heat exchanger		TM	turbomachinery
Pre-C	precompressor		UA	heat exchanger conductance

Appendix A

The cost of each component is computed as [33]:

$$\text{Cost} = a \times SP^b \times f_T$$

where SP is the scaling parameter of each component and f_T is its temperature correction factor:

$$f_T = \begin{cases} 1, & T_{max} < T_{bp} \\ 1 + c \times \left(T_{max} - T_{bp}\right) + d \times \left(T_{max} - T_{bp}\right)^2, & T_{max} \geq T_{bp} \end{cases}$$

Being T_{max} the maximum temperature the component withstands and $T_{bp} = 550\,°C$ the temperature breakpoint. W stands for the work performed by each component and must be expressed in MW units. UA unit must be W/K.

Table A1. Cost fit coefficients for each component.

Component	a	b	c	d
PHX/RHX	3500	1	0	5.4×10^{-5}
REG/LTR/Additional heat regenerator/HTR	49.45	0.7544	0.02141	0
Direct air coolers/IC	32.88	0.75	0	0
Axial turbines	182,600	0.5561	0	1.106×10^{-4}
IG centrifugal MC/Rc/Pre-C	1,230,000	0.3992	0	0
Gearbox	177,200	0.2434	0	0
Generator	108,900	0.5463	0	0
Motor MC/Rc/Pre-C–open drip-proof motor	399,400	0.6062	0	0

Table A2. Scaling parameter for each component.

Component	SP
PHX/RHX	0
REG/LTR/Additional heat regenerator/HTR	UA
Direct air coolers/IC	UA
Axial turbines	W_T
IG centrifugal MC/Rc/Pre-C	W_C
Gearbox	W_T
Generator	W_E
Motor MC	W_C
Motor Rc	W_{RC}
Motor Pre-C	W_{PCM}

Appendix B

The optimum values of pressure ratio and conductance obtained to maximize efficiency and minimize the electricity-specific cost can be found in this section.

Appendix B.1. Turbine Inlet Temperature

Figure A1 shows the pressure ratio and regenerator conductance used to maximize the cycle efficiency shown in Figure 4.

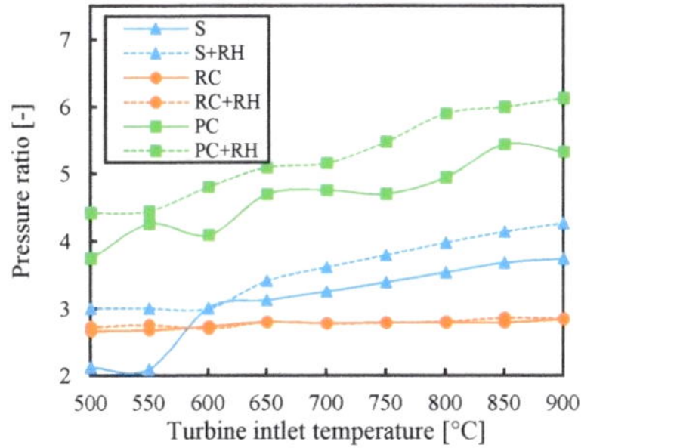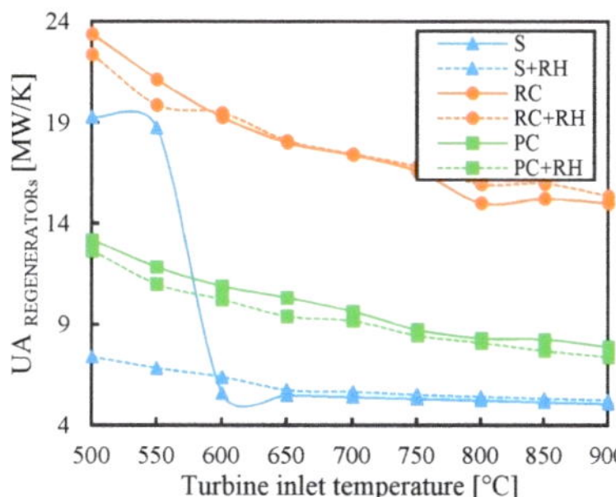

Figure A1. Pressure ratio and regenerator conductance used to maximize the cycle efficiency shown in Figure 4.

Figure A2 shows the pressure ratio and regenerator conductance used to minimize the electricity-specific cost shown in Figure 5.

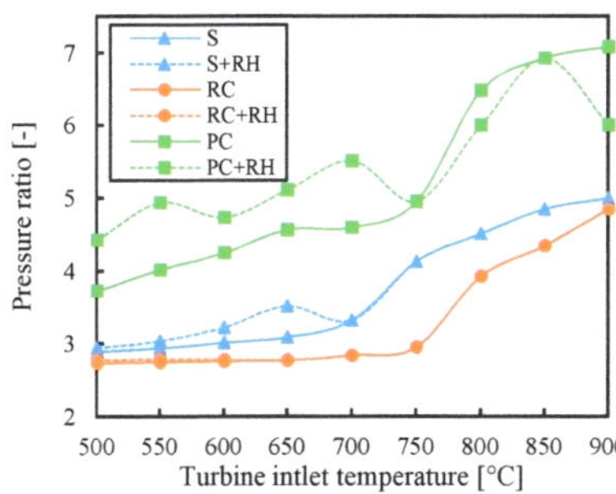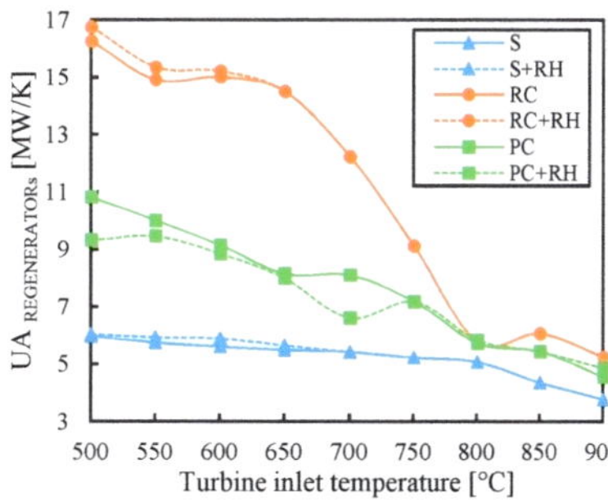

Figure A2. Pressure ratio and regenerator conductance used to minimize the electricity-specific cost shown in Figure 5.

Appendix B.2. Ambient Temperature

Figure A3 shows the pressure ratio and regenerator conductance used to maximize the cycle efficiency shown in Figure 7.

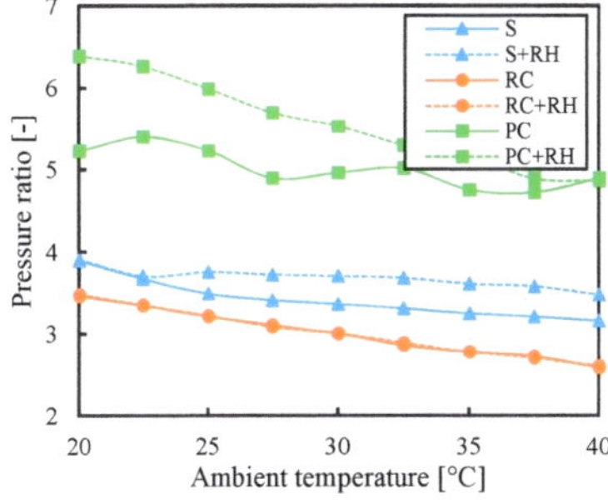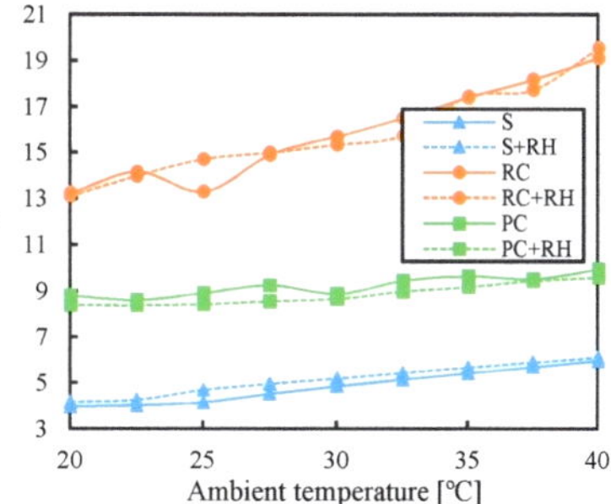

Figure A3. Pressure ratio and regenerator conductance used to maximize the cycle efficiency shown in Figure 7.

Figure A4 shows the pressure ratio and regenerator conductance used to minimize the electricity-specific cost shown in Figure 8.

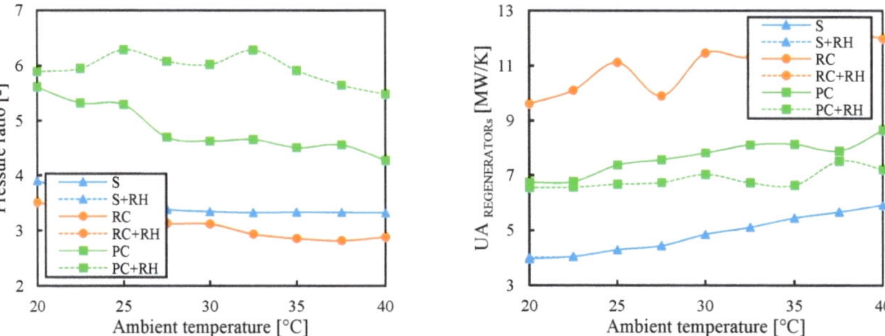

Figure A4. Pressure ratio and regenerator conductance used to minimize the electricity-specific cost shown in Figure 8.

Appendix B.3. Heating Cost

Figure A5 shows the pressure ratio and regenerator conductance used to maximize the cycle efficiency shown in Figure 9.

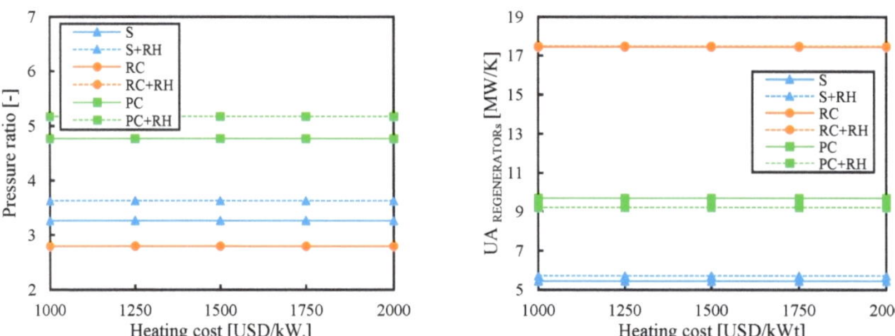

Figure A5. Pressure ratio and regenerator conductance used to maximize the cycle efficiency shown in Figure 9.

Figure A6 shows the pressure ratio and regenerator conductance used to minimize the electricity-specific cost shown in Figure 10.

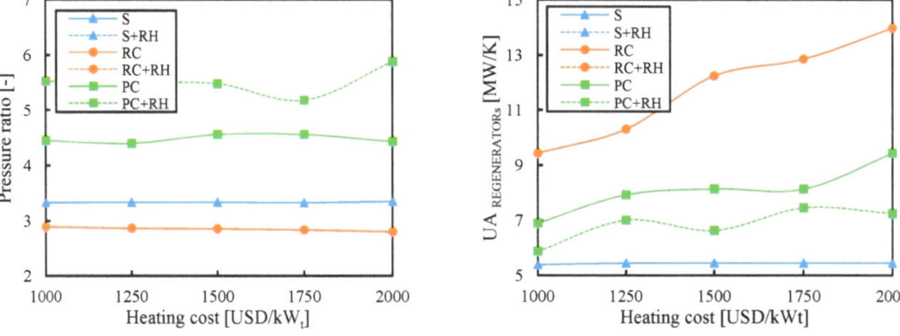

Figure A6. Pressure ratio and regenerator conductance used to minimize the electricity-specific cost shown in Figure 10.

Appendix B.4. Pressure Drop

Figure A7 shows the pressure ratio and regenerator conductance used to maximize the cycle efficiency shown in Figure 11.

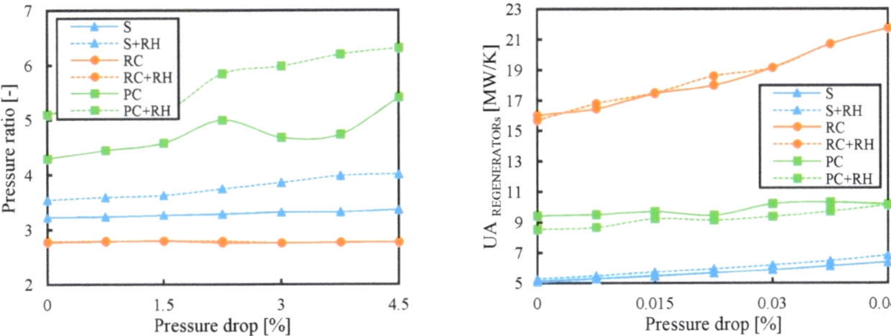

Figure A7. Pressure ratio and regenerator conductance used to maximize the cycle efficiency shown in Figure 11.

Figure A8 shows the pressure ratio and regenerator conductance used to minimize the electricity-specific cost shown in Figure 12.

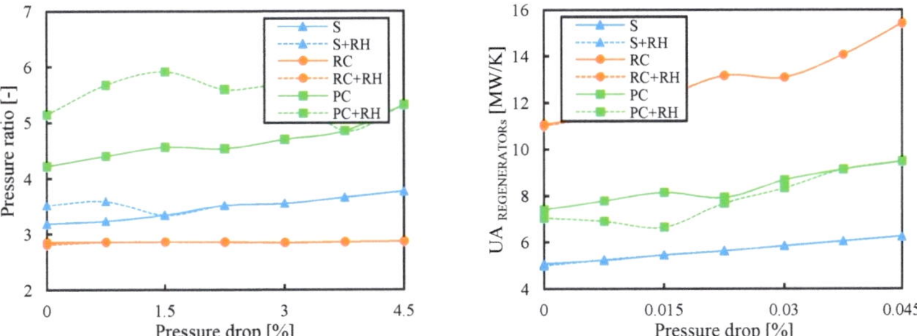

Figure A8. Pressure ratio and regenerator conductance used to minimize the electricity-specific cost shown in Figure 12.

Appendix B.5. Turbomachinery Efficiency

Figure A9 shows the pressure ratio and regenerator conductance used to maximize the cycle efficiency shown in Figure 13.

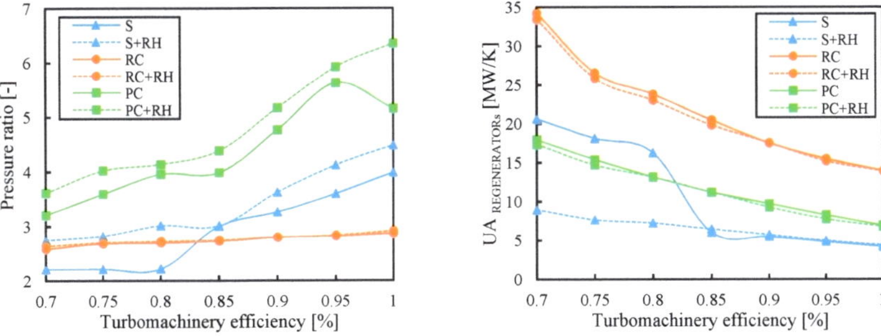

Figure A9. Pressure ratio and regenerator conductance used to maximize the cycle efficiency shown in Figure 13.

Figure A10 shows the pressure ratio and regenerator conductance used to minimize the electricity-specific cost shown in Figure 14.

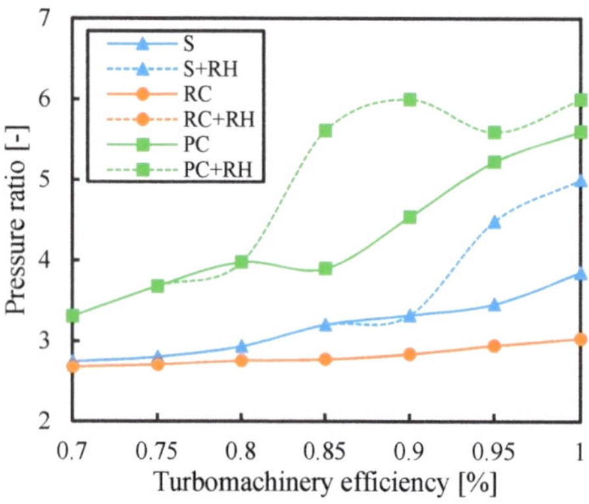
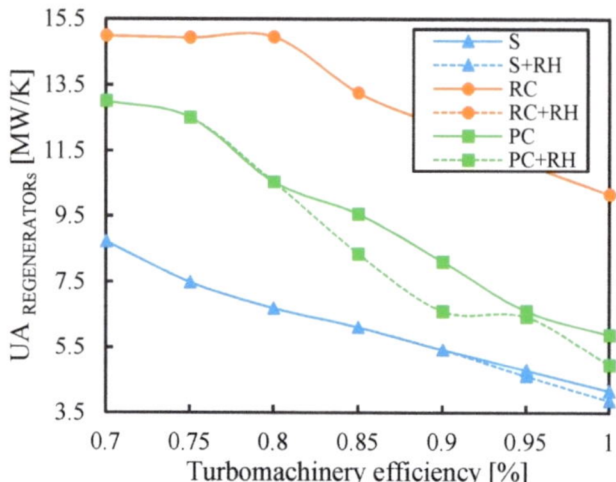

Figure A10. Pressure ratio and regenerator conductance used to minimize the electricity-specific cost shown in Figure 14.

References

1. Iverson, B.D.; Conboy, T.M.; Pasch, J.J.; Kruizenga, A.M. Supercritical CO_2 Brayton cycles for solar-thermal energy. *Appl. Energy* **2013**, *111*, 957–970. [CrossRef]
2. Dostal, V. A Supercritical Carbon Dioxide Cycle for Next Generation Nuclear Reactors. Ph.D. Thesis, Massachusetts Institute of Technology, Cambridge, MA, USA, 2004.
3. Cheang, V.T.T.; Hedderwick, R.A.A.; McGregor, C. Benchmarking supercritical carbon dioxide cycles against steam Rankine cycles for Concentrated Solar Power. *Sol. Energy* **2015**, *113*, 199–211. [CrossRef]
4. Yin, J.M.; Zheng, Q.Y.; Peng, Z.R.; Zhang, X.R. Review of supercritical CO_2 power cycles integrated with CSP. *Int. J. Energy Res.* **2020**, *44*, 1337–1369. [CrossRef]
5. Padilla, R.V.; Too, Y.C.S.; Benito, R.; McNaughton, R.; Stein, W. Thermodynamic feasibility of alternative supercritical CO_2 Brayton cycles integrated with an ejector. *Appl. Energy* **2016**, *169*, 49–62. [CrossRef]
6. Thanganadar, D.; Fornarelli, F.; Camporeale, S.; Asfand, F.; Patchigolla, K. Off-design and annual performance analysis of supercritical carbon dioxide cycle with thermal storage for CSP application. *Appl. Energy* **2021**, *282*, 116200. [CrossRef]
7. Ma, Y.; Morosuk, T.; Luo, J.; Liu, M.; Liu, J. Superstructure design and optimization on supercritical carbon dioxide cycle for application in concentrated solar power plant. *Energy Convers. Manag.* **2020**, *206*, 112290. [CrossRef]
8. Thanganadar, D.; Asfand, F.; Patchigolla, K.; Turner, P. Techno-economic analysis of supercritical carbon dioxide cycle integrated with coal-fired power plant. *Energy Convers. Manag.* **2021**, *242*, 114294. [CrossRef]
9. Yang, J.; Yang, Z.; Duan, Y. A review on integrated design and off-design operation of solar power tower system with S-CO_2 Brayton cycle. *Energy* **2022**, *246*, 123348. [CrossRef]
10. Guo, J.Q.; Li, M.-J.; He, Y.-L.; Jiang, T.; Ma, T.; Xu, J.-L.; Cao, F. A systematic review of supercritical carbon dioxide(S-CO_2) power cycle for energy industries: Technologies, key issues, and potential prospects. *Energy Convers. Manag.* **2022**, *258*, 115437. [CrossRef]
11. Chen, J.; Cheng, K.; Li, X.; Huai, X.; Dong, H. Thermodynamic evaluation and optimization of supercritical CO_2 Brayton cycle considering recuperator types and designs. *J. Clean. Prod.* **2023**, *414*, 137615. [CrossRef]
12. Yun, S.; Zhang, D.; Li, X.; Zhou, X.; Jiang, D.; Lv, X.; Wu, W.; Feng, Z.; Min, X.; Tian, W.; et al. Design, optimization and thermodynamic analysis of SCO2 Brayton cycle system for FHR. *Prog. Nucl. Energy* **2023**, *157*, 104593. [CrossRef]
13. White, C.; Gray, D.; Plunkett, J.; Shelton, W.; Weiland, N.; Shultz, T. *Techno-Economic Evaluation of Utility-Scale Power Plants Based on the Indirect sCO2 Brayton Cycle*; U.S. Department of Energy, Office of Scientific and Technical Information: Washington, DC, USA, 2017. [CrossRef]
14. Atif, M.; Al-Sulaiman, F.A. Performance analysis of supercritical CO_2 Brayton cycles integrated with solar central receiver system. In Proceedings of the IREC 2014—5th International Renewable Energy Congress, Hammamet, Tunisia, 25–27 March 2014; pp. 2–7. [CrossRef]
15. Al-Sulaiman, F.A.; Atif, M. Performance comparison of different supercritical carbon dioxide Brayton cycles integrated with a solar power tower. *Energy* **2015**, *82*, 61–71. [CrossRef]
16. Padilla, R.V.; Too, Y.C.S.; Beath, A.; McNaughton, R.; Stein, W. Effect of Pressure Drop and Reheating on Thermal and Exergetic Performance of Supercritical Carbon Dioxide Brayton Cycles Integrated with a Solar Central Receiver. *J. Sol. Energy Eng.* **2015**, *137*, 051012. [CrossRef]

17. Neises, T.; Turchi, C. A Comparison of Supercritical Carbon Dioxide Power Cycle Configurations with an Emphasis on CSP Applications. *Energy Procedia* **2014**, *49*, 1187–1196. [CrossRef]
18. Alfani, D.; Neises, T.; Astolfi, M.; Binotti, M.; Silva, P. Techno-economic analysis of CSP incorporating sCO$_2$ Brayton power cycles: Trade-off between cost and performance. In Proceedings of the SolarPACES2020, Online, 28 September–2 October 2020; p. 090001. [CrossRef]
19. Meybodi, M.A.; Beath, A.; Gwynn-Jones, S.; Veeraragavan, A.; Gurgenci, H.; Hooman, K. Techno-economic analysis of supercritical carbon dioxide power blocks. In Proceedings of the SolarPACES2016, Abu Dhabi, United Arab Emirates, 11–14 October 2016; American Institute of Physics Inc.: College Park, MD, USA, 2017; p. 060001. [CrossRef]
20. Marchionni, M.; Bianchi, G.; Tsamos, K.M.; Tassou, S.A. Techno-economic comparison of different cycle architectures for high temperature waste heat to power conversion systems using CO$_2$ in supercritical phase. *Energy Procedia* **2017**, *123*, 305–312. [CrossRef]
21. Ho, C.K.; Carlson, M.; Garg, P.; Kumar, P. Technoeconomic Analysis of Alternative Solarized s-CO$_2$ Brayton Cycle Configurations. *J. Sol. Energy Eng. Trans. ASME* **2016**, *138*, 051008. [CrossRef]
22. Crespi, F. Thermo-Economic Assessment of Supercritical CO$_2$ Power Cycles for Concentrated Solar Power Plants. Ph.D. Thesis, Universidad de Sevilla, Sevilla, Spain, 2019.
23. Crespi, F.; Sánchez, D.; Martínez, G.S.; Sánchez-Lencero, T.; Jiménez-Espadafor, F. Potential of Supercritical Carbon Dioxide Power Cycles to Reduce the Levelised Cost of Electricity of Contemporary Concentrated Solar Power Plants. *Appl. Sci.* **2020**, *10*, 5049. [CrossRef]
24. Turchi, C.S.; Ma, Z.; Neises, T.W.; Wagner, M.J. Thermodynamic Study of Advanced Supercritical Carbon Dioxide Power Cycles for Concentrating Solar Power Systems. *J. Sol. Energy Eng.* **2013**, *135*, 041007. [CrossRef]
25. Neises, T.; Turchi, C. Supercritical carbon dioxide power cycle design and configuration optimization to minimize levelized cost of energy of molten salt power towers operating at 650 °C. *Sol. Energy* **2019**, *181*, 27–36. [CrossRef]
26. Siddiqui, M.E.; Taimoor, A.A.; Almitani, K.H. Energy and exergy analysis of the S-CO$_2$ Brayton cycle coupled with bottoming cycles. *Processes* **2018**, *6*, 153. [CrossRef]
27. Yang, Z.; Kang, R.; Luo, X.; Chen, J.; Liang, Y.; Wang, C.; Chen, Y.; Xu, J. Rigorous modelling and deterministic multi-objective optimization of a super-critical CO$_2$ power system based on equation of state and non-linear programming. *Energy Convers. Manag.* **2019**, *198*, 111798. [CrossRef]
28. Mohammadi, K.; Ellingwood, K.; Powell, K. Novel hybrid solar tower-gas turbine combined power cycles using supercritical carbon dioxide bottoming cycles. *Appl. Therm. Eng.* **2020**, *178*, 115588. [CrossRef]
29. Heller, L.; Glos, S.; Buck, R. A detailed economic comparison of supercritical CO2 and steam power cycles in particle CSP plants. In Proceedings of the SolarPACES22, Albuquerque, NM, USA, 26–30 September 2022.
30. Klein, S.A. *EES: Engineering Equation Solver for the Microsoft Windows Operating System*; F-Chart Software: Fitchburg, WI, USA, 1992; Available online: https://books.google.es/books/about/EES.html?id=L0zDPwAACAAJ&redir_esc=y (accessed on 3 November 2021).
31. Lemmon, E.W.; Bell, I.H.; Huber, M.L.; McLinden, M.O. *NIST Standard Reference Database 23: Reference Fluid Thermodynamic and Transport Properties-REFPROP*; Version 10.0; National Institute of Standards and Technology: Gaithersbg, MD, USA, 2018. Available online: https://www.nist.gov/sites/default/files/documents/2018/05/23/refprop10a.pdf (accessed on 20 April 2021).
32. Turchi, C.S.; Ma, Z.; Neises, T.; Wagner, M. Thermodynamic Study of Advanced Supercritical Carbon Dioxide Power Cycles for High Performance Concentrating Solar Power Systems. In Proceedings of the ASME 2012 6th International Conference on Energy Sustainability, San Diego, CA, USA, 23–26 July 2012.
33. Weiland, N.T.; Lance, B.W.; Pidaparti, S.R. SCO$_2$ power cycle component cost correlations from DOE data spanning multiple scales and applications. In Proceedings of the ASME Turbo Expo, Phoenix, AZ, USA, 17–21 June 2019; Volume 9, pp. 1–17. [CrossRef]
34. Turchi, C.S.; Vidal, J.; Bauer, M. Molten salt power towers operating at 600–650 °C: Salt selection and cost benefits. *Sol. Energy* **2018**, *164*, 38–46. [CrossRef]
35. Tse, L.A.; Neises, T. Analysis and Optimization for off—Design Performance of the Recompression sCO2 Cycles for High Temperature CSP Applications. In Proceedings of the 5th International Symposium—sCO2 Power Cycles2, San Antonio, TX, USA, 29–31 March 2016; pp. 1–13. [CrossRef]
36. Albrecht, K.J.; Bauer, M.L.; Ho, C.K. Parametric Analysis of Particle CSP System Performance and Cost to Intrinsic Particle Properties and Operating Conditions. In Proceedings of the ASME 2019 13th International Conference on Energy Sustainability, Bellevue, WA, USA, 15–17 July 2019; American Society of Mechanical Engineers: New York, NY, USA, 2019. [CrossRef]
37. Dyreby, J. Modeling the Supercritical Carbon Dioxide Brayton Cycle with Recompression. Ph.D. Thesis, The University of Wisconsin, Madison, WI, USA, 2014.
38. Padilla, R.V.; Too, Y.C.S.; Benito, R.; Stein, W. Exergetic analysis of supercritical CO$_2$ Brayton cycles integrated with solar central receivers. *Appl. Energy* **2015**, *148*, 348–365. [CrossRef]
39. Wilk, G.; DeAngelis, A.; Henry, A. Estimating the cost of high temperature liquid metal based concentrated solar power. *J. Renew. Sustain. Energy* **2018**, *10*, 023705. [CrossRef]
40. Caraballo, A.; Galán-Casado, S.; Caballero, Á.; Serena, S. Molten Salts for Sensible Thermal Energy Storage: A Review and an Energy Performance Analysis. *Energies* **2021**, *14*, 1197. [CrossRef]

41. Ho, C.K.; Carlson, M.; Garg, P.; Kumar, P. Cost and Performance Tradeoffs of Alternative Solar-driven S-CO$_2$ Brayton Cycle Configurations. In Proceedings of the ASME 2015 Power and Energy Conversion Conference, San Diego, CA, USA, 28 June–2 July 2015; pp. 1–10. [CrossRef]
42. González-Portillo, L.F.; Albrecht, K.; Ho, C.K. Techno-Economic Optimization of CSP Plants with Free-Falling Particle Receivers. *Entropy* **2021**, *23*, 76. [CrossRef]
43. NREL. SAM (System Advisor Model) version 2020.11.29. 2020. Available online: https://sam.nrel.gov (accessed on 12 September 2023).
44. Crespi, F.; Gavagnin, G.; Sánchez, D.; Martínez, G.S. Analysis of the Thermodynamic Potential of Supercritical Carbon Dioxide Cycles: A Systematic Approach. *J. Eng. Gas Turbines Power* **2018**, *140*, 051701. [CrossRef]
45. *Gen 3 CSP Topic 1—sCO$_2$ Cycle Modeling Assumptions and Boundary Conditions*; United States Department of Energy, Office of Energy Efficiency and Renewable Energy: Washington, DC, USA, 2019.
46. Seo, H.; Cha, J.E.; Kim, J.; Sah, I.; Kim, Y.W. Design and performance analysis of a supercritical carbon dioxide heat exchanger. *Appl. Sci.* **2020**, *10*, 4545. [CrossRef]
47. Dennis, R. DOE Initiative on SCO$_2$ Power Cycles (STEP)—Heat Exchangers: A Performance and Cost Challenge. In *EPRI—NETL Workshop on Heat Exchangers for SCO$_2$ Power Cycles*; US Department of Energy: San Diego, CA, USA, 2015.
48. Reznicek, E.P. Design and Simulation of Supercritical Carbon Dioxide Recompression Brayton Cycles with Regenerators for Recuperation. Ph.D. Thesis, Colorado School of Mines, Golden, CO, USA, 2019.
49. Zhang, R.; Su, W.; Lin, X.; Zhou, N.; Zhao, L. Thermodynamic analysis and parametric optimization of a novel S–CO$_2$ power cycle for the waste heat recovery of internal combustion engines. *Energy* **2020**, *209*, 118484. [CrossRef]
50. González-Portillo, L.F.; Muñoz-Antón, J.; Martínez-Val, J.M. Thermodynamic mapping of power cycles working around the critical point. *Energy Convers. Manag.* **2019**, *192*, 359–373. [CrossRef]
51. Mehos, M.; Turchi, C.; Vidal, J.; Wagner, M.; Ma, Z.; Ho, C.; Kolb, W.; Andraka, C.; Kruizenga, A. Concentrating Solar Power Gen3 Demonstration Roadmap. 2017. Available online: www.nrel.gov/publications (accessed on 1 June 2023).
52. Linares, J.I.; Montes, M.J.; Cantizano, A.; Sánchez, C. A novel supercritical CO$_2$ recompression Brayton power cycle for power tower concentrating solar plants. *Appl. Energy* **2020**, *263*, 114644. [CrossRef]

Article

Improved Waste Heat Management and Energy Integration in an Aluminum Annealing Continuous Furnace Using a Machine Learning Approach

Mohammad Andayesh [1,2,*], Daniel Alexander Flórez-Orrego [1,*], Reginald Germanier [3], Manuele Gatti [2] and François Maréchal [1]

1   Industrial Process and Energy Systems Engineering, École Polytechnique Fédérale de Lausanne EPFL, 1950 Sion, Switzerland; francois.marechal@epfl.ch
2   Department of Energy, Politecnico di Milano, 20156 Milan, Italy; manuele.gatti@polimi.it
3   Novelis Switzerland SA, 3960 Sierre, Switzerland; reginald.germanier@novelis.adityabirla.com
*   Correspondence: mohammad.andayesh@mail.polimi.it (M.A.); daniel.florezorrego@epfl.ch (D.A.F.-O.)

Abstract: Annealing furnaces are critical for achieving the desired material properties in the production of high-quality aluminum products. In addition, energy efficiency has become more and more important in industrial processes due to increasing decarbonization regulations and the price of natural gas. Thus, the current study aims to determine the opportunities to reduce energy consumption in an annealing continuous furnace and the associated emissions. To this end, the heat transfer phenomenon is modeled and solutions for the decreasing fuel consumption are evaluated so that the overall performance of the process is enhanced. A heat transfer model is developed using the finite difference method, and the heat transfer coefficient is calculated using machine learning regression models. The heat transfer model is able to predict the heat transfer coefficient and calculate the aluminum temperature profile along the furnace and the fuel consumption for any given operating condition. Two solutions for boosting the furnace exergy efficiency are evaluated, including the modulation of the furnace temperature profiles and the energy integration by the recycling of exhaust flue gases. The results show that the advanced energy integration approach significantly reduces fuel consumption by up to 20.7%. Sensitivity analysis demonstrates that the proposed strategy can effectively reduce fuel consumption compared with the business-as-usual scenario for a range of sheet thicknesses and sheet velocities.

Keywords: annealing continuous furnace; decarbonization; computational fluid dynamics; machine learning; exergy analysis; energy integration

Citation: Andayesh, M.; Flórez-Orrego, D.A.; Germanier, R.; Gatti, M.; Maréchal, F. Improved Waste Heat Management and Energy Integration in an Aluminum Annealing Continuous Furnace Using a Machine Learning Approach. *Entropy* **2023**, *25*, 1486. https://doi.org/10.3390/e25111486

Academic Editor: Tatiana Morosuk

Received: 4 September 2023
Revised: 21 October 2023
Accepted: 24 October 2023
Published: 26 October 2023

1. Introduction

The global aluminum market is expected to grow annually by 5.8%, stimulated by an increasing demand for aluminum products, such as sheets and coils, in the automotive industry. Aluminum alloys have low density, good corrosion resistance, a high strength-to-weight ratio and good ductility [1,2]. For these reasons, aluminum is the second most used metal in the modern economy, finding applications not only in transportation sector but also in packaging and buildings [3]. Aluminum alloys are also widely used in aircraft components and structures [4]. Another advantage of the aluminum is its high recyclability, which makes it a sustainable choice for many applications. In fact, the increase in aluminum recycling rates has gained renewed interest, considering that primary (pure) aluminum production has a CO_2 emission intensity of around 17.1 t_{CO2}/t_{Al} [5].

Annealing is a critical process in the manufacture of aluminum coils, as it relieves concentrated stresses that have been introduced during the rolling process and modifies the microstructure of aluminum to improve the material strength, toughness, and corrosion resistance. In this way, it also increases its ductility, which allows the material to be formed and shaped more easily. It is achieved by heating the aluminum coil to a given temperature below the melting point, holding the temperature for a pre-defined time, and finally cooling it down either with water or air. The quality of the finished product is improved and the risk of defects is reduced. Complex microstructure evolutions including static re-crystallization, phase transformation, and a change in crystal orientation, grain morphology and size happen during the heat treatment of the aluminum coils [6]. In this regard, the precise heat transfer and band transportation processes in the annealing continuous line (ACL) ensures the efficient and reliable heating rates that comply with the expected production quality and throughput.

The carbon footprint and the production cost of the aluminum coils strongly depend on the energy efficiency of the ACL furnace. Therefore, it is necessary to develop an accurate operational model for predicting and improving the furnace performance and energy consumption. Computational fluid dynamics (CFD) has been widely applied as a powerful tool for analyzing the heat transfer and fluid flow in heat treatment furnaces [7–12]. A three-dimensional CFD model simulates the heat and mass transfer through the specified domain by numerically solving the governing equations in discrete zones called finite volumes. However, one disadvantage of the CFD simulations is the high computational time. Hajaliakbari and Hassanpour [13] applied a numerical approach based on the finite volume method to calculate the energy efficiency of an annealing continuous furnace in the steel industry. According to the authors, both of strip velocity and heating power should be carefully adjusted in each heating schedule. Strommer et al. [14] developed a first-principle model that relies on mass and energy balances to describe the dynamic behavior of the furnace. Although the model differs from measurements, it is suitable for real-time applications of control due to the moderate computational effort. Cho et al. [15] proposed a data-driven neural network MPC (model predictive controller) as a fast predictive model for the real-time control of an ACL furnace. He et al. [16] developed a first-principle model to determine the strip temperature using the heat balance method. The model inputs are the strip dimensions and zone temperature, and it provides the strip temperature distribution in the furnace. Differently from the configuration of the ACL furnace studied in the present work, the annealing furnaces of previous studies rely on temperature-resistant rolls for transporting the band and mostly use a radiant tube to supply the heat. In this work, both the forced convection and radiation heating processes and the levitation force for lifting and transporting the aluminum band are driven by the hot flue gases injected through the furnace nozzles. This contactless transportation system leads to a higher quality for the final product.

The application of machine learning methods to model and predict heat transfer phenomena in thermofluid systems has drawn the attention of researchers in order to reduce the computational time related to CFD simulations. Supervised machine learning can be used to improve the understanding of the heat transfer processes by developing accurate models for predicting heat transfer coefficients. Kwon et al. [17] the applied random forest algorithm to predict the heat transfer coefficient for convection in a cooling channel integrated with variable rib roughness. Accordingly, compared with simple analytical correlations, machine learning regressors can be much more accurate, especially for unsteady, nonlinear systems. Mehralizadeh et al. [18] developed several machine learning models to predict the boiling heat transfer coefficient of different refrigerants in finned-tube applications and compared them with existing empirical correlations. The models predict the heat transfer coefficient for the test data with good agreement. Yoo et al. [19] used machine learning to predict the heat transfer coefficient for condensation in the presence of non-condensable gas, in terms of the total pressure, mass fraction of the non-condensable gas, and wall subcooling. According to the authors, outside of the application range, the existing correlations do not accurately predict the heat transfer coefficient. Thus, a machine learning technique was applied to better predict the heat transfer coefficient for other operating conditions based on new experimental results. In the present work, four supervised learning algorithms are used to predict the heat transfer coefficient of the aluminum furnace using relevant operating conditions as the model inputs. Thus, the application of machine learning models may help in improving the operational efficiency and reliability of ACL furnaces.

In view of this, major cost savings could be achieved by implementing enhanced waste heat recovery approaches, thanks to reduced fuel consumption, lower risk perception, and mitigation of the environmental impact. Some authors have studied ways of recovering waste heat energy in the aluminum industry. Senanu et al. [20] studied the effect of flue gas recycling from aluminum electrolysis cells with a CO-to-CO_2 converter to chemically recover waste heat. Jouhara et al. [21] designed a heat-pipe heat exchanger for recovering waste heat from a thermal treatment furnace. Brough and Jouhara [22] highlighted the relevant potential for waste heat recovery in the aluminum production processes and reviewed different sources of waste heat and applicable technologies. Flórez-Orrego et al. [23] conducted a systemic study on decarbonization processes in the aluminum remelting industry to elucidate opportunities for enhanced waste heat recovery and renewable energy integration.

In contrast to previous studies, in which waste heat recovery was performed by recuperation, in the present work, two solutions are proposed and analyzed for improving the waste heat recovery and furnace efficiency. The first solution deals with the adoption of optimal temperature profiles that guarantee the lowest exergy loss for each one of the furnace zones. The adjustment of the temperature of each zone to a suitable level that still ensured the heat transfer rate proved to be a thermodynamically efficient way to distribute the energy requirement among different zones while reducing the stack loss. The second approach consists of thermally integrating the different zones of the ACL furnace, as in certain zones the roof gases may still have enough energy to preheat the aluminum band in the colder zones. Currently, each zone temperature is controlled by a number of fired heaters and the waste heat available in the stack gases is used for preheating hot water distribution networks at low temperature. In this regard, the first solution could be highly compatible with the decarbonization strategy via electrification of the heat supply, which may halve the emissions of the aluminum industry provided that electricity from renewable resources is available [24]. The second solution can be adopted in the case of biomass integrated gasification approaches, as the amount of waste heat released in the gasifier and other reactors could be harvested to preheat the combustion air of syngas-fired ACL furnaces.

In this study, computational fluid dynamics (CFD) tools are used for modeling and simulating different operating conditions of the ACL furnace and retrieve data that characterizes the thermodynamic and fluid dynamic performance. After the pre-processing of the experimental and CFD data, four machine learning models are trained and their accuracies for predicting the overall heat transfer coefficients are evaluated. Moreover, the operating model quantifies the natural gas consumption and exergy losses. Next, strategies for fuel reduction, e.g., energy integration, are proposed and analyzed to improve the energy efficiency of the ACL furnace. Sensitivity analysis is also conducted to verify the effect of aluminum sheet thickness and transport velocity on the energy integration results.

The novelty of this work relies on the development of a fast and accurate machine-learning-based tool able to predict the operation of an actual ACL (annealing continuous line) furnace. To the best authors' knowledge, previous studies deal with the application of computational fluid dynamic models for specific case studies and, thus, they cannot be applied to predict the heat transfer parameters for the other operating conditions of the furnace. This work represents a novel solution that can be applied in real plants using a model predictive control for adjusting the heat treatment process according to the real-time performance. Secondly, some strategies for fuel reduction targeting are proposed and analyzed using the developed model considering the quality of the energy flows (namely, exergy), which has not been analyzed in previous studies. In this regard, the analysis of different potential heating profiles allows us to determine the best temperature profile in the furnace that guarantees the lowest irreversibility of the whole energy system.

2. Description of the ACL Furnace Operating Principle

The unrolled aluminum coil enters the furnace at 20 °C and moves through the fourteen furnace zones (see Figure 1) at constant speed in order to achieve an annealing temperature of around 500–600 °C. The aluminum must remain at high temperature for a specific duration, which varies from recipe to recipe and is typically confidential. Each zone contains three nozzles at the top and three nozzles at the bottom that are fed by two recirculation fans on each side. Natural-gas-fired burners provide the heat duty. The aim of the nozzles is twofold, namely, to redirect the hot gases towards the heating strip of the aluminum and support its mass, so that the material never touches the furnace internals and the integrity of the treated surface is maintained. The furnace is 42 m long, and the typical values for aluminum speed range from 30 to 40 m/min. The nominal thickness can vary from 1 to 2.5 mm. For higher production throughputs, sheet velocity can be increased up to 45 m/min.

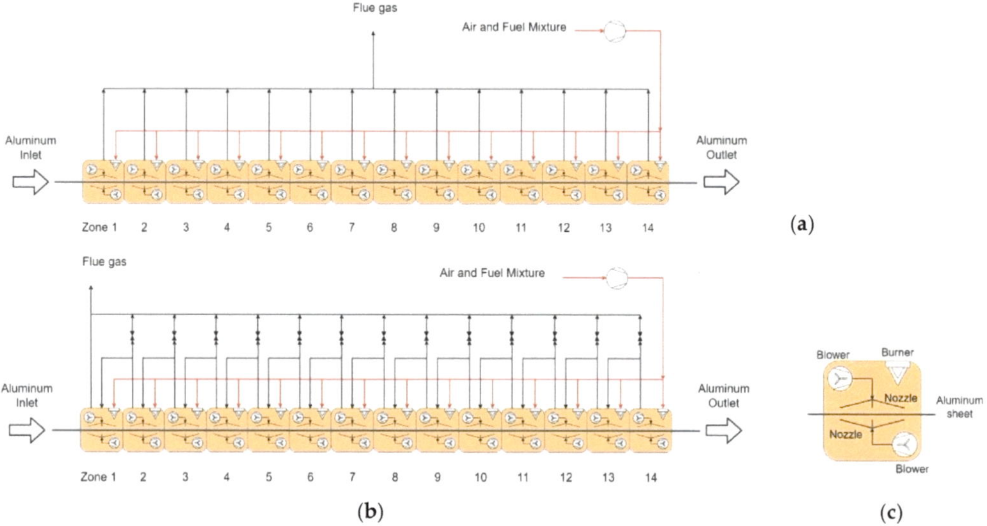

Figure 1. Process flow diagrams of ACL furnaces considering (**a**) separate exhausts and direct discharge to environment and (**b**) integrated configuration for recycling exhaust gas. (**c**) Schematic view of each zone.

Differently from the layout shown in Figure 1a, wherein the flue gases from the stack are discharged directly into the environment, in the proposed integrated configuration (see Figure 1b), the exhaust gas of a hotter zone can be sent to a colder zone in order to capitalize on the waste heat still available in the flue gases. The heat integration approach using recycling is further analyzed to quantify the reduction in fuel consumption. In order to operate within practical conditions, a limitation is considered on the volumetric flowrate of the recycled flue gas. In other words, to maintain the space velocity of the hot gases inside each zone, the maximum volumetric flowrate of each zone in the integrated configuration should be no higher than in the configuration of Figure 1a, i.e., without recycled exhaust gases. In this way, the overall effect of the energy integration will be the maximization of the waste heat recovery from hot exhaust gases to preheat the colder zones, whereas the balance of the energy requirement is achieved using conventional fired heaters.

3. Methodology

A combination of computational fluid dynamics simulations and machine learning models is used to predict the heat transfer coefficient for different operating conditions of the ACL furnace. The input parameters are selected as the velocity and the thickness of the aluminum sheet, the gas temperature, and the fan recirculation percentage. This approach allows determining important energy transport parameters with a low computational time, in comparison to the execution of a complete CFD simulation. By applying the energy balance, the temperature profile of the aluminum sheet along the whole furnace can be outlined, as well as the waste heat available in the ACL stack.

Figure 2 summarizes the procedure used to develop and apply the proposed energy integration approach and hierarchize the different temperature profiles that ensure the minimum exergy destruction rate. Measured data, whenever available, are used to validate the computational fluid dynamic simulations performed in ANSYS Fluent® 2022 R2 software. Further details on the mathematical models and solvers used in the CFD simulation are discussed in the next section. After the computational results are validated, the data is extracted and prepared to be fitted to the regression model. The regression model consists of a simplified finite different representation of the heating process of the aluminum sheet. The derivation of this simplified model is described in the subsequent sections. It is worth noting that, differently from an oversimplified lumped model, which does not consider the lag between the internal and external temperature profiles of the aluminum sheet, the proposed simplified model relying on finite difference discretization can capture the inertia of the aluminum heating and the delay in the heat diffusion from the surface to the center of the heated material. After the data is regressed on the simplified model using polynomial regression machine learning method in Tensorflow/Keras libraries of Python programming language, the model is able to predict the heat transfer coefficient and it can be used to calculate both energy and exergy balances, and perform the recycling and sensitivity analyses, as shown in Figure 2.

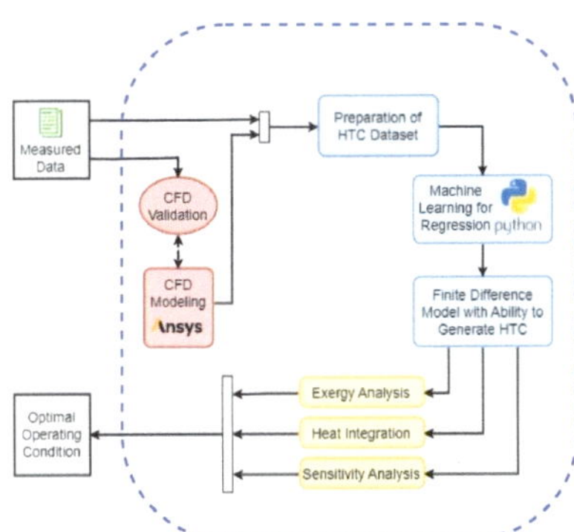

Figure 2. Schematic of the methodology used to validate the CFD calculation using experimental data, prepare the simulated dataset, perform the regression on a simplified heat transfer model, and calculate the energy integration performance in the ACL furnace.

3.1. Configuration of the Computational Fluid Dynamics Model

The control volume adopted for the CFD simulation is based on three zones of the furnace, as shown in Figure 3. Since the setup parameters used in the simulation can be adjusted to analyze either the frontend or backend zones of the furnace, a sample volume allows reducing the computational time while keeping the accuracy of the solution. A Cartesian meshing is applied with an inflation mesh near the most critical heat transfer and flow surfaces, namely, the nozzle and aluminum faces. The simulation setup considers the activation of the energy equation for coupled heat transfer between the aluminum sheet and the hot gases. A k-ω SST turbulence model is selected to represent the perturbation and eddies present in the highly non-laminar flow. Additionally, the P1 radiation model is considered. Since the aluminum coil is continuously unrolled and passed at a constant speed through the different zones in the ACL furnace, the CFD simulation considers a constant motion for the aluminum sheet. The governing equations of the simulations performed in ANSYS Fluent® [25,26] can be summarized in Equations (1)–(5):

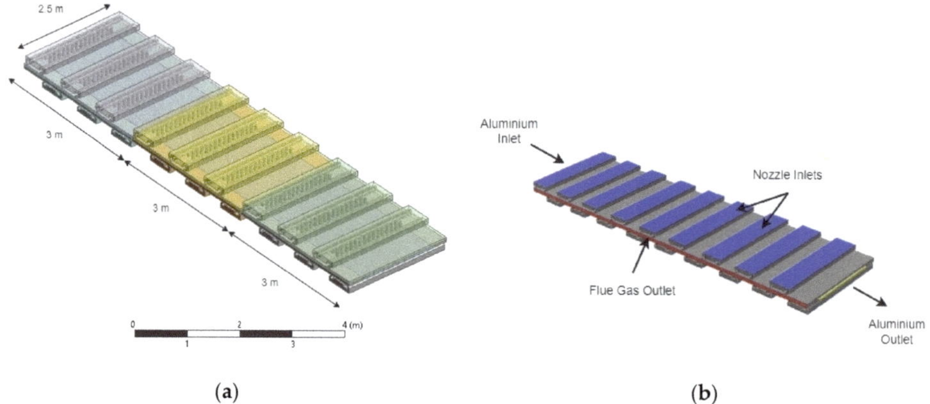

Figure 3. CFD simulation in ANSYS Fluent®. (a) Control volume, which includes three zones. (b) Boundaries conditions of the domain.

- Mass conservation equation:

$$\nabla.\left(\rho \vec{v}\right) = 0 \tag{1}$$

- where $\vec{v}$ (m/s) and ρ $\left(\frac{kg}{m^3}\right)$ are gas velocity vector and density, respectively.

- Momentum conservation equation:

$$\nabla.\left(\rho \vec{v} \vec{v}\right) = -\nabla p + \nabla.\left(\overline{\overline{\tau}}\right) + \rho \vec{g} \tag{2}$$

where p (N/m^2), $\overline{\overline{\tau}}$ (N/m^2), and $\rho \vec{g}$ (N/m^3) are static pressure, the stress tensor, and the gravitational body force, respectively.

- Transport equations for the SST k-ω model:

The turbulence kinetic energy (k) and the specific dissipation rate (ω) are obtained from the following transport equations:

$$\frac{\partial}{\partial t}(\rho k) + \frac{\partial}{\partial x_i}(\rho k u_i) = \frac{\partial}{\partial x_i}\left(\Gamma_k \frac{\partial k}{\partial x_i}\right) + \tilde{G}_k - Y_k$$
$$\frac{\partial}{\partial t}(\rho \omega) + \frac{\partial}{\partial x_i}(\rho \omega u_i) = \frac{\partial}{\partial x_i}\left(\Gamma_\omega \frac{\partial \omega}{\partial x_i}\right) + G_\omega - Y_\omega + D_\omega \tag{3}$$

In these equations, $\tilde{G}_k$ represents the generation of turbulent kinetic energy due to mean velocity gradients. G_ω is the generation of ω. Γ_k and Γ_ω represent the effective diffusivity of k and ω, respectively. Y_k and Y_ω are the dissipation of k and ω due to turbulence.

- Energy equation:

$$\nabla.\left(\vec{v}(\rho E + p)\right) = \nabla.\left(k_{eff}\nabla T - h\vec{J} + \left(\overline{\overline{\tau}}_{eff}.\vec{v}\right)\right) \tag{4}$$

where k_{eff} (W/mK) is the conductivity, T (K) is the temperature, h (J/kg) is sensible enthalpy, and $\vec{J}$ (kg/m^{2}s) is the diffusion flux. The three terms on the right-hand side of the equation correspond to energy transfer due to conduction, species diffusion, and viscous dissipation, respectively.

- The P-1 model equations:

$$-\nabla . q_r = \alpha G - 4\alpha n^2 \sigma T^4 \tag{5}$$

where α is the absorption coefficient (-), G (W/m^2) is the incident radiation, and σ is the Stefan–Boltzmann constant (5.67×10^{-8} W m^{-2} K^{-4}). The expression for $-\nabla . q_r$ is directly substituted into the energy equation to account for heat sources due to radiation.

3.2. The Simplified Heat Transfer Model Used to Apply Supervised Learning Techniques

A simplified model of the heat transfer process in the ACL furnace was developed in order to apply supervised learning regression techniques to the data obtained from the CFD simulations and validated using experimental runs. In this way, a metamodeling approach allows calculation of the heat transfer coefficient given a number of operating conditions and, consequently, determining the temperature profile of the aluminum along the zones of the ACL furnace. The exergy loss and fuel consumption can be also determined based on the overall energy balance of the system, including stack losses.

Using the finite differences method (FDM), the simplified model concept can be devised in such a way that the gas–solid and the internal solid heat transfer phenomena is represented in one superficial and one inner point of the aluminum material, respectively (Figure 4). In other words, it is assumed that the heat diffusion towards and the energy accumulation inside the aluminum body ($_o$) occur within a given time lapse thanks to continuous radiative and convective heat transfer from the hot gas ($_{inf}$) to the aluminum surface ($_s$). In this way, the temperature variation of the internal mass can be differentiated from that of the aluminum surface, which is contrary to other approaches that impose lumped models [27] with a given time constant and consider the internal aluminum temperature as equal to the superficial temperature. According to Figure 4, the explicit finite differences-based discretization of the differential energy balances given in Equations (6) and (8) for the aluminum inner body (T_o) and surface (T_s), respectively, results in Equations (7) and (9). The aluminum band is discretized along the length of the furnace and Equations (7) and (9) are applied for each cell to determine the aluminum temperature (T_o) profile along the length of the furnace. Controlling this has a significant impact on the achievement of the heat treatment requirements.

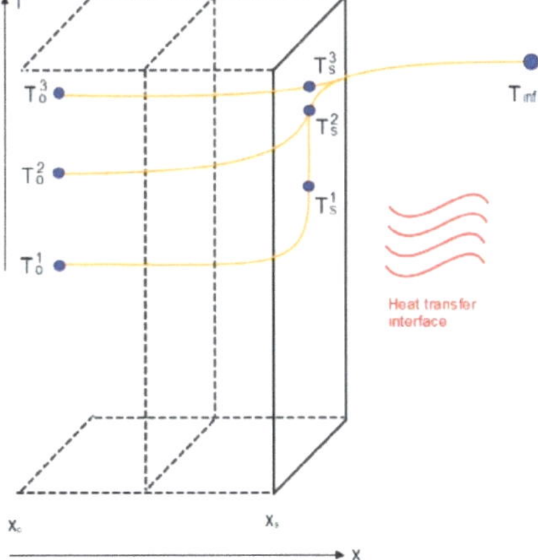

Figure 4. Schema of the derivation of the explicit finite difference-based discretization approach used to apply supervised machine learning to the experimental and simulation data.

For the aluminum inner body (at T_O):

$$kA\frac{dT}{dx} = \rho A dx C_p \frac{\partial T}{\partial t} \tag{6}$$

$$kA\frac{(T_S(t_1) - T_O(t_1))}{(x_S - x_O)} = \rho A \frac{(x_S - x_O)}{2} C_p \frac{(T_O(t_2) - T_O(t_1))}{(t_2 - t_1)} \tag{7}$$

For the aluminum surface (at T_s):

$$-kA\frac{dT}{dx} + HTC\, A(T_\infty - T_S) = \rho A dx C_p \frac{\partial T}{\partial t} \tag{8}$$

$$-kA\frac{(T_S(t_1) - T_O(t_1))}{(x_S - x_O)} + HTC\, A(T_\infty(t_1) - T_S(t_1)) = \rho A \frac{(x_S - x_O)}{2} C_p \frac{(T_S(t_2) - T_S(t_1))}{(t_2 - t_1)} \tag{9}$$

Since the unknown total heat transfer coefficient is required to determine the temperatures along the ACL furnace in a transient regime and for various operational conditions; supervised learning techniques are used to regress the data gathered from CFD simulations and experimental runs and to predict the heat transfer coefficient [28]. To this end, the explicit finite difference-based discretization model is used to fit the known operating conditions, such as gas temperature, fan power percentage, and aluminum temperature, to the unknown heat transfer coefficient (HTC) in Equation (9). Figure 5 depicts two cases of the FDM model fitting on experimental data where the inner aluminum temperature (T_o) profile along the length of ACL furnace is fitted on the measured temperatures to determine the corresponding HTC. This procedure is conducted on all the experimental data to form a dataset of HTCs. Afterwards, four types of supervised learning algorithms, namely linear, polynomial, decision tree, and random forest are trained based on the dataset to predict the HTC for arbitrary operating conditions.

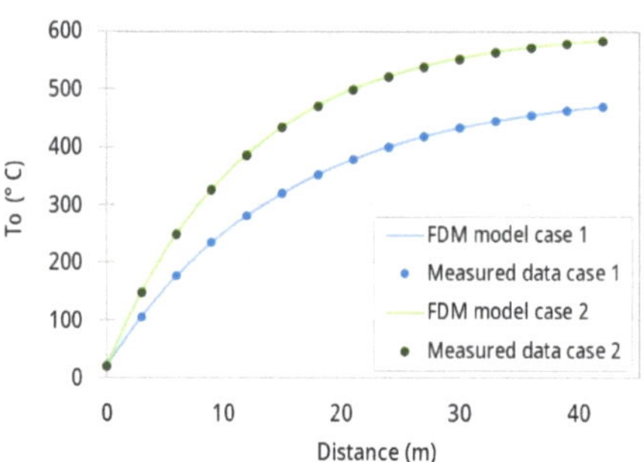

Figure 5. Simplified finite difference-based model fitting on the experimental data points of aluminum temperature (T_o) along the ACL furnace length.

3.3. Exergy Loss Calculation Based on the Energy Balance on the ACL Furnace

In order to determine the fuel consumption in the whole ACL furnace, a zone-wise energy balance can be calculated, as shown in Equation (10):

$$Q_{Al,z} + Q_{FG,z} + Q_{wall,z} + Q_{leakage,z} = Q_{fuel,z} + Q_{recirculation,z} \quad \text{for } z = 1\text{--}14 \tag{10}$$

where $Q_{Al,z}$ is the amount of energy that is effectively absorbed by the aluminum sheet (kW); $Q_{FG,z}$ is the energy leaving the ACL furnace with the flue gas (kW); and $Q_{wall,z}$ and $Q_{leakage,z}$ are the heat dissipation through the furnace walls and the leakage losses (e.g., hot gas leakage), respectively. $Q_{fuel,z}$ and $Q_{recirculation,z}$ are, respectively, the energy input with the fuel consumed and with the heat recovered from recycled flue gases produced at a downstream (hotter) zone (e.g., for heat integration). All the terms in the energy balance of Equation (10) can be explicitly represented as in Equation (11) if no combustion air preheating is adopted

$$Q_{Al,z} + (\alpha + 1)\cdot \dot{m}_{F,z}\cdot C_{p,FG,z}\cdot \left(T_{FG,z} - T_{ref}\right) + U\cdot A\cdot (T_{wall,z} - T_{amb}) + \xi_z\cdot \dot{m}_{F,z}\cdot LHV \\ = \dot{m}_{F,z}\cdot LHV + \dot{m}_{RC}\cdot C_{P,FG,z}\cdot (T_{FG,z+1} - T_{FG,z}) \qquad \text{for } z = 1\text{--}14 \qquad (11)$$

where α is the mass air-to-fuel ratio (kg_{air}/kg_{fuel}); U is the heat transfer coefficient at the furnace walls (W/m^2K); ξ is the percentage of energy loss due to hot gas leakage (-); LHV is the lower heating value of the fuel (kJ/kg); $C_{p,FG}$ is the heat capacity of the flue gases (kJ/kgK); m_F and m_{RC} are the mass flows of the fuel and the recycled gases from a hotter zone (kg/s); T_{wall} and A are the temperature (K) and the external surface area (m^2) of the furnace walls; T_{FG}, T_{ref}, and T_{amb} are the flue gases (K), reference (298 K), and ambient temperatures (298 K); and $T_{FG,z}$ and $T_{FG,z+1}$ are the temperatures of the recycled hot gases at the current and a next (hotter) zones. Rearranging Equation (11), the rate of fuel consumed per zone (kg/s) can be calculated using Equation (12):

$$\dot{m}_{F,z} = \frac{Q_{Al,z} + U\cdot A\cdot (T_{wall,z} - T_{amb}) - \dot{m}_{RC,z}\cdot C_{P,FG,z}\cdot (T_{FG,z+1} - T_{FG,z})}{(1 - \xi_z)\cdot LHV - (\alpha + 1)\cdot C_{P,FG,z}\cdot \left(T_{FG,z} - T_{ref}\right)} \qquad (12)$$

Different mechanisms of exergy destruction occur inside the ACL furnace. Expectedly, combustion is the most irreversible phenomenon; however, its impact can only be mitigated either by reducing the amount of fuel consumption (e.g., better heat recovery and isolation) or avoiding highly irreversible diffusion and heat transfer mechanisms between the combustion gas species. The latter is technically challenging, unless electrical heating powered by an ideal van 't Hoff fuel cell supersedes combustion technology. Another important source of exergy destruction is the heat transfer rate at a finite temperature difference between the hot gases and the aluminum sheet. This contribution to the exergy destruction can be calculated using Equation (13) for each zone:

$$Ex_{dest-HT,z} = Q_{Al,z}\cdot \left(1 - \frac{T_{amb}}{\overline{T}_{FG,z}}\right) - Q_{Al,z}\cdot \left(1 - \frac{T_{amb}}{\overline{T}_{Al,z}}\right) = Q_{Al,z}\cdot \left(\frac{T_{amb}}{\overline{T}_{Al,z}} - \frac{T_{amb}}{\overline{T}_{FG}}\right) \qquad (13)$$

where T_{amb} is the dead state temperature (298 K); T_{Al} and T_{FG} are aluminum and hot gases temperatures (K), respectively; and Q_{al} is the heat transferred from the hot gas to the aluminum sheet (kW). The other irreversibility mechanisms are the losses associated with the hot gas leakage, the flue gases leaving the stack at hot temperatures (e.g., if heat integration is not or only partially implemented), and the exergy destruction via wall heat losses. These exergy destruction rates can be calculated based on Equations (14)–(16), respectively:

$$Ex_{dest,Leakage,\ z} = \xi_z\cdot \dot{m}_{F,z}\cdot \varphi\cdot LHV \qquad (14)$$

$$Ex_{dest,StackGas,z} = [(\alpha + 1)\cdot \dot{m}_{F,z}\cdot C_{p,FG,z}\cdot (T_{FG,z} - T_{ref}) - \dot{m}_{RC}\cdot C_{P,FG,z} \\ \cdot (T_{FG,z} - T_{FG,z-1})]\cdot (1 - \tfrac{T_{amb}}{T_{FG,z}}) \qquad (15)$$

$$Ex_{dest,Wall,z} = UA(T_{wall,z} - T_{amb})\left(1 - \frac{T_{amb}}{\overline{T}_{FG,z}}\right) \qquad (16)$$

where φ is the ratio between the chemical exergy and the LHV of the fuel (b^{CH}/LHV ~ 1.02) and $\overline{T}_{FG}$ is the logarithmic mean temperature of the flue gases calculated as $\overline{T}_{FG} = (T_{FG} - T_{amb})/\ln(T_{FG}/T_{amb})$. Note that in Equation (15), the energy available in the flue gases of the current zone (z) can still be recycled and used to preheat the aluminum sheet in a previous zone ($z - 1$), thus reducing the total amount of energy rejected in the flue gases of the current zone. It is worth noticing that although the calculation of the heat flowing through the ACL furnace walls uses the furnace wall temperature, the exergy loss must be calculated using the actual temperature of the hot gases inside the furnace. This approach aims to include the exergy destruction along the isolation layer.

Finally, a zone-wise and total exergy destruction can be calculated in the furnace considering the exergy inflows and outflows (Figure 6):

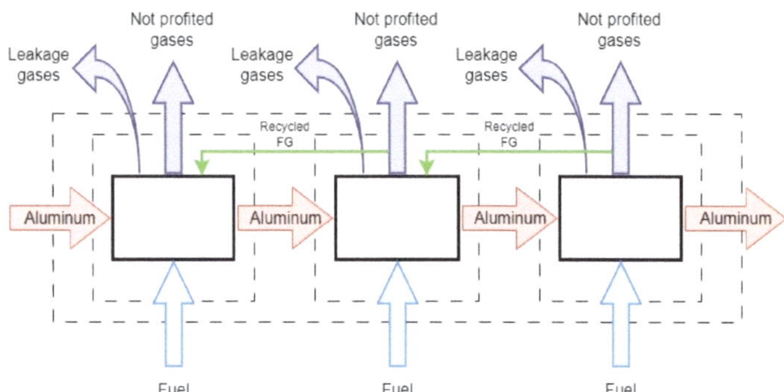

Figure 6. Schema of the calculation of the zone-wise exergy destruction and the total exergy destruction in the ACL. The dash lines depict the zones and total control volumes.

For the total exergy destruction $Ex_{dest,total}$, Equation (17):

$$Ex_{Al,\,in} - Ex_{Al,\,out} + \sum_{1..14} Ex_{fuel,\,z} = Ex_{dest,\,total} \tag{17}$$

For the zone-wise exergy destruction $Ex_{dest,z}$, Equation (18):

$$Ex_{Al,\,in,z} - Ex_{Al,\,out,z} + Ex_{fuel,\,z} - \dot{m}_{RC} \cdot C_{P,FG,z} \cdot (T_{FG,z} - T_{FG,z-1}) \left(1 - \frac{T_{amb}}{\overline{T}_{FG,z}}\right) = Ex_{dest,z} \tag{18}$$

where the exergy supplied by the fuel and the exergy recovered by the aluminum are calculated by using Equations (19) and (20), respectively:

$$Ex_{fuel,\,z} = \dot{m}_{F,z} \cdot \varphi \cdot LHV \tag{19}$$

$$Ex_{Al,\,in,z} - Ex_{Al,\,out,z} = Q_{Al,z} \cdot \left(1 - \frac{T_{amb}}{\overline{T}_{Al,z}}\right) \tag{20}$$

4. Results and Discussion

In this section, the CFD model validation is presented. The performance evaluation of the machine learning algorithms is discussed in the light of statistical indicators that measure the goodness of regression. Next, improvements for waste heat management and energy integration are analyzed by energy and exergy analysis in the ACL furnace. Finally, a sensitivity analysis is applied to estimate the variation in the fuel consumption as a function of the aluminum sheet velocity and thickness.

4.1. Validation of CFD Model

A computational fluid dynamic modelling and simulation of the ACL furnace is applied in the current study and this model is calibrated and validated using different experimental data available for certain operating conditions ($T_{hot\,gas}$ = 500 °C). Then, after the calibration, the error between the CFD and experimental data are calculated for two cases of $T_{hot\,gas}$ = 400 and 600 °C. The maximum error of the temperature profile is equal to 0.9%. Figure 7 depicts the aluminum temperature at three first zones for three constant profiles of hot gas temperatures. The temperature variations predicted using the CFD model (dashed lines) show good agreement with the measured data (solid lines).

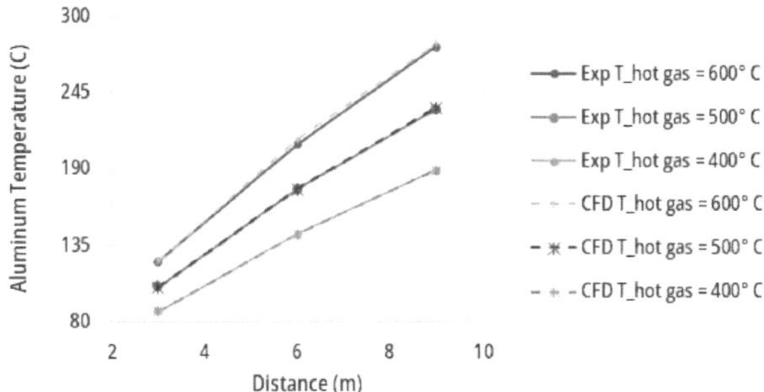

Figure 7. Validation of the simulated (CFD) temperature profiles along the ACL furnace.

4.2. Implementation of Supervised Learning Algorithms to Predict the Heat Transfer Coefficient

After the CFD and experimental data of the ACL furnace are processed, the simplified model based on the finite differences discretization approach is used to regress the heat transfer coefficient (h) as the output variable for different input operating conditions. To this end, four machine learning regression models, namely linear, polynomial, decision tree, and random forest, are applied to the dataset, which is divided into two subsets, namely training and testing data. The performance of the regression algorithms is checked by calculating different statistical metrics, such as the mean squared error (MSE) and coefficient of determination (R^2). The results of the metrics evaluation are presented in Figure 8. Accordingly, the polynomial regression model shows the best performance when predicting the heat transfer coefficient. Next, the heat transfer coefficient is predicted and used together with the simplified finite difference-based discretized model to calculate the aluminum temperature profiles for any given operating condition. The model inputs arethe aluminum thickness and band speed, percentage of recirculation fan, and furnace gas temperature.

4.3. Energy Input and Exergy Destruction as Functions of the Temperature Profile in the ACL Furnace

As expected, the higher the temperature of the hot gases, the higher the energy loss via stack and leakage gases and through the furnace walls. In addition, the higher the zone temperatures, the larger the internal exergy losses due to increased finite temperature differences between the hot gases and the aluminum. Thus, rational selection of the roof temperatures may avoid exergy losses (both internal and to the environment) being exacerbated. The zone temperatures could be controlled by using electrically heated elements, by consuming less fuel, or even by recycling hot gases from hotter zones to achieve the heating rates without incurring excessive avoidable exergy losses. This opportunity of optimizing the temperature profiles inside the ACL furnaces is further analyzed.

Figure 9 shows the constant hot gas temperature and the increasing aluminum temperature profiles in the business-as-usual (BAU) scenario, in which no temperature variation or waste heat recovery is implemented along the different zones of the ACL furnace. The operational constraints, such as the temperature set point for the continuous annealing process (~500–600° C) and the maintaining time of the treatment process, often defined by planning and material engineers as a recipe, should be observed. The sheet thickness and velocity are set to 1 mm and 30 m/min, respectively. According to Figure 9, there is a large driving force at the initial ACL furnace zones that reduces as the aluminum set point temperature is reached. During the initial zones, the temperatures of the stack and leakage gases are consequently higher, entailing not only larger exergy losses to environment but also avoidable exergy losses due to high finite temperature differences between the hot gases and the aluminum sheet.

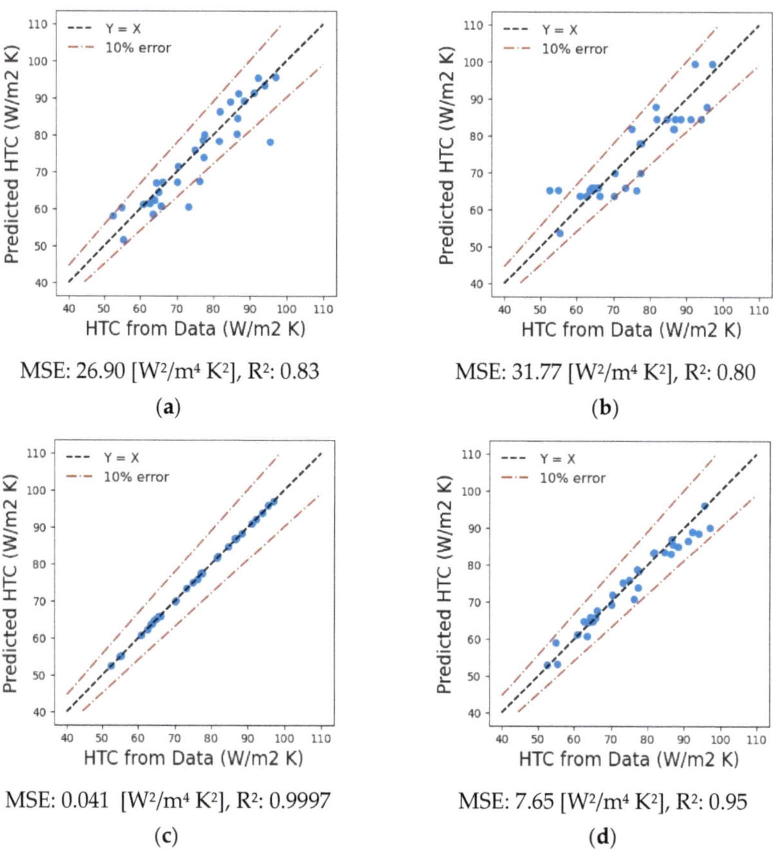

MSE: 26.90 [$W^2/m^4\ K^2$], R^2: 0.83

(**a**)

MSE: 31.77 [$W^2/m^4\ K^2$], R^2: 0.80

(**b**)

MSE: 0.041 [$W^2/m^4\ K^2$], R^2: 0.9997

(**c**)

MSE: 7.65 [$W^2/m^4\ K^2$], R^2: 0.95

(**d**)

Figure 8. Performance metrics of the test predictions for the heat transfer coefficient using (**a**) linear, (**b**) decision tree, (**c**) polynomial, and (**d**) random forest regression models.

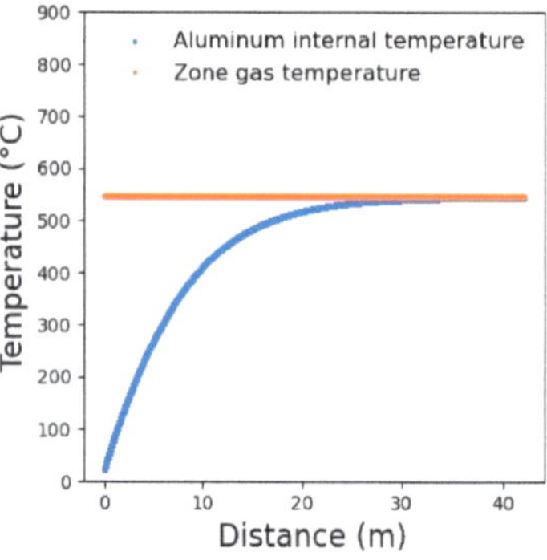

Figure 9. Hot gases and aluminum temperature profiles along the ACL furnace for the business-as-usual scenario (i.e., constant zonal temperatures).

The fuel consumption and the total exergy losses (internal and to the environment) in the ACL furnace are calculated for the business-as-usual scenario as 26.5 m^3_{NG}/tAl and 248.4 kWh/tAl, respectively. Those values are also determined for other temperature profiles, such as those shown in Figure 10, including different linear and polynomial temperature profiles in the ACL furnace. It is worth noting that energy integration via hot gas recycling is not yet applied, thus the total fuel consumption and process exergy destruction in the ACL furnace (Figure 11a,b, respectively) are initially calculated only as a function of the variation of the temperature profiles over the zones. Accordingly, the profile (o) demonstrates the best outcome, with a decrease in NG consumption of 8.5%, where exergy destruction reduces by 11.0% as the temperature profiles vary along the ACL furnace.

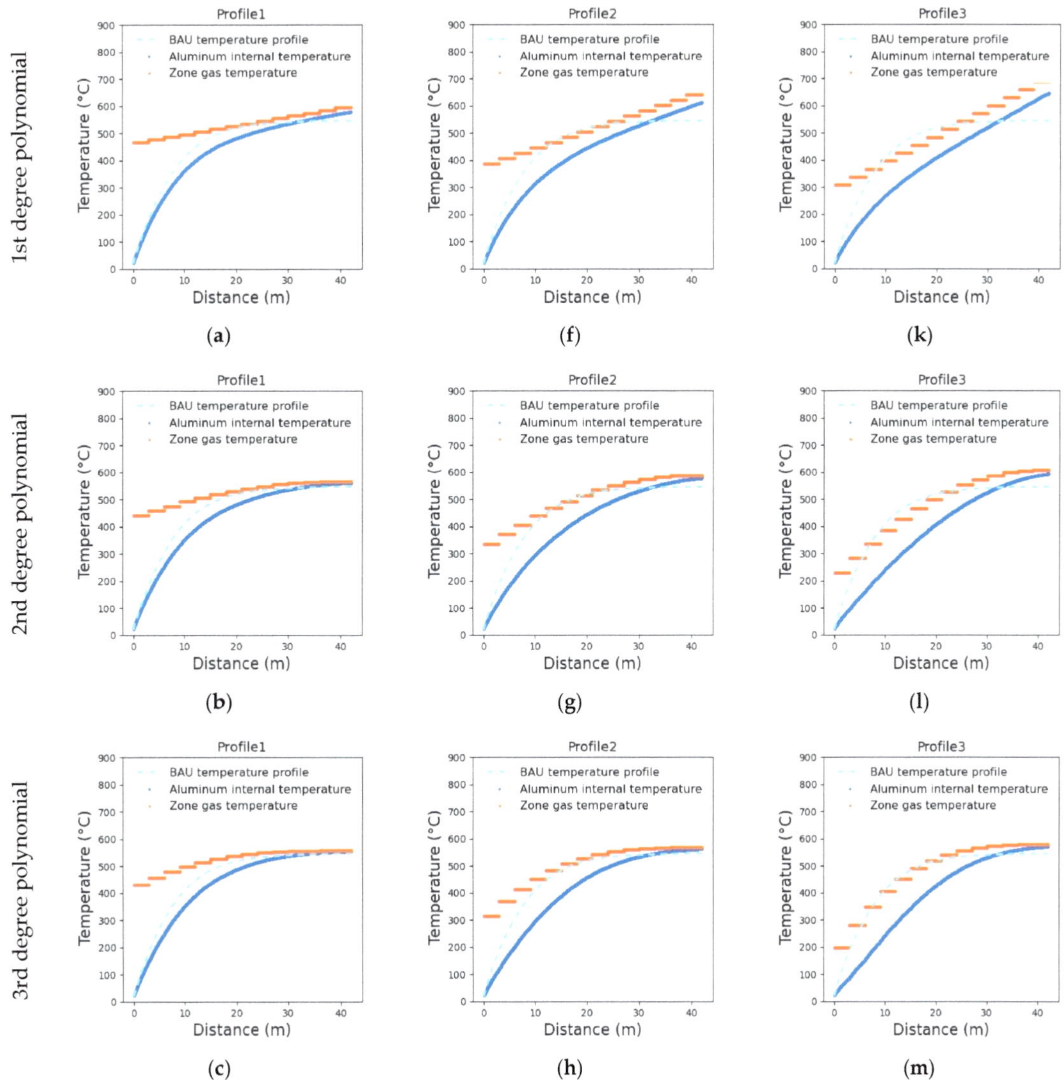

Figure 10. *Cont.*

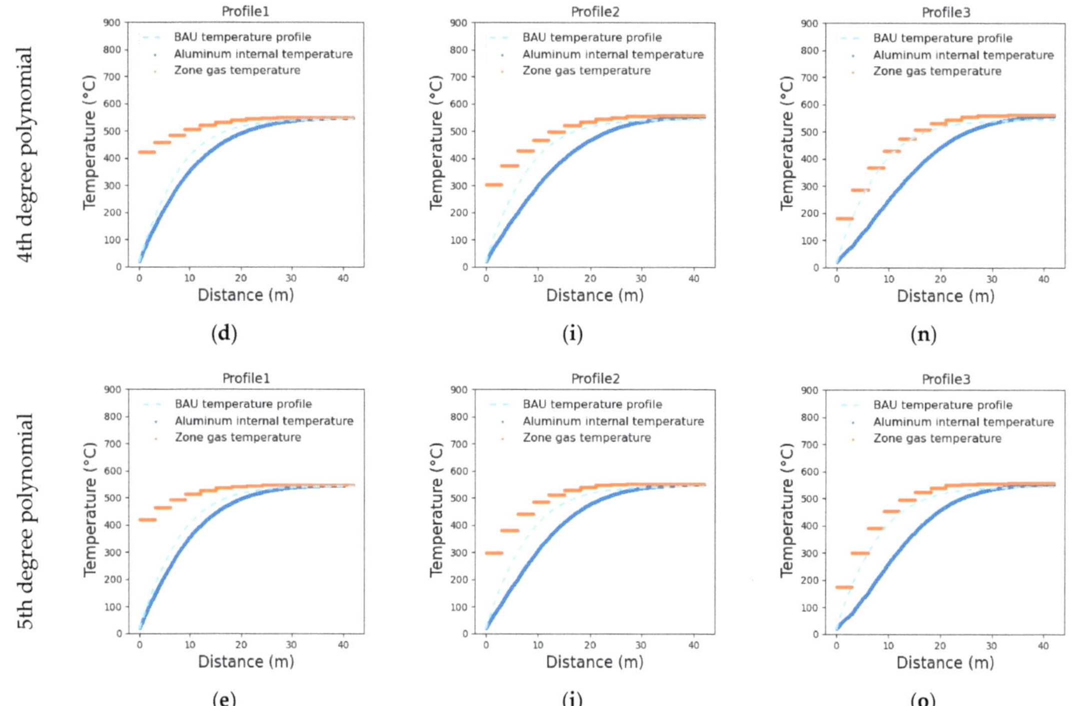

Figure 10. Simulated aluminum (solid blue curves) and roof gas (orange curve) temperature profiles along the ACL furnace zones considering linear and polynomic roof gas temperature profiles in subfigures (**a**–**o**). The aluminum temperature profile for the BAU scenario is shown as a dashed light blue curve. The profiles' equations (T = a_0+$a_1 \times$ x+$a_2 \times$ x^2+$a_3 \times$ x^3+$a_4 \times$ x^4+$a_5 \times$ x^5) are provided in Appendix A.

The exergy balance is depicted in Grassmann diagrams in Figure 12 for the two cases of the BAU scenario (Figure 9) and for profile (o) (Figure 10) without gas recycling. The evident variation in the profiles entails a major impact in terms of stack exergy loss reduction (29.5%) since the exhaust gas temperature is much lower in the initial zones of profile (o). For the same reason, considering that the energy and exergy losses are a function of the zone temperatures, the wall exergy destruction decreases by 14.6% when the variation in the hot gas profile is implemented. Moreover, a reduction of 8.6% is observed in exergy loss from leakages. Internal exergy destruction, which includes exergy losses from combustion, heat transfer, and other internal losses, decreases by 4.1% when employing the operational strategy suggested by profile (o) (Figure 10). A lower finite temperature difference between the hot gases and the aluminum reduces the irreversibility associated with the large driving force of the heat transfer phenomena.

Clearly, if the temperature of stack hot gases leaving the ACL furnace zones is limited by the maximum temperature attainable by the aluminum sheet, a large share of the hot gases' exergy may still be available at relatively high temperatures before it is rejected to the environment. Thus, together with the temperature variation over the zones, heat integration via hot gas recirculation can also be a suitable solution for reducing the irreversibility rates by recovering heat from the hot gases leaving hotter, downstream zones to heat the colder zones. This fact, in turn, reduces the fuel consumption. Obviously, waste heat recovery via the recycling of hot stack gases is not as interesting for the BAU scenario as for the variable zone temperature scenario. In the BAU case, energy could still be used to preheat the combustion air, but it will not produce the same effect in terms of decreasing the fuel input, since the exergy input necessary at the highest temperature will always need an additional consumption of fuel to reduce the exergy balance. On the other hand, when the exhaust gas of a zone is sent to a previous zone for heat recovery purposes (see Figure 1b), the potential for waste heat recovery is limited by the maximum temperature of the aluminum sheet; however, the total energy available at a high temperature may still be thermodynamically sufficient to provide the entire heat to the aluminum load. Thus, the amount of energy, but more importantly the quality thereof, plays an important role in the rational energy use and may help issuing recommendations based on the second principle of thermodynamics. The results of the energy consumption (m^3/tAl) and exergy destruction (kWh/tAl), calculated for the same profiles in Figure 10 but considering the energy integrated approach, are shown in Figure 13a,b, respectively. Those figures could be contrasted with Figure 11a,b to find that the exergy losses decrease when variable temperature profiles along the ACL furnace zones are adopted. Interestingly, it can be argued that the linear temperature profiles of the hot gases may indicate the minimum temperature necessary to achieve the heating process; thus, it could be an ideal candidate for the temperature set points in the zones. However, it can be also observed from Figure 10 that those profiles also impose heating rates that may delay the attainment of the annealing temperatures and thus represent shorter maintaining times at those conditions. However, depending on the recipe adopted by the materials engineers, the profiles will provide the required heating and maintaining rates. In this regard, other temperature profiles are analyzed in the light of the energy integration analysis, so that the effect of those temperature profiles on the reduction in natural gas consumption and exergy losses can be elucidated. For the sake of comparison, heating the aluminum sheet by using a constant temperature profile for hot gases may demand as much as 26.5 m^3 of natural gas per ton of aluminum, whereas the adoption of a polynomial profile such as that shown in Figure 10o and with heat recovery would only require 21.0 m^3/tonAl. Thus, the latter profile can save 20.7% of the required fuel and decrease 25.8% of the total exergy losses. Figure 13c shows the mass flowrate of recirculation and exhaust streams for profile (o). A limitation is set on the volumetric flowrate of the recycled flue gas to prevent the stream from flow choking. Due to this limitation, the exhaust gases from zones 3–6 are partially sent to their previous zones and some fractions are discharged through the stack. Lower zonal temperatures (thus, higher gas density) entail higher mass flowrates corresponding to a volumetric limit. As shown in Figure 13c, the recirculation mass flowrate increases from zones 6 to 3.

Additionally, the Grassmann diagrams with the exergy flows for both the BAU scenario and the scenario with temperature profile (o) (Figure 10), considering the energy integration approach (i.e., recycling enabled), are depicted in Figure 14a,b, respectively. For the heat integrated configuration, a higher fraction of the total exergy input flows into the aluminum load. Reductions in exergy losses are observed for all the types. The heat recovery from exhaust gases reduces the fuel consumption by 20.7%. As a result, the internal irreversibility due to combustion, leakages, and stack loss is sharply decreased.

4.4. Sensitivity Analysis to the Aluminum Sheet Thickness and Velocity

In the previous section, the energy consumption and the exergy destruction in the ACL furnace are calculated to show the effect of the variable profiles of the set-point temperatures for the hot gases and the advantages of the waste heat recovery from the zone stacks. The results showed a significant reduction in fuel consumption for specified values of aluminum sheet thickness (1 mm) and velocity (30 m/min) throughout the ACL furnace. However, the ACL operating conditions may vary depending on customers' requests and production throughput. Thus, the performance of the proposed energy-saving solutions should be also discussed for a range of aluminum sheet velocities and thicknesses. The specific fuel consumption (m^3/tAl) and the reduction percentages (%) with respect to the BAU scenario are recalculated for aluminum thickness ranging from 1 to 2.2 mm (Figure 15).

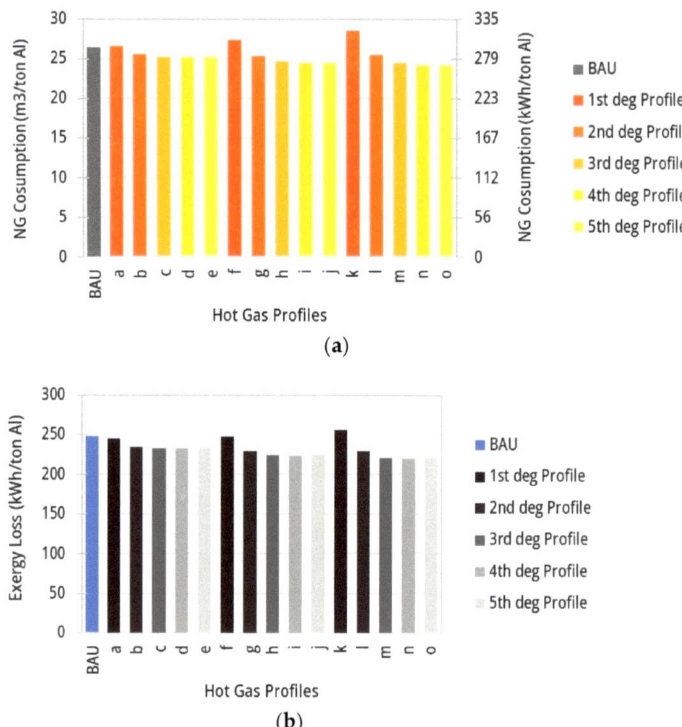

Figure 11. (**a**) Fuel consumption for different temperature profiles of hot gases along the ACL zones and for separate flue gas discharge (no gas recycling). (**b**) Total exergy loss for different temperature profiles of hot gas along the ACL zones and for separate flue gas discharge (no gas recycling).

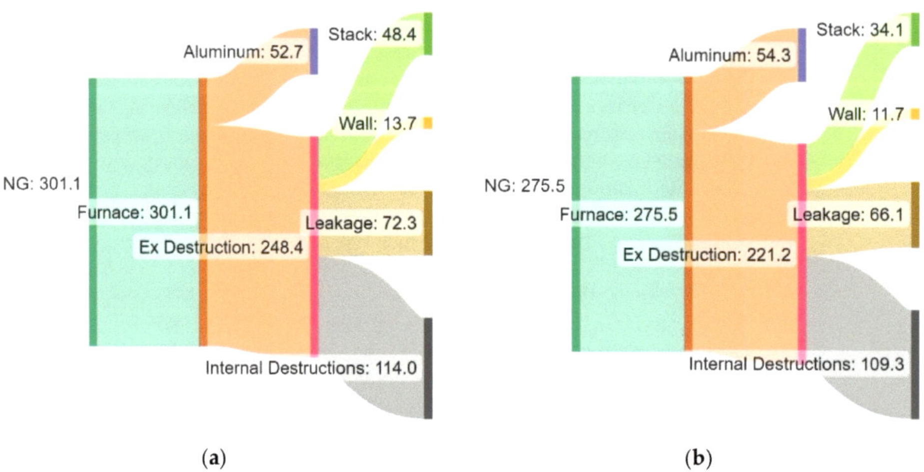

Figure 12. Grassmann diagrams of exergy flows in the ACL furnace (in kWh/tAl) for two representative cases: (**a**) BAU scenario and (**b**) profile (o) (Figure 10), i.e., with separate flue gas discharge (no recycling).

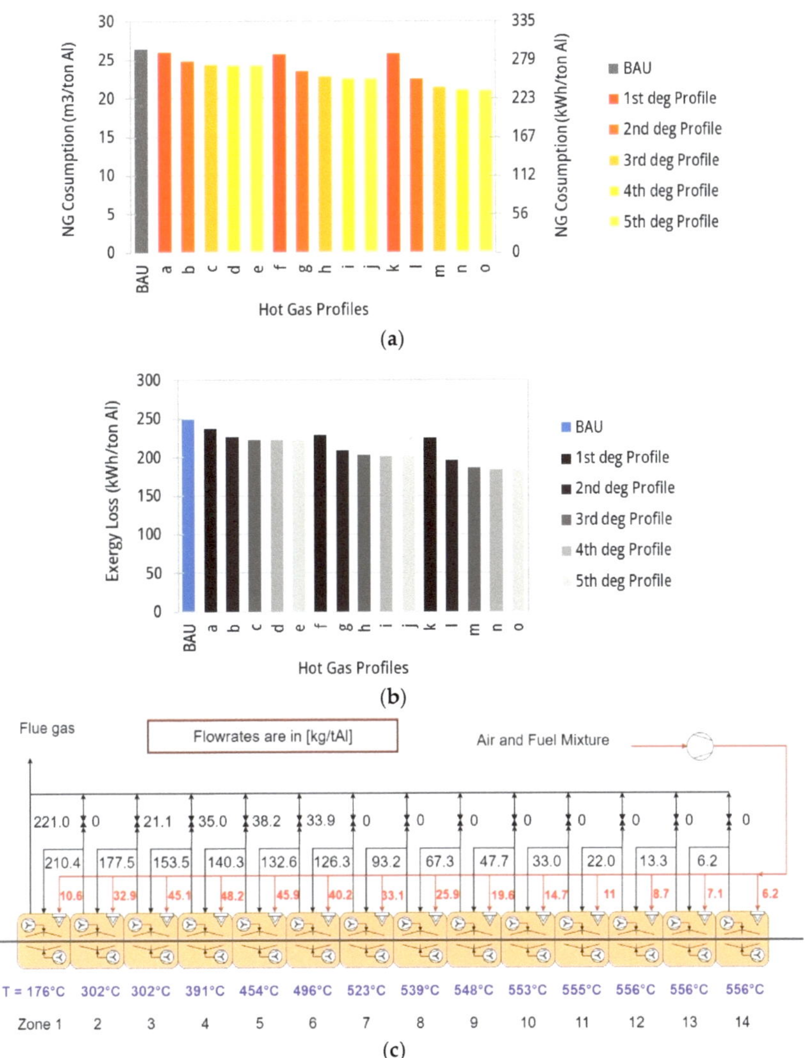

Figure 13. (**a**) Fuel consumption for different temperature profiles of hot gases (Figure 10) along the ACL zones for the energy integrated configuration (recycling enabled). (**b**) Total exergy losses for the different temperature profiles of hot gases (Figure 10) along the ACL zones for the energy integrated configuration (recycling enabled). (**c**) Mass flowrates of the flue gases (kg/tAl) per zone for temperature profile (o) (Figure 10) when the energy integration approach is adopted (recycling enabled).

According to Figure 15a, fuel consumption expectedly increases by increasing the sheet thickness. For lower thicknesses, a higher surface-to-volume ratio allows the aluminum to more easily and quickly achieve the treatment temperature and the maintaining time along the ACL furnace. By adopting the constant hot gas profile (BAU scenario), the fuel consumption ranges from 26.5 to 30.2 m^3/tAl for sheet thicknesses between 1 and 2.2 mm, respectively. On the other hand, the application of a non-constant profile (i.e., profile (o) (Figure 10)) together with enhanced energy integration decreases the fuel consumption by up to 21.0 and 26.8 m^3/tAl for the same range of thicknesses. Figure 15b depicts the reduction percentages of NG consumption due to the implementation of profile (o) (Figure 10) with heat integration in comparison with the BAU profile. As can be seen, fuel consumption reduces by 20.7 to 11.3% for sheet thicknesses of 1 to 2.2 mm, respectively.

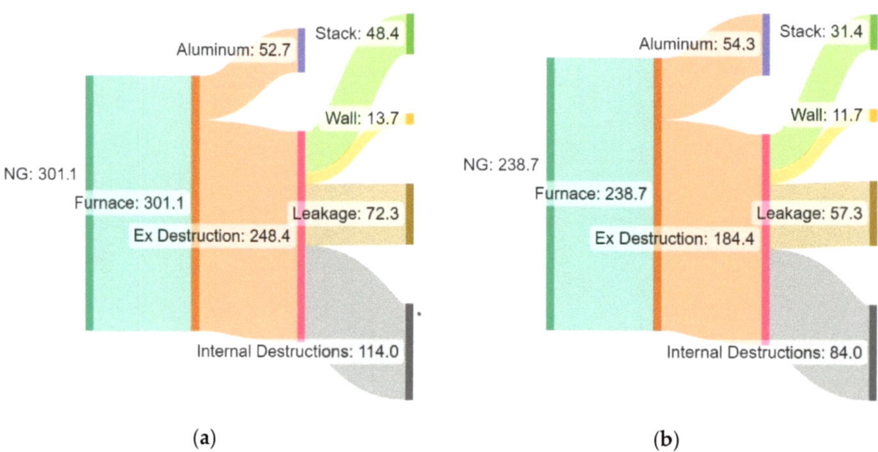

Figure 14. Grassmann diagrams of exergy flows (in kWh/tAl) in the ACL furnace for (**a**) the BAU scenario (same as Figure 12a) and (**b**) profile (o) (Figure 10), i.e., when the energy integration approach is adopted (recycling enabled).

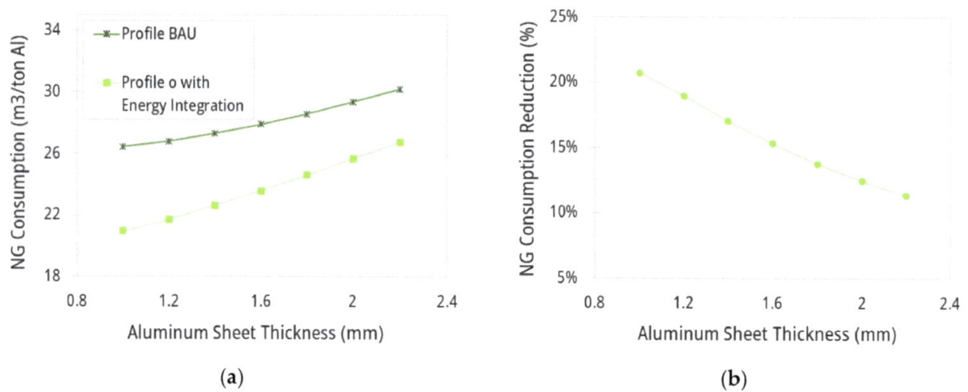

Figure 15. Plots of (**a**) specific fuel consumption (m^3/tAl) and (**b**) reduction percentages (%) of the scenario with temperature profile (o) (Figure 10), with heat integration (recycling enabled), in comparison with the BAU scenario as a function of the aluminum sheet thickness.

Figure 16 shows the effect of the aluminum sheet velocity on the energy consumption in the ACL furnace, considering the BAU temperature profile and profile (o) (Figure 10). Naturally, fuel consumption increases by increasing the band velocity, as a larger amount of mass of aluminum needs to be processed per unit of time. In cases of higher velocity, it is required that the furnace operates at higher temperatures to provide a higher heat flux for reaching the heat treatment temperatures on time and providing the maintaining times. For temperature profile (o) (Figure 10), the natural gas (NG) consumption varies between 21.0 to 25.7 m^3/tAl for aluminum velocities of 30 to 45 m/min, respectively. Implementation of improved temperature profiles and energy integration approaches (hot gas recycling) may lead to a 20.7 to 10.8% reduction in fuel consumption for band velocities from 30 to 45 m/min.

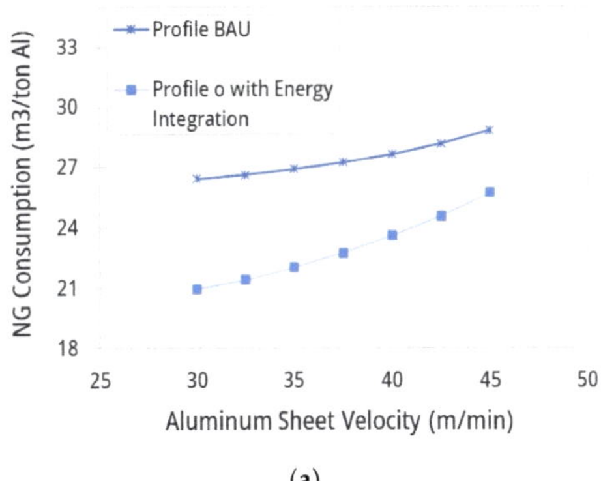

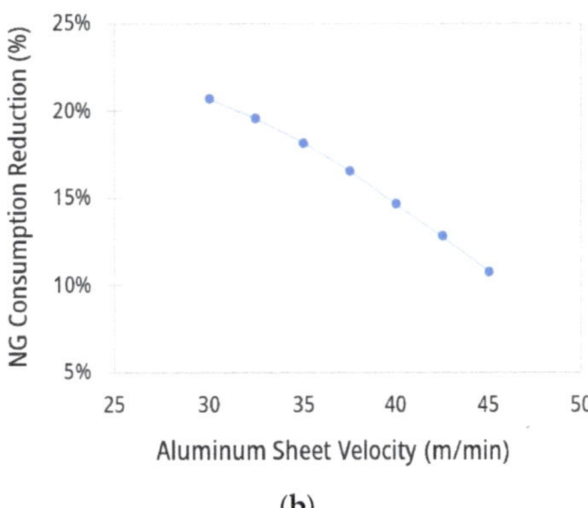

(a)   (b)

Figure 16. Plots of (**a**) specific fuel consumption (m^3/tAl) and (**b**) reduction percentages (%) for the scenario with temperature profile (o) (Figure 10), with heat integration (recycling enabled), in comparison with the BAU scenario as a function of the aluminum sheet velocity.

5. Conclusions

ACL furnaces are utilized for the heat treatment of aluminum sheets in the rolling industry. In the current study, an energy model of this furnace is developed using the finite difference method (FDM) and machine learning (ML) approaches trained on the basis of the experimental data available and the computational fluid dynamics (CFD) simulations in order to characterize the thermal performance of the system and propose solutions for waste heat management. Four ML models are evaluated for regression of the heat transfer coefficient (HTC). The polynomial model shows the best performance, with an MSE of 0.06 (W^2/m^4 K^2) and a coefficient of determination (R^2) equal to 0.9997. An average error of 0.43% is observed, which is more precise than the models in the open literature (around 5% reported by [13]). This operational model calculates the aluminum temperature profile along the ACL furnace, as well as the fuel consumption, and allows achieving an exergy analysis based on arbitrary operating conditions. A low computational time makes the model suitable for real-time controlling and optimization applications.

Solutions for improvement of the energy performance are also assessed, including the variation in the furnace temperature profiles and the energy integration via the partial recycling of the hot flue gases. The results demonstrate that the heat integration significantly increases the efficiency of the operating conditions for non-constant temperature profiles and also decreases the fuel consumption by up to 20.7% compared with the business-as-usual (BAU) scenario. The sensitivity analysis on ACL fuel consumption for aluminum sheet thickness variation from 1 to 2.2 mm shows an increase in natural gas consumption of 27.6%. Additionally, an increase in the band speed from 30 to 45 m/min leads to a 22.4% increase in fuel consumption. By increasing both parameters, more fuel is required, but the proposed solutions with non-constant temperature profiles in the ACL furnace with energy integration (hot gas recycling) still effectively reduce the fuel consumption and therefore the associated environmental impact.

Author Contributions: Conceptualization, M.A. and D.A.F.-O.; methodology, M.A. and D.A.F.-O.; software, M.A.; validation, M.A., D.A.F.-O. and R.G.; formal analysis and investigation, M.A., D.A.F.-O. and R.G.; data curation, M.A. and R.G.; writing—original draft preparation, M.A. and D.A.F.-O.; writing—review and editing, M.G. and D.A.F.-O.; visualization, M.A. and D.A.F.-O.; supervision and project administration, F.M.; funding acquisition, F.M. and R.G. All authors have read and agreed to the published version of the manuscript.

Funding: The authors would like to thank the Swiss Federal Office of Energy and Suisse Energie for funding this research work in the context of the Net Zero Lab consortium, with the project "Enhanced waste heat recovery approach for the reduction of fuel consumption in the aluminium industry" grant No. SuisseEnergie 2022-09-30.

Institutional Review Board Statement: Not applicable.

Data Availability Statement: Not applicable.

Acknowledgments: The authors would like to acknowledge the Net Zero Lab team for their insights on the final version of this manuscript.

Conflicts of Interest: The authors declare no conflict of interest. The funders had no role in the design of the study; in the collection, analyses, or interpretation of data; in the writing of the manuscript; or in the decision to publish the results.

Nomenclature

Symbols

ρ	Density (kg/m^3)
$\vec{v}$	Gas velocity vector (m/s)
p	Static pressure (Pa)
$\bar{\bar{\tau}}$	Stress tensor (N/m^2)
$\vec{g}$	Gravitational acceleration (m/s^2)
k	Turbulence kinetic energy (J/kg)
ω	Specific dissipation rate (m^2/s^3)
$\tilde{G}_k$	Generation of turbulent kinetic energy (J/kg)
G_ω	Generation of specific dissipation rate (m^2/s^3)
Γ_k	Effective diffusivity of turbulent kinetic energy (m^2/s)
Γ_ω	Effective diffusivity of specific dissipation rate (m^2/s)
Y_k	Dissipation of turbulent kinetic energy (J/kg)
Y_ω	Dissipation of specific dissipation rate (m^2/s^3)
k_{eff}	Conductivity (W/m K)
T	Temperature (°C)
h	Sensible enthalpy (J/kg)
$\vec{J}$	Diffusion flux (kg/m^2 s)
α	Absorption coefficient (-)
G	Incident radiation (W/m^2)
σ	Stefan–Boltzmann constant, 5.67×10^{-8} (W/m^2 K^4)
C_p	Aluminum specific heat capacity (J/kg K)
A	Cross-sectional area of aluminum (m^2)
α	Mass air-to-fuel ratio (kg$_{\text{air}}$/kg$_{\text{fuel}}$)
U	Heat transfer coefficient of furnace walls (W/m^2 K)
ξ	Percentage of energy loss due to hot gas leakage (-)
Ex	Exergy (J)
φ	Ratio of the chemical exergy to LHV of the fuel (-)
MSE	Mean squared error
R^2	Coefficient of determination (-)

Subscripts

o	Aluminum inner body
s	Aluminum surface
∞	Hot gas
FG	Flue gases
F	Fuel
Al	Aluminum
RC	Recycled gases from a hotter zone
wall	Furnace walls
ref	Reference
amb	Ambient

z	Current zone
z+1	Next (hotter) zone
dest	Destruction
HT	Heat transfer
Abbreviations	
FDM	Finite difference method
ML	Machine learning
HTC	Heat transfer coefficient
NG	Natural gas
LHV	Lower heating value
BAU	Business-as-usual

Appendix A

The equations of zone temperature profiles shown in Figure 10 are summarized in Table A1.

Table A1. The equations of the profiles in Figure 10 ($T = a_0 + a_1 \times x + a_2 \times x^2 + a_3 \times x^3 + a_4 \times x^4 + a_5 \times x^5$). T in °C, x in m.

Profile	Curve Equation
BAU	$T = 544.3$
a	$T = 3.25x + 459.18$
b	$T = -0.0833x^2 + 6.75x + 427.81$
c	$T = 0.0021x^3 - 0.2596x^2 + 10.514x + 412.56$
d	$T = -5.0E-5x^4 + 0.0089x^3 - 0.5392x^2 + 14.558x + 402.6$
e	$T = 1.0E-06x^5 - 0.0003x^4 + 0.023x^3 - 0.9332x^2 + 18.898x + 394.73$
f	$T = 6.5x + 375.25$
g	$T = -0.1667x^2 + 13.5x + 311.63$
h	$T = 0.0043x^3 - 0.5192x^2 + 21.029x + 280.41$
i	$T = -0.0001x^4 + 0.0178x^3 - 1.0784x^2 + 29.117x + 261.59$
j	$T = 3.0E-6x^5 - 0.0006x^4 + 0.0461x^3 - 1.8665x^2 + 37.796x + 246.05$
k	$T = 9.75x + 290.43$
l	$T = -0.25x^2 + 20.25x + 195.44$
m	$T = 0.0064x^3 - 0.7788x^2 + 31.543x + 149.67$
n	$T = -0.0002x^4 + 0.0266x^3 - 1.6176x^2 + 43.675x + 120.09$
o	$T = 4.0E-6x^5 - 0.0009x^4 + 0.0691x^3 - 2.7997x^2 + 56.694x + 97.179$

References

1. Dong, H.-R.; Li, X.-Q.; Li, Y.; Wang, Y.-H.; Wang, H.-B.; Peng, X.-Y.; Li, D.-S. A review of electrically assisted heat treatment and forming of aluminum alloy sheet. *Int. J. Adv. Manuf. Technol.* **2022**, *120*, 7079–7099. [CrossRef]
2. Lee, J.; Bong, H.J.; Kim, D.; Lee, Y.-S.; Choi, Y.; Lee, M.-G. Mechanical properties and formability of heat-treated 7000-series high-strength aluminum alloy: Experiments and finite element modeling. *Met. Mater. Int.* **2020**, *26*, 682–694. [CrossRef]
3. Majeau-Bettez, G.; Krey, V.; Margni, M. What future for primary aluminium production in a decarbonizing economy? *Glob. Environ. Change* **2021**, *69*, 102316.
4. Zhou, B.; Liu, B.; Zhang, S. The advancement of 7xxx series aluminum alloys for aircraft structures: A review. *Metals* **2021**, *11*, 718. [CrossRef]
5. Deng, L.; Johnson, S.; Gencer, E. Environmental-Techno-Economic analysis of decarbonization strategies for the Indian aluminum industry. *Energy Convers. Manag.* **2022**, *274*, 116455. [CrossRef]
6. Gao, P.; Ren, Z.; Zhan, M.; Xing, L. Tailoring of the microstructure and mechanical properties of the flow formed aluminum alloy sheet. *J. Alloys Compd.* **2022**, *928*, 167139. [CrossRef]
7. Mayrhofer, M.; Koller, M.; Seemann, P.; Prieler, R.; Hochenauer, C. CFD investigation of a vertical annealing furnace for stainless steel and non-ferrous alloys strips—A comparative study on air-staged & MILD combustion. *Therm. Sci. Eng. Prog.* **2022**, *28*, 101056.
8. Arkhazloo, N.B.; Bazdidi-Tehrani, F.; Jadidi, M.; Morin, J.-B.; Jahazi, M. Determination of temperature distribution during heat treatment of forgings: Simulation and experiment. *Heat Transf. Eng.* **2022**, *43*, 1041–1064. [CrossRef]
9. Arkhazloo, N.B.; Bazdidi-Tehrani, F.; Morin, J.-B.; Jahazi, M. Optimization of furnace residence time and loading pattern during heat treatment of large size forgings. *Int. J. Adv. Manuf. Technol.* **2021**, *113*, 2447–2460. [CrossRef]

10. Jóźwiak, P.; Hercog, J.; Kiedrzyńska, A.; Badyda, K.; Olevano, D. Thermal Effects of Natural Gas and Syngas Co-Firing System on Heat Treatment Process in the Preheating Furnace. *Energies* **2020**, *13*, 1698. [CrossRef]
11. Nave, O. Modification of semi-analytical method applied system of ODE. *Mod. Appl. Sci.* **2020**, *14*, 75. [CrossRef]
12. Dou, R.; Zhao, H.; Zhao, P.; Wen, Z.; Li, X.; Zhou, L.; Zhang, R. Numerical model and optimization strategy for the annealing process of 3D coil cores. *Appl. Therm. Eng.* **2020**, *178*, 115517. [CrossRef]
13. Hajaliakbari, N.; Hassanpour, S. Analysis of thermal energy performance in continuous annealing furnace. *Appl. Energy* **2017**, *206*, 829–842. [CrossRef]
14. Strommer, S.; Niederer, M.; Steinboeck, A.; Kugi, A. A mathematical model of a direct-fired continuous strip annealing furnace. *Int. J. Heat Mass Transf.* **2014**, *69*, 375–389. [CrossRef]
15. Cho, M.; Ban, J.; Seo, M.; Kim, S.W. Neural network MPC for heating section of annealing furnace. *Expert Syst. Appl.* **2023**, *223*, 119869. [CrossRef]
16. He, F.; Wang, Z.-X.; Liu, G.; Wu, X.-L. Calculation Model, Influencing Factors, and Dynamic Characteristics of Strip Temperature in a Radiant Tube Furnace during Continuous Annealing Process. *Metals* **2022**, *12*, 1256. [CrossRef]
17. Kwon, B.; Ejaz, F.; Hwang, L.K. Machine learning for heat transfer correlations. *Int. Commun. Heat Mass Transf.* **2020**, *116*, 104694. [CrossRef]
18. Mehralizadeh, A.; Shabanian, S.R.; Bakeri, G. Investigation of boiling heat transfer coefficients of different refrigerants for low fin, Turbo-B and Thermoexcel-E enhanced tubes using computational smart schemes. *J. Therm. Anal. Calorim.* **2020**, *141*, 1221–1242. [CrossRef]
19. Yoo, J.M.; Lee, D.H.; Hong, D.J.; Jeong, J.J. Application of machine learning technique in predicting condensation heat transfer coefficient and droplet entrainment rate. In Proceedings of the Transactions of the Korean Nuclear Society Virtual Spring Meeting, Jeju, Republic of Korea, 10–11 May 2007.
20. Senanu, S.; Solheim, A.; Lødeng, R. Gas Recycling and Energy Recovery. Future Handling of Flue Gas from Aluminium Electrolysis Cells. In *Light Metals*; Springer: Berlin/Heidelberg, Germany, 2022; pp. 1004–1010.
21. Jouhara, H.; Nieto, N.; Egilegor, B.; Zuazua, J.; González, E.; Yebra, I.; Igesias, A.; Delpech, B.; Almahmoud, S.; Brough, D.; et al. Waste heat recovery solution based on a heat pipe heat exchanger for the aluminium die casting industry. *Energy* **2023**, *266*, 126459. [CrossRef]
22. Brough, D.; Jouhara, H. The aluminium industry: A review on state-of-the-art technologies, environmental impacts and possibilities for waste heat recovery. *Int. J. Thermofluids* **2020**, *1*, 100007. [CrossRef]
23. Orrego, F.; Alexander, D.; Dareen, D.; Reginald, G.; François, M. A Systemic Study for Enhanced Waste Heat Recovery and Renewable Energy Integration towards Decarbonizing the Aluminium Industry. In Proceedings of the 36th International Conference on Efficiency, Cost, Optimisation, Simulation and Environmental Impact of Energy Systems—ECOS 2023, Canary Islands, Spain, 25–30 June 2023.
24. Teske, S.; Niklas, S.; Talwar, S. Decarbonisation pathways for industries. In *Achieving the Paris Climate Agreement Goals: Part 2: Science-Based Target Setting for the Finance Industry—Net-Zero Sectoral 1.5 °C Pathways for Real Economy Sectors*; Springer: Berlin/Heidelberg, Germany, 2022; pp. 81–129.
25. Fluent, A. *Ansys Fluent Theory Guide*; Ansys Inc.: Canonsburg, PA, USA, 2011; Volume 15317, pp. 724–746.
26. Zhao, J.; Ma, L.; Zayed, M.E.; Elsheikh, A.H.; Li, W.; Yan, Q.; Wang, J. Industrial reheating furnaces: A review of energy efficiency assessments, waste heat recovery potentials, heating process characteristics and perspectives for steel industry. *Process Saf. Environ. Prot.* **2021**, *147*, 1209–1228. [CrossRef]
27. Kumbhar, S.V.; Sonage, B.K. Unsteady-state lumped heat capacity system design for tube furnace for continuous inline wire annealing process. *Heat Transf.* **2019**, *48*, 874–884. [CrossRef]
28. Hughes, M.T.; Kini, G.; Garimella, S. Status, challenges, and potential for machine learning in understanding and applying heat transfer phenomena. *J. Heat Transf.* **2021**, *143*, 120802. [CrossRef]

Article

Simultaneous Optimization and Integration of Multiple Process Heat Cascade and Site Utility Selection for the Design of a New Generation of Sugarcane Biorefinery

Victor Fernandes Garcia [1] and Adriano Viana Ensinas [2,*]

1   Center of Engineering, Modeling and Social Science Applied, Federal University of ABC, Santo André 09210-580, Brazil; v.garcia@ufabc.edu.br
2   Department of Engineering, Federal University of Lavras, Lavras 37000-200, Brazil
*   Correspondence: adriano.ensinas@ufla.br

Abstract: Biorefinery plays a crucial role in the decarbonization of the current economic model, but its high investments and costs make its products less competitive. Identifying the best technological route to maximize operational synergies is crucial for its viability. This study presents a new superstructure model based on mixed integer linear programming to identify an ideal biorefinery configuration. The proposed formulation considers the selection and process scale adjustment, utility selection, and heat integration by heat cascade integration from different processes. The formulation is tested by a study where the impact of new technologies on energy efficiency and the total annualized cost of a sugarcane biorefinery is evaluated. As a result, the energy efficiency of biorefinery increased from 50.25% to 74.5% with methanol production through bagasse gasification, mainly due to its high heat availability that can be transferred to the distillery, which made it possible to shift the bagasse flow from the cogeneration to gasification process. Additionally, the production of DME yields outcomes comparable to methanol production. However, CO_2 hydrogenation negatively impacts profitability and energy efficiency due to the significant consumption and electricity cost. Nonetheless, it is advantageous for surface power density as it increases biofuel production without expanding the biomass area.

Keywords: biorefinery; MILP superstructure; carbon credit; process integration; biofuels; optimization; heat integration

Citation: Garcia, V.F.; Ensinas, A.V. Simultaneous Optimization and Integration of Multiple Process Heat Cascade and Site Utility Selection for the Design of a New Generation of Sugarcane Biorefinery. *Entropy* **2024**, *26*, 501. https://doi.org/10.3390/e26060501

Academic Editors: Daniel Flórez-Orrego, Meire Ellen Ribeiro Domingos and Rafael Nogueira Nakashima

Received: 4 April 2024
Revised: 4 June 2024
Accepted: 5 June 2024
Published: 8 June 2024

1. Introduction

The economic development of a region is linked to the increase in its energy consumption. At present, the world's energy matrix is mainly composed of non-renewable sources; therefore, in a global scenario of worsening climate change, the development of renewable energy sources that do not cause greater greenhouses gas emissions is fundamental for sustainable development and the creation of a low-carbon economy. In this sense, biorefineries play a crucial role in sustainable economic development by enabling the recovery of waste that would otherwise be discarded. A biorefinery is a collection of processes that can sustainably convert biomass into marketable products such as bioplastics, biofuels, and chemical intermediates. Several biorefineries have been developed and implemented, including those for sugarcane, wood, microalgae, and municipal solid waste. However, the presence of biofuels and other products from renewable resources is still low due to technological and economic barriers. To achieve competitive improvement, a biorefinery must operate in an optimized manner, making sustainable use of all available resources. However, the diversity of resources, processes, and products makes it difficult to identify this optimal configuration, making the development of a biorefinery a complex and difficult task to solve. In this sense, superstructure modeling and optimization, a computational tool used to generate and evaluate systematically all the configurations that element sets can present, have been successfully used in process synthesis. This approach uses mathematical programming to identify the ideal combination of a set of alternatives that should be adopted to achieve a given objective. In process synthesis engineering, the superstructure is widely used in heat exchange network synthesis, conversion route evaluation, and supply chain networks. A superstructure typically consists of an objective function used to evaluate and compare different outcomes, along with a set of constraints, variables, and parameters. Optimization by superstructure is a viable solution to this problem, as evidenced by the results of several researchers, as will be presented. Based on thermodynamic laws, process heat integration (HI) combined with pinch analysis (PA) is an essential tool that can increase the economic viability of a biorefinery while reducing its carbon dioxide (CO_2) emissions from the utility system [1]. This method identifies opportunities for heat exchange between heat flows within the same or different processes, thereby reducing fuel consumption in the utility system. When the method is applied considering different processes, it is known as total site heat integration (TSHI) [2]. Various works in the literature apply TSHI concepts to superstructures [3–6]. Due to the combinatorial nature of the problem, most studies implement sequential strategies to solve it. However, this approach lacks the guarantee of finding a global optimal solution and may overlook more suitable options for the specific problem. This approach can have a high computational cost, making its application impractical in certain cases. Limitations are also present in other works, as they offer specific formulations for certain problems.

The sugar and alcohol sector is a crucial industry for the Brazilian economy, with Brazil currently ranking second in the world in bioethanol production. Traditional bioethanol production in the sugarcane industry is a consolidated process in which bioethanol is produced from the sugars present in sugarcane juice in seven stages: cleaning, preparation, and extraction; processing; concentration; sterilization and cooling; fermentation; distillation; and dehydration. First, the sugarcane arrives at the distillery, where it undergoes a process of cleaning, cutting, and then grinding. The extracted juice is sent to the treatment stage, where impurities are removed through a coagulation and decantation process. The bagasse produced is used as fuel in the utility system. Next, the treated broth is sent to an evaporator system where it is concentrated by an evaporator system until it reaches 19°Brix (a unit of measurement used in the industry to express the mass of sugars in a solution) [7]. In the sterilization stage, the already concentrated broth is heated to 130 °C and then cooled to 32 °C. During fermentation, the sugars present in the broth are consumed and converted into bioethanol by the action of yeasts of the genus Saccharomyces cerevisiae. It also produces carbon dioxide, which is released into the atmosphere after purification. To recover the bioethanol, the wine produced is sent to a distillation column system made up of four columns (A, A1, D, and B-B1). At this stage of the process, hydrated bioethanol (92.6–93.8% by mass of bioethanol) and a stream of vinasse [8] are obtained, a dark brown effluent, acidic in nature and with a high pollution potential. For each liter of bioethanol, 10 to 15 L of vinasse are produced [9]. To achieve a concentration of 99.3% by mass for bioethanol, the hydrated bioethanol undergoes a dehydration process in which excess water is removed using a solvent such as monoethylene glycol (MEG) or cyclohexane [10].

A standard distillery uses a cogeneration system to meet its energy needs, generating both electrical and thermal energy for all stages of the process. This system consists of a boiler, steam turbine, and electrical generator that form a steam cycle. Bagasse is used as a fuel to heat water and produce superheated steam. The resulting steam is sent to the steam turbine to expand and generate electricity. The steam then moves to the process, provides the required energy, and finally returns to the cogeneration system. In certain cases, there may be an excess of steam, which is then transported to thermoelectric facilities. The excess electricity produced by these plants is then sold to the power grid. According to Albarelli [10], improvements in energy integration of the process with investments in heat recovery technology can make a large amount of bagasse available as feedstock for other processes, such as gasification integrated with methanol production. According to Fuess et al. [11], the integration of new processes, in addition to increasing energy efficiency, would diversify the products obtained, improving the biorefinery nature of the sugarcane industry. In this sense, the biodigestion of vinasse is a possibility to be evaluated. It is usually used for the fertigation of sugarcane crops, but its polluting properties limit its use in the soil. In addition, the large quantity produced makes its proper disposal a problem for the distillery. Thus, biodigestion of vinasse, in addition to reducing its pollution potential, would produce biomethane, a biofuel considered strategic for the energy transition [12]. Methanol is a product that can be blended with gasoline, is used in the production of biodiesel or directly in fuel cells, and is commonly obtained from natural gas; the conversion of carbon dioxide to methanol is a process that has received attention [13,14].

As mentioned earlier, determining the configuration of a biorefinery can be a complex task due to the numerous processes that can be integrated. To optimize biorefineries and identify the best production route, several authors have adopted the use of superstructure as an alternative. Infante et al. [15] presented a MILP formulation to evaluate a microalgae biorefinery considering the production of different biofuels in Colombia. Its formulation found that microalgae liquefaction was the most viable route, while bagasse was used as process fuel and pellet production. Fonseca et al. [16] formulated a superstructure to evaluate the best strategy and the economic feasibility of integrating a second-generation ethanol process into an existing distillery. The results suggested that all bagasse was allocated to hydrolysis, while sugarcane straw, lignin, and biogas were directed to a Rankine Cycle. Huynh et al. [17] used a superstructure aiming to maximize the biodiesel production profit. The study by Kenkel et al. [18] employs a bicriteria superstructure, a superstructure with two objective functions, to investigate the conversion of CO_2 to methanol in Germany. The authors found that the price of electricity significantly influences the selection of technologies, impacting directly the production costs and emissions. The study suggests that synthetic methanol production from renewable energy sources could become competitive with natural gas in the future if its cost were reduced. Pyrgakis and Kokossis [19] employed a superstructure with a bipartite graphical representation and a modified total site cascade to study a real lignocellulosic biorefinery. The proposed formulation identified operational synergies between the thermal currents of the processes, which reduced the demand for hot and cold utilities by 9% and 14%, respectively. Celebi et al. [20] proposed a multi-objective superstructure for comparing sugar and syngas biorefinery platforms, ranking thirty-four configurations, with the lowest cost configuration integrating DME production with succinic acid. With its superstructure formulation, Galanopoulos et al. [21] reduce the biodiesel production costs by up to 80% in an integrated algae biorefinery using wastewater and CO_2 emissions. As highlighted above, the approach taken by the works considers sequential procedures to solve problems, so it lacks the guarantee of finding a globally optimal solution and may ignore more appropriate options for the specific problem. To address this issue, this paper presents a new superstructure formulation that utilizes mixed integer linear programming (MILP) for biorefinery optimization. The presented model's innovation lies in its ability to perform the process selection with scale adjustment, simultaneously to utility selection, and heat recovery by heat cascade integration. Unlike other approaches, this method results in the optimal configuration of the biorefinery, ensuring that the presented configuration is the best possible.

2. Methodology

The formulation presented in this paper consists of general mass and energy balance and is based on the previous work of Kantor et al. [22], while the constraints to perform heat cascade constraint is based on Bagajewicz and Rodera [23]. To achieve this, black box models representing each technology are inserted into the biorefinery. Each of the inserted models is based on previous works and describes the conversion, input, and output flows of each process, as well as their thermal stream, which are used to perform the heat cascade integration. Therefore, the superstructure receives information related to process and resource economics, including process operation, maintenance, and investment costs, as well as resource acquisition costs and market prices. The superstructure was modeled as a MILP model and implemented in LINGO software v.21 [24]. Figure 1 shows the flow of information in the superstructure. Next, the formulation of the superstructure is presented.

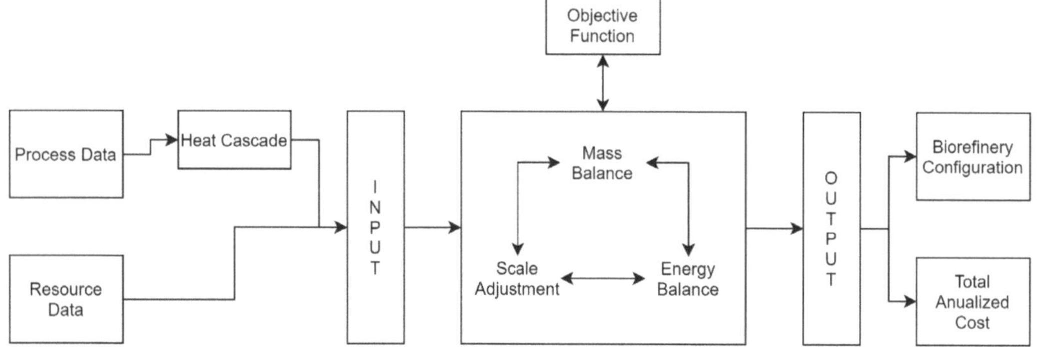

Figure 1. Schematic representation of superstructure flow information.

2.1. Main Sets Definitions

The formulation considers different sets, subsets, and their combinations. This structure makes it possible to perform the inclusion and exclusion of processes and their parameters in a faster and more organized way, in addition to allowing the superstructure to be extended by including new concepts in sets. There are two main sets, RESOURCE (R) and UNIT (U). An r element ($r \in R$) represents anything that can be transported, e.g., a biomass or a utility, while a u element ($u \in U$) represents a unit that can transform one resource into another, e.g., a distillery that transforms sugarcane into bioethanol (EtOH), vinasse, bagasse, and CO_2. Each unit is inserted into the superstructure as a black box model. In the PROCESS (PU) subset (PU$\subset$U), the elements represent units that have at least one heat stream available for heat integration. UTILITIES (UT) subset (UT$\subset$R) contains the elements that represent thermal utilities, while HUT (HUT$\subset$UT) and CUT (CUT$\subset$UT) contain the hot and cold utilities, respectively. The elements in the sets LRA (LRA$\subset$R) and LRB (LRB$\subset$R) are the resources that can be acquired and commercialized. Figure 2 shows a schematic representation of the main sets and subsets in the superstructure.

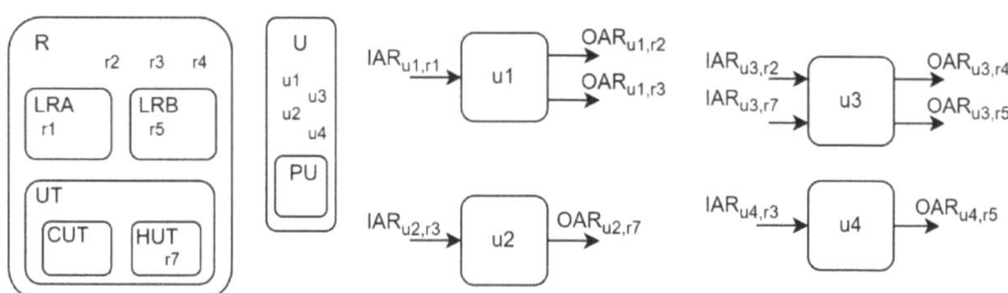

Figure 2. Superstructure main sets and subsets representation, and the inlets and outlets units.

2.2. Objective Function

As an objective function presented in Equation (1), the superstructure considers the minimization of the total annual cost (TAC), which considers the annualized unit capital cost (UCCu), the resource acquisition cost (RCr), the product commercialization revenue (PCr), and the carbon credits (CCs) resulting from the replacement of fossil fuels by their respective renewable energy sources, as expressed in Equation (1).

$$\text{TAC} = \sum_{u} \text{UCC}_u + \sum_{r} \text{RC}_r - \sum_{r} \text{PC}_r - \text{CC} \tag{1}$$

2.3. Unit Selection and Scale Adjustment

The selection and scale adjustment of each element u, is carried out by Equation (2), where $CapMin_u$ and $CapMax_u$ are parameters that represent the maximum and minimum scale adjustment that a unit can have, y_u is a binary variable that represents the existence of that unit, and w_u is a continuous variable responsible for the unit scale adjustment. When selected, a unit has its y_u equal to one, and its w_u is limited by $CapMin_u$ and $CapMax_u$. If not selected, y_u assume value zero, resulting in a w_u equal to zero.

$$CapMin_u y_u \leq w_u \leq CapMax_u y_u \quad \forall u \in U, \forall l \in L \tag{2}$$

2.4. Mass Balance

As mentioned above, the superstructure uses the concept of scale adjustment. In this concept, each element u is inserted as a black box model with a specific scale and its input flow ($IAR_{u,r}$) and output flow ($OAR_{u,r}$) for each resource, and it is relative to the specific scale. Depending on the situation, the scale of this unit can be adjusted, increased, or decreased. To do this, a continuous variable w_u is multiplied by each one of these parameters, adjusting them linearly and proportionally to the required scale. Thus, the produced and consumed quantities of a resource are calculated by Equations (3) and (4) respectively, where $prod_{u,r}$ is the production of resource r by unit u and $cons_{u,r}$ is the consumption of resource r by unit u.

$$OAR_{u,r} w_u - prod_{u,r} = 0 \quad \forall u \in U, \forall r \in R \tag{3}$$

$$IAR_{u,r} w_u - cons_{u,r} = 0 \quad \forall u \in U, \forall r \in (R - UT) \tag{4}$$

As the superstructure does not consider the accumulation of resources, every resource produced or bought needs to be consumed or sold. To represent this condition, Equations (5)–(7) were developed, where $bought_r$, $sold_r$, and fop represents the amount bought and sold of a resource r, and the fop is the hours of operation in a year. Equations (5) and (6) are applied to the features contained in sets LRA and LRB, respectively. Equation (7) is applied to features that are not present in the LRA and LRB subsets.

$$bought_r = \sum_u cons_{u,r} fop \quad \forall r \in LRA \tag{5}$$

$$\sum_u prod_{u,r} fop = \sum_u cons_{u,r} fop + sold_r \quad \forall r \in LRB \tag{6}$$

$$\sum_u prod_{u,r} fop = \sum_u cons_{u,r} fop \quad \forall r \in R - (LRA + LRB) \tag{7}$$

The Equations (8) and (9), where $avail_r$ and $demand_r$ represents the amount avail and demanded of resource r, ensure that every resource bought is available, and every resource sold is demanded.

$$bought_r \leq avail_r \quad \forall r \in LRA \tag{8}$$

$$sold_r \leq demand_r \quad \forall r \in LRB \tag{9}$$

As the superstructure enables the selection of utilities simultaneously with total site heat integration, the consumption of a thermal utility (element contained in UT) by a process (element contained in PU) cannot be defined as a parameter, as it can vary depending on the scale of the process and whether it is energetically integrated with other processes. Therefore, for a pu element, Equation (4) is rewritten as Equation (10), where $massUtility_{pu,ut}$ is the value of the mass flow of utility ut consumed by the pu process.

$$massUtility_{pu,ut} - cons_{pu,r} = 0 \quad \forall pu \in PU, \forall r \in UT \tag{10}$$

2.5. Multiple Cascade Heat Integration and Utility Selection

The energy balance and energy integration constraints between cascades were used and applied to all *pu* elements to perform utility selection and energy integration. These constraints identify regions of heat exchange between processes, as well as between processes and utilities. For this, it is necessary that the superstructure receives the heat cascade formulation, that is, the number of stages (s) and the inlet (Te_s) and outlet (Ts_s) temperature of each one, the minimum amount of energy required (MER_{pu}), and the minimum consumption of cold utility (UF_{pu}) of each *pu* element, in addition to the inlet ($Tin_{pu,n}$) and outlet ($Tout_{pu,n}$) temperature for each stream and its respective thermal capacity $\left(MCp_{Ppu,n}\right)$.

Equations (11) and (12) were used to perform the selection of utilities and heat integration simultaneously, which relate the minimum amount of energy that a process needs to receive/transfer with its possible sources/receivers. Equation (11) expresses that the heat demanded by a process is equal to the heat received in the form of utilities ($Qu_{pu,ut}$) or by direct integration with another process (Qin_{pu}). Since it is possible that there is more than one utility that can supply heat, the heat received from utilities is placed is special summations that limits the utilities that can exchange heat. Equation (12) express that the cold utility consumed by a process is equal to the heat transferred to a cold utility ut ($Qu_{pu,ut}$) or transferred to another process ($Qout_{pu}$).

$$MER_{pu}w_{pu} = \sum_{ut} Qu_{pu,ut} + Qin_{pu} \quad \forall u \in PU, \ \forall(ut \in HUT \wedge UTout_{ut} \geq Tpinch_{ut}) \quad (11)$$

$$UF_{pu}w_{pu} = \sum_{ut} Qu_{pu,ut} + Qout_{pu} \quad \forall u \in PU, \ \forall(ut \in CUT \wedge UTout_{ut} \leq Tpinch_{ut}) \quad (12)$$

When a unit is scaled, its heat demand must be adjusted proportionally as its scale increases or decreases. This is done by multiplying its minimum energy requirement of hot utility (MER_{pu}) and cold utility (UF_{pu}) by its scaling variable (w_{pu}). The heat supplied by a utility to a unit is determined by the unit pinch temperature ($Tpinch_{pu}$) Equations (13)–(15), where $Qh_{pu,s}$ is the heat available by the hot streams in stage *s* by process *pu* and $Qc_{pu,s}$ is the heat demanded by the cold streams in stage s by process pu and is limited by the interval temperature of the heat cascade stage and the heat demand of that stage. Figure 3 shows a representation of the utility placement in heat cascade as a general example.

$$Qu_{pu,ut} \leq \sum_{s| \ Ts_s \geq Tpinch_{pu} \wedge Te_s \leq UTout_{ut}} \left(Qh_{pu,s} - Qc_{pu,s}\right)w_{pu} \quad \forall puPU, \ \forall \ ut \in HUT \quad (13)$$

$$Qh_{pu,s} = \sum_{n| \ Tin_{pu,n} \geq Te_s \wedge Tout_{pu,n} \leq Ts_s} MCp_{pu,n}(Te_s - Ts_s) \quad \forall pu \in PU; \ \forall s \quad (14)$$

$$Qc_{pu,s} = \sum_{n| \ Te_s \leq Tout_{pu,n} \wedge Ts_s \geq Tin_{pu,n}} MCp_{pu,n}(Te_s - Ts_s) \quad \forall pu \in PU; \ \forall s \quad (15)$$

Equations (16) and (17) are responsible for converting the heat demand of a utility into its mass flow, connecting mass balance and energy balance.

$$Qu_{pu,ut} = massUtility_{pu,ut}hv_{ut} \quad \forall u \in PU, \ \forall ut \in HUT \quad (16)$$

$$Qu_{pu,ut} = massUtility_{pu,ut}hs_{ut} \quad \forall u \in PU, \ \forall ut \in CUT \quad (17)$$

To consider the heat entering and leaving one HC stage to another HC, the variables Qf and Qs have been inserted, and this represents the inlet heat into unit pu and stage s and Outlet heat into unit pu and stage s, respectively. For the stages below the pinch temperature, Qf has a value of zero, while for the stages above the pinch, Qs has a value of zero. These considerations are made to ensure that there is no heat input into the region below the pinch or heat loss above the pinch. This consideration is expressed in Equations (18) and (19), as well as the limitation of the amount of heat that can enter or leave a stage of the thermal cascade.

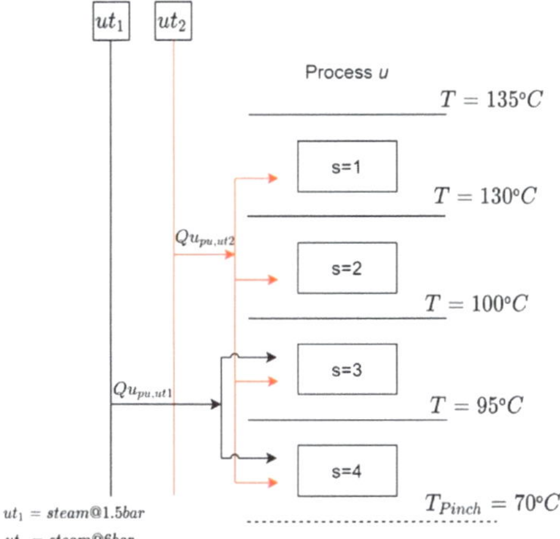

Figure 3. Representation of the utility placement in heat cascade.

$$Qf_{pu,s} \begin{cases} = 0 \\ \leq \left(Qc_{pu,s} - Qh_{pu,s} \right) w_{pu} \end{cases} \tag{18}$$

$$Qs_{pu,s} \begin{cases} = 0 \\ \leq \left(Qh_{pu,s} - Qc_{pu,s} \right) w_{pu} \end{cases} \tag{19}$$

By preventing heat from leaving the region above the pinch or entering the region below the pinch of an HC, heat transfer from one HC to another HC is limited to the region between the pinches of the two HCs, with heat leaving the one with the higher pinch temperature and entering the lowest pinch temperature. This region is shown in Figure 4.

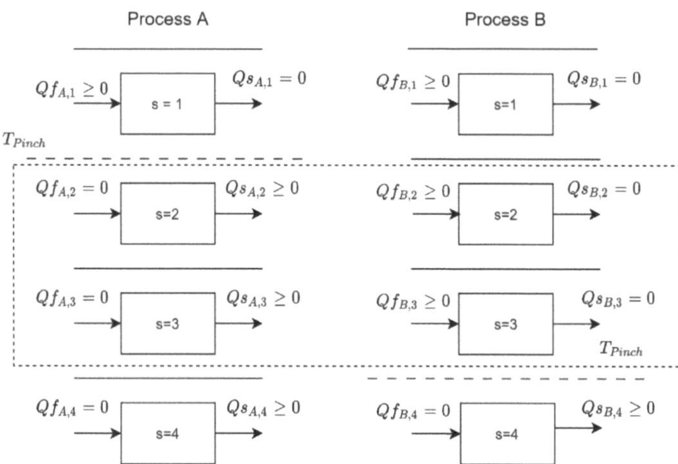

Figure 4. Region between pinch process, which allows heat integration.

The total amount of heat transferred from one process to another is calculated by Equation (20). Equation (21) expresses the total heat received from other processes. Equation (22) limits the heat released by a process to the total heat received by others in the region bounded by the pinch. Equation (23) states that the heat received by a unit must be less than or equal to the total heat transferred by the others in the region between the pinches.

$$Qout_{pu} = \sum_{s|\ Te_s \leq Tpinch_{pu}} Qs_{pu,s} \tag{20}$$

$$Qin_{pu} = \sum_{s|\ Ts_s \geq Tpinch_{pu}} Qf_{pu,s} \tag{21}$$

$$Qout_{pu} \leq \sum_{pu'|\ Tpinch_{pu'} > Tpinch_{pu}} \sum_{S|\ Te_s \leq Tpinch_{pu}\ \&Ts_s \geq Tpinch_{pu'}} Qf_{pu',s} \tag{22}$$

$$Qin_{pu} \leq \sum_{pu'|\ Tpinch_{pu'} < Tpinch_{pu}} \sum_{Te_s \leq Tpinch_{pu'}\ \&Ts_s \geq Tpinch_{pu}} Qs_{pu',s} \tag{23}$$

Equation (24) guarantees that the total heat output of the processes is equal to the total heat input of the processes.

$$\sum_{pu} Qin_{pu} = \sum_{pu} Qout_{pu} \tag{24}$$

2.6. Unit Capital Cost and Investment Cost Linearization

Since each unit inserted in the superstructure has a reference scale, in addition to the resource consumption and production values, each of them has a capital cost related to the scale considered. Process capital costs tend to vary nonlinearly with scale, so to maintain the model linearity, a piecewise linearization of the capital cost function was performed for each process. First, Equation (25) was used to obtain the cost curve as a function of the scaling factor (w_u), where C_u is the adjusted capital cost for unit u, $C0_u$ is the annualized capital cost at the reference scale for unit u, and se is a scaling exponent. As recommended by Peters et al. [25] it was considered that the unit u could be reduced or increased by up to 10 times the reference value. In this way, the curve obtained starts with 10% of the unit's reference value and ends with 1000% of the unit's reference value. The curve is then divided into three levels limited by a minimum and maximum value, $CapMin_{u,l}$ and

$\text{CapMax}_{u,l}$, respectively, as shown in Figure 5. The linearized cost curve coefficients for each process are present in the Supplementary Materials.

$$C_u = C0_u w_u^{se} \tag{25}$$

When a unit is selected and scaled, one of the levels must also be selected. Therefore, Equation (2) is rewritten as Equation (26). Since only one level can be selected when a unit is selected, Equation (27) guarantees that only one level is selected. If a level is not selected, its binary variables ($yl_{u,l}$) will have a value of zero, so the local scaling factor variable $wl_{u,l}$ will also have a value of zero, so Equation (28) guarantees that w_u will be equal to the value of $wl_{u,l}$ of the selected level.

$$\text{CapMin}_{u,l} y_{u,l} \leq wl_{u,l} \leq \text{CapMax}_{u,l} y_{u,l} \tag{26}$$

$$\sum_{l} yl_{u,l} \leq 1 \tag{27}$$

$$\sum_{l} wl_{u,l} = w_u \tag{28}$$

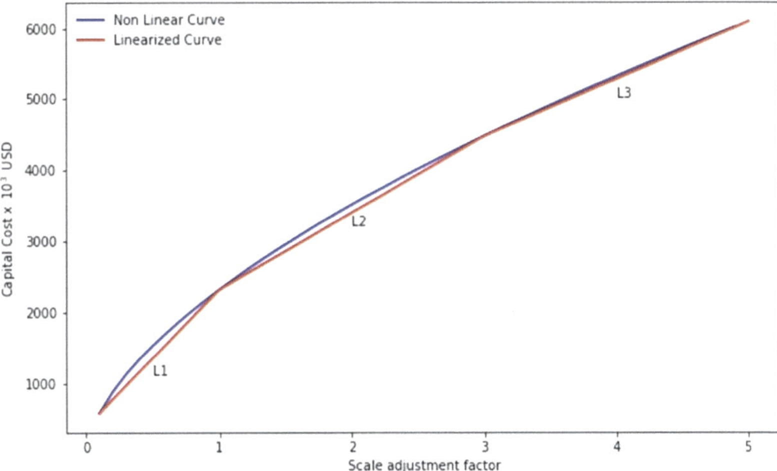

Figure 5. Piecewise linearization of the investment cost function of a process.

The capital cost of a unit can be determined by Equation (29), where $a_{u,l}$ and $b_{u,l}$ are the angular and linear coefficients of each linearized segment, respectively, MC, OC, AF and LC are maintenance cost, operation cost, annualization factor, and other cost, which are fixed as 6%, 8.6%, 0.086, and 10%, respectively. Equations (30)–(32) calculate the resource acquisition cost, product commercialization revenue, and carbon credits revenue, respectively, where ResCost_r is the cost of resource r, MPr is the market price of resource r, PC_r is the commercialization revenue of resource r, and CarbVal is the carbon credit price.

$$UCC_u = \sum_{l} \left(a_{u,l} wl_{u,l} + b_{u,l} y_{u,l} \right) . AF . (1 + MC + OC + LC) \tag{29}$$

$$RC_r = \text{Bought}_r \text{ResCost}_r \tag{30}$$

$$PC_r = \text{Sold}_r MP_r \tag{31}$$

$$CC = \left(\sum_r Sold_r MP_r - \sum_r Bought_r ResCost_r \right) CarbVal \tag{32}$$

The capital cost of each process was corrected using the CEPCI index. To annualize the process, before linearization, each curve was multiplied by the annualization factor expressed by Equation (33), considering a plant lifetime (n) of 25 years for all units and an interest rate (i) of 7%.

$$AF = \frac{i(1+i)^n}{(1+i)^n - 1} \tag{33}$$

3. Sugarcane Biorefinery Case Studies Description

In this paper, a conventional Brazilian sugarcane distillery with a typical processing capacity of 2,640,000 tons of sugarcane per year is considered, as previously described and performed by others authors [8]. For the proposed study, different technologies are considered to compare their impact on the performance of the biorefinery. In this study, several cases have been evaluated where different technologies have been integrated to improve the performance of the biorefinery by utilizing the wastes generated during the production of bioethanol from sugarcane juice. The wastes considered were vinasse, sugarcane bagasse, and carbon dioxide generated during the fermentation process. This study aimed to convert vinasse through the biodigestion process and bagasse through the gasification process to produce methanol or a bagasse power plant to produce and export electricity. For the carbon dioxide stream, the CO_2 catalytic hydrogenation process was introduced, which also produces methanol. This study also included methanol catalytic dehydration (MCD) technology, which converts methanol to DME. Because the processes require electricity and consume utilities to operate, various production technologies were included. Three cogeneration systems were evaluated to produce hot utilities, each producing saturated steam at different pressures. In addition to the cogeneration systems, the possibility of importing electricity from a photovoltaic panel system was investigated. These technologies were integrated into the superstructure to collect data from existing published work. Additional information and process descriptions are provided in the following sections. Table 1 provides a summary of the technologies used for each case evaluated, and Figure 6 summarizes the combined process in the superstructure.

Table 1. Biorefineries technology cases.

Case	Route
1	Distillery + Photovoltaic Power Station (PPS)
2	Distillery + PPS + Vinasse Biodigestion (VBD)
3	Distillery+ PPS + VBD + Bagasse Gasification (BG)
4	Distillery+ PPS + VBD + BG + Methanol Catalytic Dehydration (MCD)
5	Distillery+ PPS + VBD + BG + Catalytic CO_2 Hydrogenation (CCH)
6	Distillery+ PPS + VBD + BG + MCD + CCH

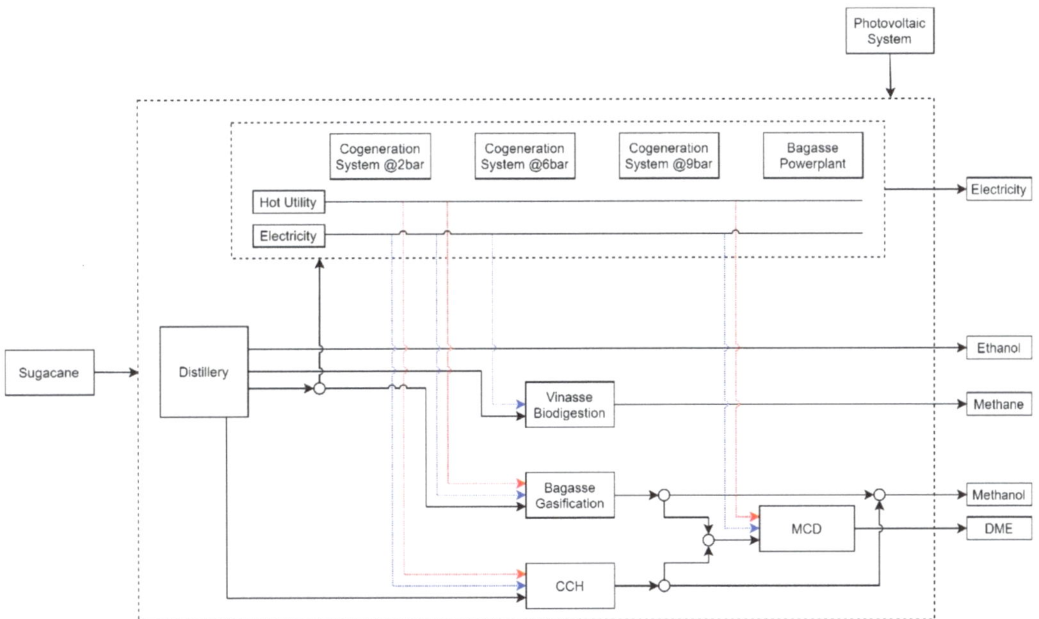

Figure 6. Biorefinery process superstructure.

Four additional indicators were included to facilitate the evaluation and comparison of technologies: payback, energy efficiency, total CO_2 avoided, and surface power density. The payback calculation considers the total investment and cash flows, considering the purchase and sale of resources, commercialization of carbon credits, and operating costs, as shown in Equation (34). For energy efficiency, Equation (35), energy input and output flows were considered in terms of resources, which were obtained based on their respective lower heating values (LHVs), as shown in Table S1. The calculation of the surface power density considers the energy produced in the form of biofuels per area of sugarcane cultivated, assuming a productivity of 76.8 tons of sugarcane/hectare. In order to calculate the total avoided CO_2 and consequently the generation of carbon credits, the avoided CO_2 for sugarcane [26], bioethanol [26], biomethanol [27], bioDME [26], biomethane [26], and electricity [28] are provided in the Supplementary Materials. To commercialize carbon credits, a sales price of USD 65.00 per credit was considered. The supplemental material includes the IAR and OAR values for each unit, as well as the linearized cost curves.

$$\text{Payback} = \frac{\text{Total Investment}}{\text{Cash flow}} \tag{34}$$

$$\text{Energy Efficiency} = \frac{\text{Output Energy}}{\text{Input Energy}} \tag{35}$$

Technologies Description

Next, the technologies considered are described, as well as the ancillary processes for providing other resources, such as electricity and hydrogen. Figure 7 shows a representation of the superstructure formulation for this study.

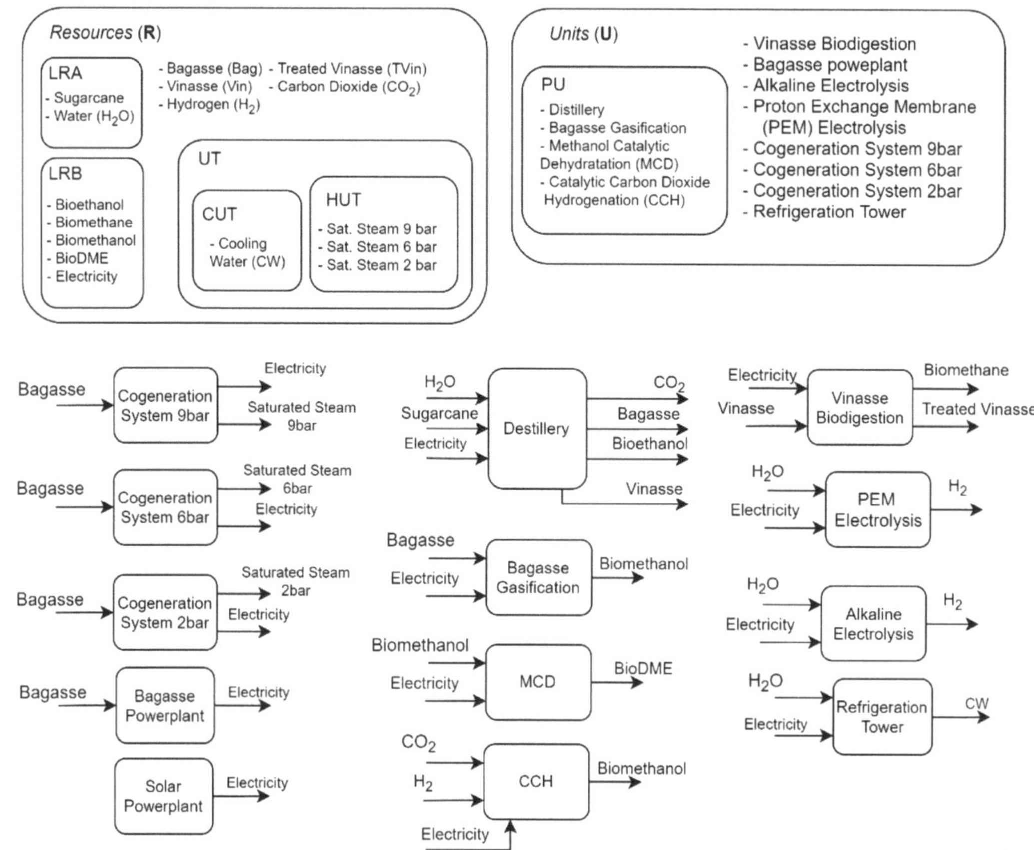

Figure 7. Representation of the main sets and models used in the case study.

Distillery: A Brazilian conventional autonomous distillery, with a typical milling capacity of 2,640,000 tons of cane per year and a crushing rate of 500 tons of sugarcane per hour is considered to produce bioethanol. This process includes the following steps: cleaning, preparation, and extraction system; cane juice treatment; juice concentration; sterilization and must cooling; fermentation; distillation and rectification; and dehydration. This unit receives sugarcane, water, electricity, and utilities as resources and produces bioethanol as the main product and bagasse, vinasse, and CO_2 as by-products. Process and investment data were taken from [7,10]. Figure 8 shows a diagram of the sugarcane distillery.

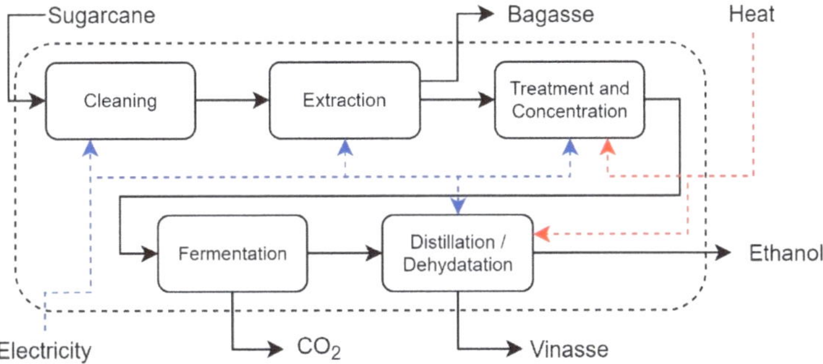

Figure 8. Distillery flowsheet of the sugarcane distillery.

Vinasse Biodigestion: The process consists of two steps, anaerobic digestion and biogas purification. In the first, the vinasse is fed directly into an anaerobic biodigester where microorganisms consume part of the organic material and produce biogas, a gas mixture of CO_2, methane (CH_4), and hydrogen sulfide (H_2S). In the second step, H_2S is removed from the produced biogas by micro-aeration of the biogas and then CO_2 is removed by the pressure swing absorption (PSA) process, resulting in a high CH_4 purity (>98%) [29], as shown in Figure 9.

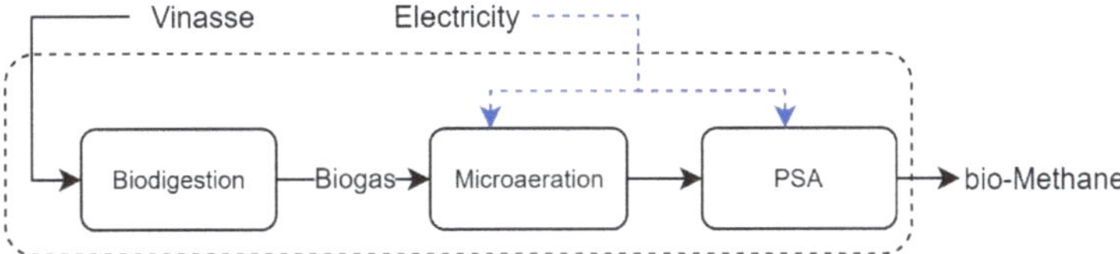

Figure 9. Representation of the vinasse biodigestion process.

Bagasse Gasification: In the first, the production of biomethanol by bagasse gasification takes place in five steps: bagasse pretreatment, gasification, syngas conditioning, methanol synthesis, and upgrading (Figure 10). In the pretreatment stage, the bagasse is dried in an air dryer and fed into the steam gasification reactor. Syngas conditioning removes major impurities such as particulates and tar from the produced gas. The composition of the syngas is adjusted with hydrogen to achieve a stoichiometric ratio s, defined by Equation (36), of 2.05, as recommended for methanol synthesis [30]. The adjusted syngas is sent to the methanol synthesis where its pressure is adjusted to 50 bar, it is mixed with unreacted syngas, and it is preheated to 225 °C before entering the reactor where its temperature is adjusted. The methanol synthesis reactor is a fixed bed reactor containing a copper/zinc oxide/alumina catalyst. The reactor effluent is decompressed and degassed. The liquid methanol is cooled to 43.3 °C and sent to a distillation column where it is purified [31]. Process and economic data were obtained from the literature [31].

$$s = \frac{H_2 - CO_2}{CO + CO_2} \tag{36}$$

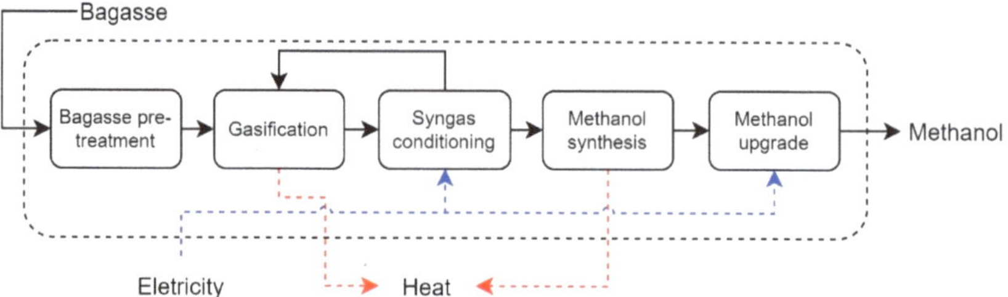

Figure 10. Main steps presented in bagasse sugarcane gasification integrated to methanol production.

Catalytic CO$_2$ Hydrogenation (CCH): CCH is a three-step process, gas compression, methanol synthesis, and upgrading, as shown in Figure 11. First, CO$_2$ is compressed to 48 bar in a 4-stage compressor and then mixed with hydrogen. Before entering the preheater, the CO$_2$-H$_2$ mixture receives a recycle stream of unreacted gases. The final mixture is compressed and preheated to reactor conditions (220 °C, 83 bar). The CO$_2$ hydrogenation reactor is a fixed bed reactor with a Cu/ZnO/Al$_2$O$_3$ catalyst. The reactor effluent is cooled and decompressed; 95% of the unreacted gases are recycled and 5% is purged. The liquid methanol is cooled to 43.3 °C and fed to a distillation column as in the bagasse gasification unit. The process configuration and conditions are based on the previous work of [32,33].

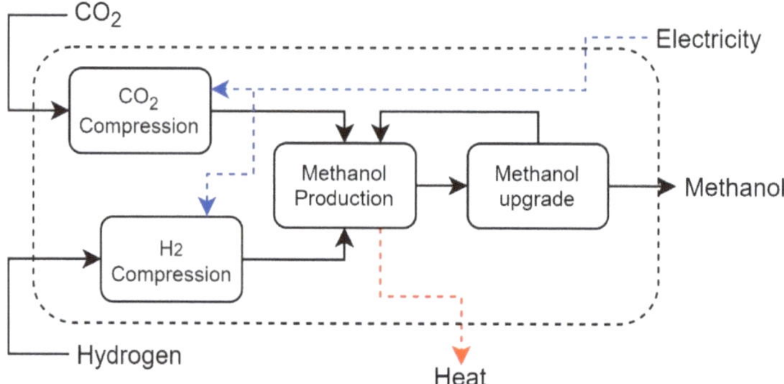

Figure 11. Catalytic carbon dioxide hydrogenation representation.

Methanol Catalytic Dehydration (MCD): The process can be divided into two steps, production and purification, as shown in Figure 12. In the first, fresh methanol is mixed with recycled reactant and evaporated before entering the reactor. After cooling, the effluent is sent to the purification stage where the product is recovered by column distillation. The unreacted methanol is recovered by another column and recycled. The process and economic data, including information on the process flows, were obtained by Aspen plus simulation, following the work of Shim et al. [34], Dutta et al. [31], and Turton et al. [35].

Hydrogen Production: Since hydrogen is required to produce methanol and DME, two different technologies are considered in the superstructure, alkaline electrolysis and proton exchange membrane electrolysis. The first is the most mature technology, and the electrodes are immersed in an aqueous KOH solution, allowing water electrolysis and thus hydrogen production. In the second, a proton exchange membrane is placed in the center of the cell to conduct the protons produced in the anode to the cathode, where they are reduced to produce hydrogen [36].

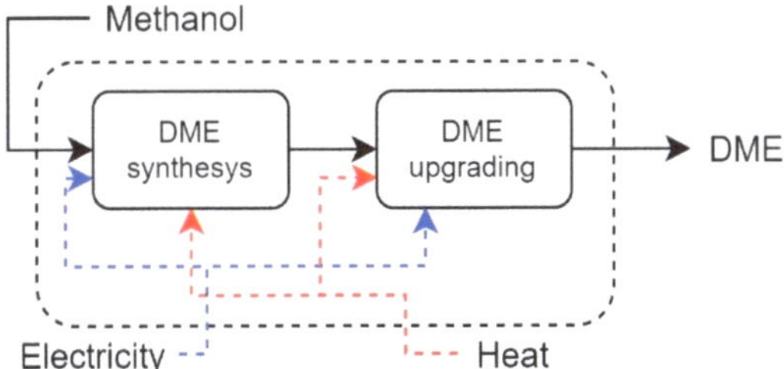

Figure 12. Main steps of ethanol catalytic dehydration.

Utilities Production: For hot utilities, the superstructure considers three different cogeneration schemes, modeled as a steam-based cycle with steam turbines and sugarcane bagasse as fuel. Each model considers a different level of turbine output saturated steam pressure: 2.2, 6, and 9 bar. For bagasse consumption and hot utilities and electricity production, a model was developed in EES software, version 10, considering a cogeneration efficiency of 85% and a net calorific value of 7500 kJ/kg for bagasse. The operating temperatures of sugarcane ethanol distilleries are relatively low, approaching 115 °C. Consequently, the distilleries rely on cogeneration systems powered by sugarcane bagasse. To provide a more realistic representation of the process, similar systems were selected for this work. For the cold utilities, cooling water is considered [31].

Electricity Production: In conventional distilleries, the electricity demand is met by the cogeneration system or, in some cases, by a biomass power plant. In this sense, for electricity generation, the superstructure has three alternatives: cogeneration units (described above), bagasse, and import from a solar photovoltaic supplier.

4. Results and Discussion

Six different cases were evaluated after inserting the data into the superstructure. All cases considered the existence of a sugarcane distillery and the possibility of heat exchange between processes. For each of the cases, different technologies are integrated into the existing plant, making it possible to obtain a different biorefinery configuration and result. Tables 2 and 3 provide a summary of the cost analysis results of the optimization problem solved for the cases under consideration in this paper, where the total investment represents the investment with process, resource bought represents the total expenses with resources like sugarcane. Biofuels and carbon credit revenues represent the income with biofuel and carbon credit commercialization respectively, while total avoided CO_2 and labor cost represent the CO_2 total mass that was no longer emitted and the cost associated with process labor cost, respectively.

Table 2. Results of the optimal biorefineries configurations cost analysis.

Parameter	Case 1 [1]	Case 2 [2]	Case 3 [3]	Case 4 [4]	Case 5 [5]	Case 6 [6]
TAC [$\times 10^6$ USD/year]	−91.23	−90.89	−113.85	−101.65	−48.76	−42.17
Total Investment [$\times 10^6$ USD]	233.24	246.02	325.95	333.89	408.03	418.18
Resource bought [$\times 10^6$ USD·y^{-1}]	30.29	30.29	41.54	41.60	115.05	115.13
Biofuels Revenues [$\times 10^6$ USD·y^{-1}]	161.03	163.01	205.27	195.54	237.15	222.93
Carbon Credit Revenue [$\times 10^6$ USD·y^{-1}]	24.84	26.05	41.45	41.24	50.06	49.74
Labor Cost [$\times 10^6$ USD·y^{-1}]	57.37	60.52	80.18	82.13	100.37	102.87
Payback [y^{-1}]	2.38	2.50	2.64	2.99	6.69	7.65

[1] Destillery; [2] Destillery + Vinasse Biodigestion (VBD); [3] Destillery + VBD + Bagasse Gasification (BG); [4] Destillery + VBD + BG + Methanol Catalytic Dehydration (MCD); [5] Destillery + VBD + BG + Catalytic CO_2 Hydrogenation (CCH); [6] Destillery + VBD + BG + MCD + CCH.

Table 3. Energy balance results of the optimal biorefinery cases.

Parameter	Case 1	Case 2	Case 3	Case 4	Case 5	Case 6
Energy Consumed [$\times 10^9$ MJ$\cdot$y^{-1}]	11.71	11.71	12.36	12.36	16.76	16.76
Energy Produced [$\times 10^9$ MJ$\cdot$y^{-1}]	5.88	6.256	9.21	9.33	11.14	11.31
Energy Efficiency [%]	50.25	53.43	74.49	75.43	66.46	67.46
Surface Power Density [GJ$\cdot$ha^{-1}]	171.17	182.00	253.74	257.12	288.83	293.44

For Case 1, the configuration obtained is very similar to that found in bioethanol distilleries in Brazil, which essentially consist of a distillery and a cogeneration system. The biorefinery has the potential to produce 171,072 tons of bioethanol, 729,907 tons of bagasse, 2,364,595.2 tons of vinasse (used as fertilizer), and 161,040 tons of CO_2 per year. Figure 13 shows the initial GCC of the biorefinery with the indication of the hot utility supplied by the cogeneration system. Although the distillation column systems, from the bioethanol distillation section, are the largest consumers of utilities and require heat at a temperature close to 110 °C, the temperature of the extraction and recovery columns directly affects the selection of the utility level, causing the cogeneration system to supply heat at higher pressure levels. Thus, the cogeneration system produces saturated steam at 6.5 bar and uses 45.6 tons of bagasse per hour to produce hot utilities and electricity. Excess bagasse is diverted to a bagasse power plant, which exports excess electricity to the grid. Energetically, the biorefinery can produce 5.88×10^9 MJ/year of energy with an initial energy efficiency of 50.25%.

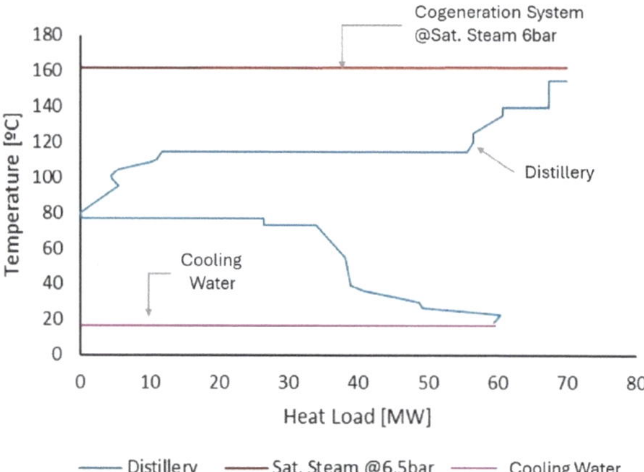

Figure 13. Grand composite curve with steam placement for Case 1.

If the biodigestion of vinasse is integrated into the biorefinery, as in Case 2, the configuration remains unchanged. However, the increased consumption of the biorefinery causes a reduction in the exported electricity. The new process allows the biorefinery to produce 7075.24 tons of biomethane per year, and the biodigested vinasse is used for fermentation. Although this new unit resulted in increased TAC and payback due to higher total investment and operating costs, the production of biomethane improved environmental performance by reducing total avoided CO_2. In addition, the increase in biofuel production from vinasse biodigestion resulted in a small increase in the energy efficiency of the biorefinery, as shown in Figure 14. In a carbon credit valorization scenario, the revenues from biomethane production can exceed the investment costs, making the process more economically viable. Figure 15a,b shows the biorefinery final configurations for Cases 1 and 2, respectively.

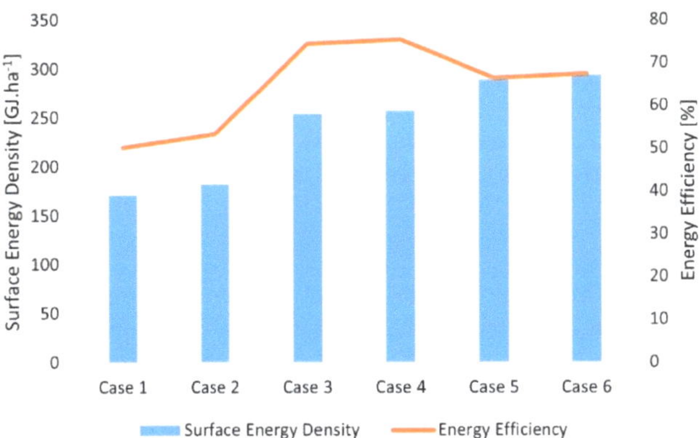

Figure 14. Energy efficiency comparison among all cases evaluated.

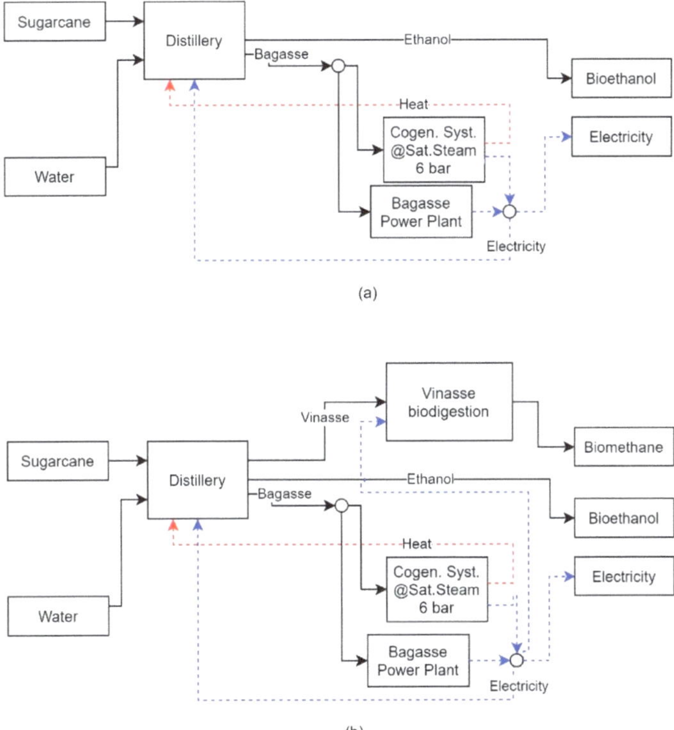

Figure 15. Biorefinery configuration for Case 1 (**a**) and Case 2 (**b**).

In Case 3, the integration of the gasification unit allowed the biorefinery to produce 209,795.6 tons of methanol per year, resulting in a 25% increase in revenue from biofuel sales. The presence of the BG unit allows heat exchange with another process, eliminating the use of utilities. The high-temperature characteristics of the BG unit result in a pinch temperature higher than that of the distillery. Therefore, when heat is exchanged, the BG unit serves as a source of thermal energy for the distillery. In addition, the heat from the BG unit was sufficient to meet the distillery's needs. As a result, the CHP unit was not needed and was eliminated from the biorefineries' optimal design.

Figure 16a shows the GCC of the bagasse gasification unit. Unlike other processes, the gasification unit does not receive heat from external sources and uses a portion of the syngas produced to supply the energy required by the gasification process. As a result, two streams with a high thermal load are produced, with the first consisting of the combustion gas produced and the second consisting of the synthesis gas that has not been consumed and must be cooled before being sent to the next stages. Therefore, this unit can act as a heat source for other units present, meaning that other heat sources are not necessary. Figure 16b shows the grand composite curve of the biorefinery, and it is possible to observe that even after the integration of the distillery, there is still a large amount of heat available that can be used in other processes. It is also possible to visualize the region of the curve where it was possible to recover part of the heat present in the gasification. In this sense, when bagasse is sent to gasification, it is no longer used as fuel but instead generates revenue for the distillery while still providing heat for the processes. Thus, through energy integration, the gasification unit significantly increases the biorefinery energy efficiency, as can be observed in Figure 14.

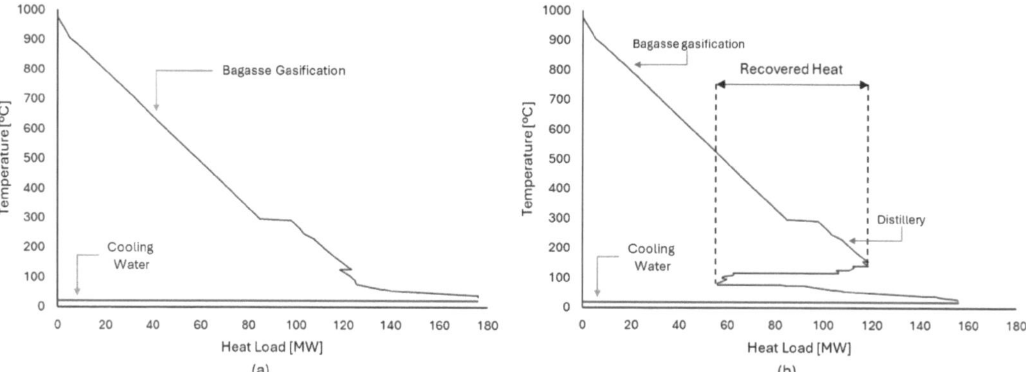

Figure 16. (**a**) Grand composite curve for methanol production by bagasse gasification process. (**b**) Grand composite curve for an integrated biorefinery in Case 3.

Due to the delivery of the bagasse stream to the gasification unit, the biorefinery required the importation of 180,805.20 MWh of electricity from a photovoltaic system, resulting in a 37.15% increase in resource acquisition costs. Although it had a higher payback than the previous setups, Case 3 had a lower TAC, suggesting that this configuration was more beneficial over time than the others. A significant increase in this metric can be seen by examining avoided emissions. The conversion of bagasse to methanol results in the retention of more carbon in the form of biofuel, thereby increasing the amount of CO_2 avoided. In addition, the imported electricity is supplied by a photovoltaic system, which eliminates any increase in CO_2 emissions and results in a greater total amount of CO_2 avoided than in the previous cases. Figure 17 shows the biorefinery configuration provided by the superstructure in this scenario.

In Case 4, the integration of the MCD process enabled the biorefinery to produce 148,756.60 tons of DME per year by converting all 209,795 tons of methanol produced through the bagasse gasification unit. The MCD integration required an additional investment of USD 7.94×10^6 compared to Case 3. Although DME has a higher market price than methanol, the increase in operating costs and process losses led to a decrease in revenue, which negatively affected the payback of the biorefinery. The heat demand of the MCD unit can be met through energy integration with the biorefinery. Figure 18a shows the superposition of the GCC of the MCD and the biorefinery, where it is possible to observe the availability of heat that can be transferred from the biorefinery to the MCD unit. Figure 18b shows the GCC of the biorefinery after the integration of the MCD.

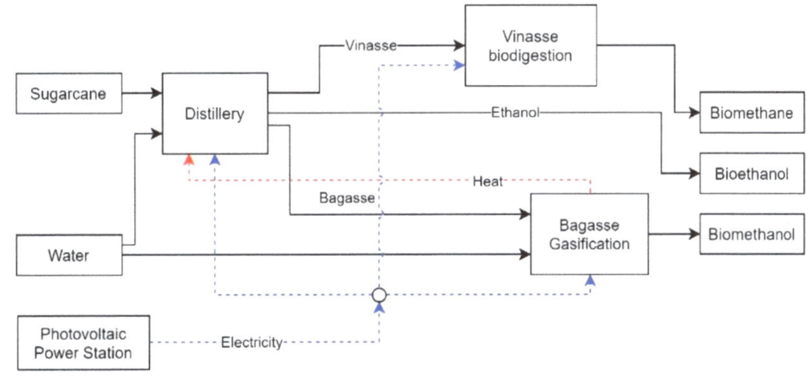

Figure 17. Biorefinery optimal configuration for Case 3.

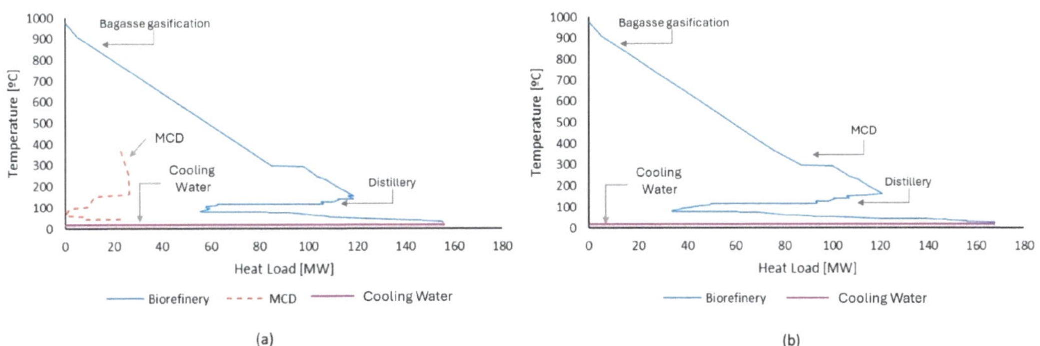

Figure 18. (**a**) Overlay of the biorefinery GCC (blue line) and MCD (red dotted line). (**b**) Biorefinery GCC after biorefinery integration.

Of the cases evaluated, Cases 3 and 4 presented the highest energy efficiency values. Comparing Cases 2 and 3, a significant increase in the energy efficiency of the biorefinery is observed when the gasification process is included, due to its heat transfer from gasification to the other process. Despite the higher investment required, the transfer of the bagasse flow to the gasification unit made a large amount of heat available while increasing the biofuel production, directly increasing the revenue and energy efficiency of the biorefinery. Observing Cases 3 and 4, the MCD process introduction did not have a negative impact on the energy efficiency of the biorefinery. In fact, there was a slight increase. As before, the biorefinery imported all the electricity it consumed. It received a supply of 181,759.82 MWh of electricity. Figure 19 shows the main flows in this biorefinery configuration.

In Case 5, the integration of the CCH process enabled the biorefinery to produce 306,400 tons of methanol per year, a 46% increase over Case 3. While biofuel production and carbon credit revenues increased, the payback period also increased, primarily due to the significant investment in the plant, its operating costs, and resource purchases. To convert CO_2 into methanol, the CCH unit requires hydrogen, which must be produced by the biorefinery. Two technologies were evaluated for this purpose: alkaline electrolysis and PEM. Alkaline electrolysis was selected because of its lower capital cost. Previously, as in many distilleries, the CO_2 produced during fermentation was vented to the atmosphere. By converting it to biofuels, the carbon capture is significantly increased, resulting in a higher total avoided CO_2. Since the biorefinery's CO_2 emissions are in sugarcane cultivation and transportation, the amount of carbon credits obtained also increases. Overall, the total avoided CO_2 increased by 21.37% compared to Case 4 and by 20.74% compared to Case 3. Figure 20 shows the biorefinery configuration and its main flows for Case 5.

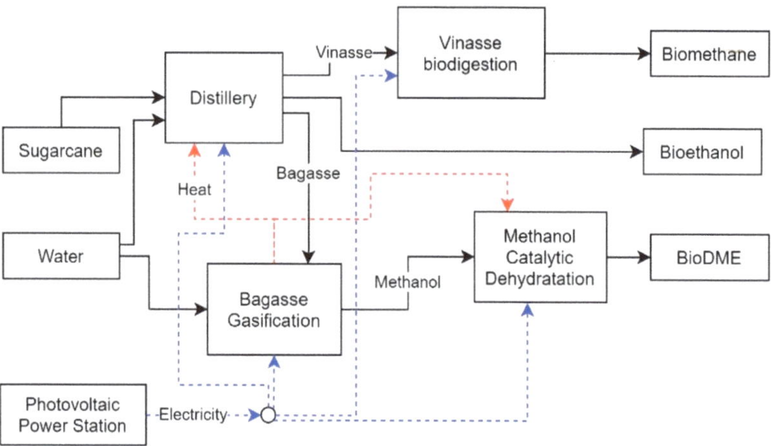

Figure 19. Biorefinery optimal configuration for Case 4.

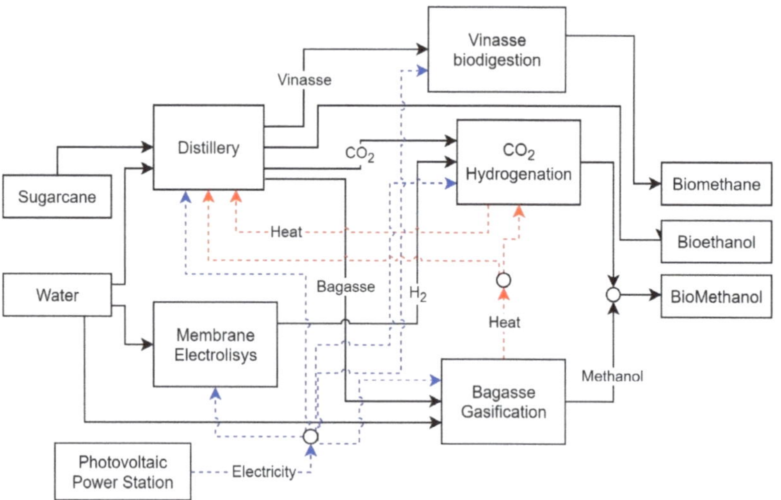

Figure 20. Biorefinery optimal configuration for Case 5.

The CCH unit uses hot and cold utilities, allowing energy integration with other processes. By having a pinch temperature higher than that of the distillery, the CCH can transfer some of its excess heat to the distillery, as shown in the GCC of the two processes in Figure 21. At the same time, the CCH also receives heat from the gassing, thus acting as a source and sink of heat for different processes. In this sense, the biorefinery recovered 427,865 MWh of heat per year through heat integration. In Case 5, the biorefinery imported 1,403,629 MWh of electricity to power its processes. This significantly increased resource acquisition costs due to the high electricity consumption of the electrolyzer. In addition, the increased electricity consumption reduced the biorefinery's energy efficiency to 66.46%.

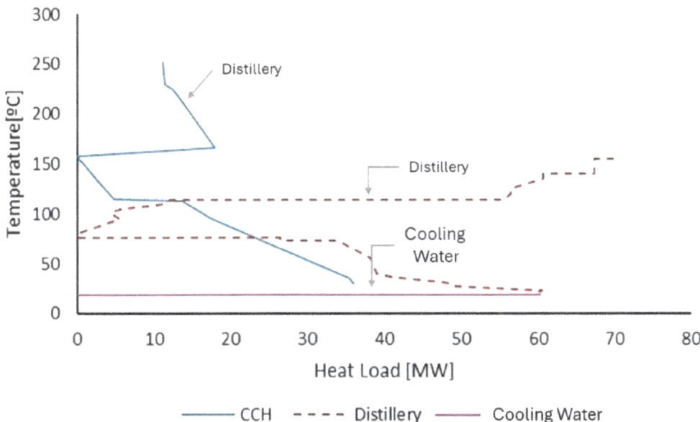

Figure 21. Representation of the overlap of the distillery GCC and CCH.

As noted above, the process of converting CO_2 into methanol using CCH requires 0.22 tons of hydrogen for each ton of methanol produced. The hydrogen must be supplied by the biorefinery. Considering all the CO_2 conversion produced by the distillery, as well as the production of hydrogen through alkaline electrolysis, the biorefinery needed to import 1,403,629 MWh of electricity. With an energy consumption of 343.34 MWh, as hydrogen form, CCH produces 36.47 tons of methanol, resulting in an energy efficiency of 70.31% (HHV) for the CCH process. However, considering the efficiency of the electrolyzes, the overall efficiency of converting CO_2 to methanol is 51.28%, which justifies the decrease in energy efficiency of the biorefinery. The CCH process model has a ratio of 4 moles of H_2 to 1.13 moles of methanol, which is very close to the stoichiometric value of the reaction of 3:1. This suggests that it is essential to improve H_2 production, electricity acquisition, or cost reduction to improve the energy efficiency of the process and the biorefinery.

In Case 6, the biorefinery had a TAC of -31.35×10^6 USD/year and required a total investment of USD 418.18×10^6, resulting in an annual production of 217,260.7 tons of DME. The implementation of heat integration allowed the recovery of 603.82 GJ of energy per year. In this new configuration, the biorefinery has a 14.3% higher payback compared to the previous configuration, and a total reduction of 0.6% in avoided CO_2 emissions. Figure 22 shows the biorefinery configuration for Case 6. Although the gasification process is present, Cases 5 and 6 show a reduction in the biorefinery's energy efficiency. Even with heat recovery between processes, the high electricity consumption of the electrolyzers and the unavailability of resources for their production meant that the biorefinery would have to import much more electricity than in other cases, severely penalizing its energy efficiency. However, an increase in the energy produced per area of sugarcane cultivated can be observed, as in other cases where there has been an increase in biofuel production.

Comparing Cases 6 and 4, the integration of the CCH unit has led to a worsening of the economic and energy indicators, requiring more payback time to recover the investments made, and a decrease in energy efficiency. As mentioned above, the biorefinery, by producing hydrogen for the CCH, significantly increases its electricity imports and, consequently, its expenditure on this resource. This situation can also be observed when comparing Cases 3 and 5, indicating that the integration of the CCH process, despite having a strong positive impact on the generation of carbon credits, proved to be detrimental to the performance of the biorefinery. In both Cases 5 and 6, for an annual production of 21,225.6 tons of H2, the annualized cost of the electrolyzer was USD 15,437,508.52. Thus, hydrogen production has a cost of USD 4.69 per kg.

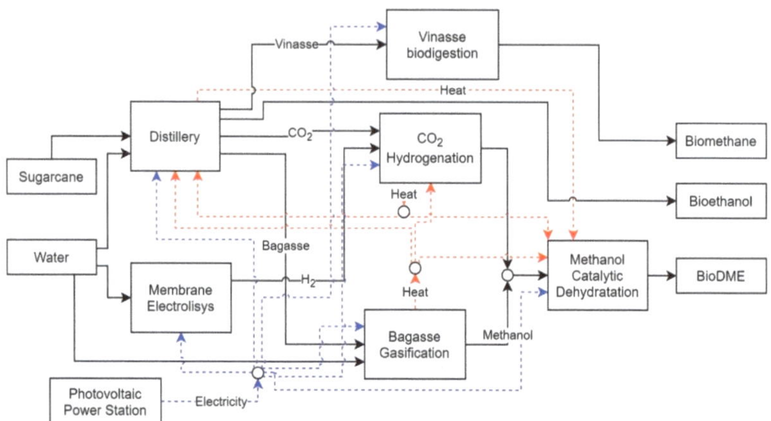

Figure 22. Biorefinery configuration for Case 6.

A comparison of the results obtained with other studies reveals that the values found are comparable to those reported. For a first-generation distillery, which produces ethanol from sugarcane juice, Albarelli [30] achieved an energy efficiency of close to 43%. By integrating methanol production through the gasification of bagasse and sugarcane straw, the energy efficiency of the biorefinery varied between 50% and 65%, depending on the configuration and mixture evaluated. It is crucial to emphasize that in Albarelli's work, the authors considered a range of biorefinery configurations, including the production of second-generation ethanol utilizing a portion of straw and bagasse. Nevertheless, as in this work, the authors concluded that the increase in the energy efficiency of the biorefinery is a result of the increase in biofuel productivity. However, this is accompanied by an increase in investments and the complexity of the technologies present in the biorefinery. Bressanin [37] evaluated the production of biofuels using the Fischer–Tropsch synthesis process from two different types of sugarcane, obtaining efficiency values between 45.4% and 57.7%.

To evaluate the impact of the electricity price on the payback of the biorefinery, Cases 3 to 6 were simulated again considering different electricity prices, 60, 45, and 30 USD·MWh^{-1}. By reducing the cost of electricity, it is possible to observe a positive impact on the payback values of all cases, as shown in Figure 23. Since the electricity imports are much higher in Cases 5 and 6 than in Cases 3 and 4, the reduction of the payback time was more significant. When the cost of electricity is 45 USD·MWh^{-1}, Cases 5 and 6 show a reduction of 34.8% and 27.8%, respectively. When the cost is 30 USD·MWh^{-1}, the reduction is 46.5% and 43.6%, respectively, compared to the first case. These results suggest that the price of electricity is crucial to increase the competitiveness of biofuels and thus improve the viability of new generations of biorefineries as proposed in this paper. Furthermore, it is also possible to observe the impact that hydrogen production can have on the performance of a biorefinery, raising the hypothesis that the development and improvement of technologies is a point of great relevance for the development of biorefineries.

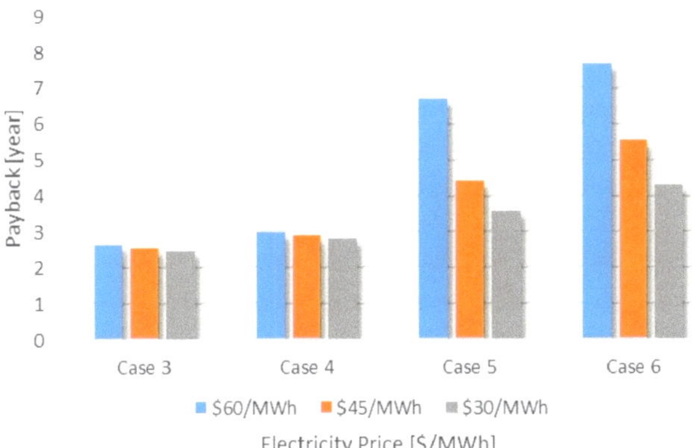

Figure 23. Obtained payback for different electricity prices considered.

5. Conclusions

This study presents a novel superstructure that uses a MILP formulation to optimize and evaluate biorefineries. In this formulation, the selection and scaling of each process are performed simultaneously with the selection of utilities and heat integration between processes. The selection and scaling of processes for biorefinery composition were governed by mass balance constraints in addition to demand constraints and feedstock availability. For heat integration between processes, the concept of process heat cascade integration is used. This approach allows heat exchange in the region between the pinch points of these processes, facilitating heat transfer integration and reducing energy consumption. The simultaneous solution is achieved by linking the mass and energy balance constraints through the calculation of the utility mass required by the processes. This eliminates the need for complex computational structures and iterative problem-solving, provided that all possible process combinations have been considered. The study evaluated the integration of different technologies to improve diversification and biofuel production in a sugarcane biorefinery. This sector is of great importance to the Brazilian economy and is considered essential for the sustainable development of a low-carbon economy. The results presented show that the integration of energy from the gasification process allowed the biorefinery to simultaneously generate revenue and energy from bagasse. Since it led to a significant improvement in the energy, economic, and environmental performance of the biorefinery, the production of methanol through bagasse gasification can be considered a key process for the expansion of the biorefinery. On the contrary, the conversion of carbon dioxide into methanol, while increasing the generation of carbon credits, has a significant negative impact on biorefinery energy efficiency and economic viability. This is due to the significant increase in electricity cost acquisition. The results also show that the price of electricity is critical to the economic viability of the biorefinery due to its high consumption of electrolyzers. Furthermore, the results indicate that the incorporation of the bagasse gasification process may be a viable technological alternative to conventional cogeneration systems. This is due to its demonstrated ability to meet the entire heat demand of the biorefinery.

Supplementary Materials: The following supporting information can be downloaded at https://www.mdpi.com/article/10.3390/e26060501/s1, Table S1: Main economic assumptions adopted for economic evaluation; Table S2. Values of CO_2 emitted and avoided used to generate carbon credits. Table S3: Linearized cost curve coefficients and their respective levels; Table S4: Steam parameters considered; Table S5: Heat streams considered for each process; Table S6: IAR values used for each model considered in the cases; Table S7: OAR values used for each model considered in the cases.

Author Contributions: All authors contributed to the analysis of the results and to writing the manuscript. V.F.G. developed and implemented the mathematical optimization model in Lingo, collected and analyzed data, and wrote the first draft of the manuscript. All authors provided feedback on the content and revised the final draft. A.V.E. conceived and supervised the research. All authors have read and agreed to the published version of the manuscript.

Funding: This research was funded by CNPq, grant number 303909/2019-6, and the Research Support Foundation of the State of Minas Gerais (FAPEMIG), grant number 37738768/2021.

Institutional Review Board Statement: Not applicable.

Data Availability Statement: Data is contained within the article.

Conflicts of Interest: The authors declare no conflicts of interest.

Nomenclature

Sets

R	Resource set
U	Unit set
PU	Process subset
UT	Utilities subset
HUT	Hot utility subset
CUT	Cold utility subset
LRA	Subset for available resources
LRB	Subset for demanded resources

Subscripts

u	Unit set index
r	Resource set index
pu	Process subset index
ut	Utility subset index
n	Stream index
l	Unit level index

Abbreviations and formulas

CO_2	Carbon Dioxide
CCH	Catalytic CO_2 Hydrogenation
CH_4	Methane
DME	Dimethyl Ether
HC	Heat Cascade
HI	Heat Integration
H_2S	Hydrogen Sulfide
MCD	Methanol Catalytic Dehydration
MeOH	Methanol
PA	Pinch Analysis
TSI	Total Site Integration

Variables and Parameters

$a_{u,l}$	Angular coefficient of linearized segment l of unit u
AF	Annualization factor
$avail_r$	Available amount of resource r
$b_{u,l}$	Linear coefficient of linearized segment l of unit u
$bought_r$	Amount bought of resource r
$CapMin_{u,l}$	Minimum capacity of unit u in level l
$CapMax_{u,l}$	Maximum capacity of unit u in level l
CarbVal	Value of carbon credit
CC	Carbon credit revenue
$cons_{u,r}$	Consumption of resource r by unit u
Cu	Adjusted capital cost for unit u
$C0_u$	Annualized capital cost at the reference scale for unit u
$demand_r$	Amount demanded of resource r

fop	Hours of operation in a year
hs_{ut}	Available heat per mass of cold utility ut
hv_{ut}	Available heat per mass of hot utility ut
$IAR_{u,r}$	Inlet flow rate of resource r in unit u
LC	Other cost
$massUtility_{pu,ut}$	Mass of utility ut consumed by unit pu
MC	Maintenance cost
$MCp_{pu,n}$	Thermal capacity of stream n of process pu
MP_r	Market price of resource r
MER_{pu}	Minimum energy requirement of hot utility ut
$OAR_{u,r}$	Outlet flow rate of resource r in unit u
OC	Operational cost
PC_r	Commercialization revenue of resource r
PC_r	Revenue commercialization from product r
$prod_{u,r}$	Production of resource r by unit u
$Qc_{pu,s}$	Heat demanded by the cold streams in stage s by process pu
$Qf_{pu,s}$	Inlet heat into unit pu and stage s
$Qh_{pu,s}$	Heat available by the hot streams in stage s by process pu
Qin_{pu}	Heat received from another unit by unit pu
$Qout_{pu}$	Heat supplied to another unit by unit pu
$Qu_{pu,ut}$	Heat consumed by unit pu of utility ut
$Qs_{pu,s}$	Outlet heat into unit pu and stage s
RC_r	Acquisition cost of resource r
$ResCost_r$	Cost of resource r
se	Scaling exponent
$sold_r$	Sold amount of resource r
TAC	Total annualized cost
Te_s	Inlet temperature of stage s
$Tin_{pu,n}$	Inlet temperature of stream n of unit pu
$Tout_{pu,n}$	Outlet temperature of stream n of unit pu
$Tpinch_{pu}$	Pinch temperature of unit pu
Ts_s	Outlet temperature of stage s
UCC_u	Capital cost of unit u
UF_{pu}	Minimum energy requirement of cold utility ut
$UTout_{ut}$	Outlet temperature of utility ut
$y_{u,l}$	Binary variable that selects a unit u in a linearized segment l
w_u	Scale adjustment variable of unit u
$wl_{u,l}$	Local scaling adjustment variable of unit u in level l

References

1. Linnhoff, B.; Sahdev, V. Pinch Technology. In *Ullmann's Encyclopedia of Industrial Chemistry*; John Wiley & Sons: Hoboken, NJ, USA, 2000.
2. Smith, R. *Chemical Process: Design and Integration*; John Wiley & Sons: Hoboken, NJ, USA, 2005; ISBN 0-470-01191-2.
3. Pedrozo, H.A.; Casoni, A.I.; Ramos, F.D.; Estrada, V.; Diaz, M.S. Simultaneous Design of Macroalgae-Based Integrated Biorefineries and Their Heat Exchanger Network. *Comput. Chem. Eng.* **2022**, *164*, 107885. [CrossRef]
4. Restrepo-Flórez, J.M.; Maravelias, C.T. Advanced Fuels from Ethanol-a Superstructure Optimization Approach. *Energy Environ. Sci.* **2021**, *14*, 493–506. [CrossRef]
5. Rhee, G.; Lim, J.Y.; Hwangbo, S.; Yoo, C. Evaluation of an Integrated Microalgae-Based Biorefinery Process and Energy-Recovery System from Livestock Manure Using a Superstructure Model. *J. Clean. Prod.* **2020**, *293*, 125325. [CrossRef]
6. Yang, Z.; Zhang, N.; Smith, R. Enhanced Superstructure Optimization for Heat Exchanger Network Synthesis Using Deterministic Approach. *Front. Sustain.* **2022**, *3*, 976717. [CrossRef]
7. Pina, E.A.; Palacios-Bereche, R.; Chavez-Rodriguez, M.F.; Ensinas, A.V.; Modesto, M.; Nebra, S.A. Reduction of Process Steam Demand and Water-Usage through Heat Integration in Sugar and Ethanol Production from Sugarcane—Evaluation of Different Plant Configurations. *Energy* **2017**, *138*, 1263–1280. [CrossRef]
8. Palacios Bereche, R. Modelagem e Integracao Energetica Do Processo de Producao de Etanol a Partir de Biomassa de Cana-de-Acucar. Ph.D. Thesis, Universidade Estadual de Campinas, Campinas, Brazil, 2011.

9. Fuess, L.T.; Garcia, M.L.; Zaiat, M. Seasonal Characterization of Sugarcane Vinasse: Assessing Environmental Impacts from Fertirrigation and the Bioenergy Recovery Potential through Biodigestion. *Sci. Total Environ.* **2018**, *634*, 29–40. [CrossRef] [PubMed]
10. Albarelli, J.Q. Produção de Açúcar e Etanol de Primeira e Segunda Geração: Simulação, Integração Energética e Análise Econômica. Ph.D. Thesis, Universidade Estadual de Campinas, Campinas, Brazil, 2013.
11. Fuess, L.T.; Cruz, R.B.C.M.; Zaiat, M.; Nascimento, C.A.O. Diversifying the Portfolio of Sugarcane Biorefineries: Anaerobic Digestion as the Core Process for Enhanced Resource Recovery. *Renew. Sustain. Energy Rev.* **2021**, *147*, 111246. [CrossRef]
12. Safari, A.; Das, N.; Langhelle, O.; Roy, J.; Assadi, M. Natural Gas: A Transition Fuel for Sustainable Energy System Transformation? *Energy Sci. Eng.* **2019**, *7*, 1075–1094. [CrossRef]
13. Chiou, H.H.; Lee, C.J.; Wen, B.S.; Lin, J.X.; Chen, C.L.; Yu, B.Y. Evaluation of Alternative Processes of Methanol Production from CO_2: Design, Optimization, Control, Techno-Economic, and Environmental Analysis. *Fuel* **2023**, *343*, 127856. [CrossRef]
14. Khamhaeng, P.; Laosiripojana, N.; Assabumrungrat, S.; Kim-Lohsoontorn, P. Techno-Economic Analysis of Hydrogen Production from Dehydrogenation and Steam Reforming of Ethanol for Carbon Dioxide Conversion to Methanol. *Int. J. Hydrogen. Energy* **2021**, *46*, 30891–30902. [CrossRef]
15. Infante, J.E.; Garcia, V.F.; Ensinas, A.V. Optimal Superstructure Model of Sugarcane-Microalgae Based Biorefinery. In *WASTES: Solutions, Treatments and Opportunities IV*; CRC Press: Boca Raton, FL, USA, 2023; ISBN 978-1-00-334508-4.
16. Fonseca, G.C.; Costa, C.B.B.; Cruz, A.J.G. Economic Analysis of a Second-Generation Ethanol and Electricity Biorefinery Using Superstructural Optimization. *Energy* **2020**, *204*, 117988. [CrossRef]
17. Huynh, T.A.; Rossi, M.; Raeisi, M.; Franke, M.B.; Manenti, F.; Zondervan, E. Promising Future for Biodiesel: Superstructure Optimization from Feed to Fuel. *Comput. Aided Chem. Eng.* **2022**, *51*, 595–600. [CrossRef]
18. Kenkel, P.; Wassermann, T.; Rose, C.; Zondervan, E. A Generic Superstructure Modeling and Optimization Framework on the Example of Bi-Criteria Power-to-Methanol Process Design. *Comput. Chem. Eng.* **2021**, *150*, 107327. [CrossRef]
19. Pyrgakis, K.A.; Kokossis, A.C. Total Site Analysis as a Synthesis Model to Select, Optimize and Integrate Processes in Multiple-Product Biorefineries. *Chem. Eng. Trans.* **2016**, *52*, 913–918. [CrossRef]
20. Celebi, A.D.; Ensinas, A.V.; Sharma, S.; Maréchal, F. Early-Stage Decision Making Approach for the Selection of Optimally Integrated Biorefinery Processes. *Energy* **2017**, *137*, 908–916. [CrossRef]
21. Galanopoulos, C.; Kenkel, P.; Zondervan, E. Superstructure Optimization of an Integrated Algae Biorefinery. *Comput. Chem. Eng.* **2019**, *130*, 106530. [CrossRef]
22. Kantor, I.; Robineau, J.L.; Bütün, H.; Maréchal, F. A Mixed-Integer Linear Programming Formulation for Optimizing Multi-Scale Material and Energy Integration. *Front. Energy Res.* **2020**, *8*, 49. [CrossRef]
23. Bagajewicz, M.; Rodera, H. Energy Savings in the Total Site Heat Integration across Many Plants. *Comput. Chem. Eng.* **2000**, *24*, 1237–1242. [CrossRef]
24. Lingo Systems Inc. *LINGO*; Version 21; Lingo Systems Inc.: Portland, OR, USA, 2022.
25. Peters, M.S.; Timmerhaus, K.D.; West, R.E. *Plant Design and Economics for Chemical Engineers*; ACS: Cambridge, MA, USA, 2003; ISBN 0-07-239266-5.
26. Matsuura, M.D.S.; Seabra, J.E.A.; Chagas, M.F.; Scachetti, M.T.; Morandi, M.A.B.; Moreira, M.M.R.; Novaes, R.M.L.; Ramos, N.P.; Cavalett, O.; Bonomi, A. RenovaCalc: A Calculadora Do Programa RenovaBio. *Embrapa Meio Ambiente-Artig. Em An. Congr. (ALICE)* **2018**, 162–167.
27. Frischknecht, R.; Jungbluth, N.; Althaus, H.-J.; Doka, G.; Dones, R.; Heck, T.; Hellweg, S.; Hischier, R.; Nemecek, T.; Rebitzer, G. The Ecoinvent Database: Overview and Methodological Framework (7 Pp). *Int. J. Life Cycle Assess.* **2005**, *10*, 3–9. [CrossRef]
28. Empresa de Pesquisa Energética. *Balanço Energético Nacional 2022*; Empresa de Pesquisa Energética: Rio de Janeiro, Brazil, 2022; p. 292.
29. Fuess, L.T.; de Araújo Júnior, M.M.; Garcia, M.L.; Zaiat, M. Designing Full-Scale Biodigestion Plants for the Treatment of Vinasse in Sugarcane Biorefineries: How Phase Separation and Alkalinization Impact Biogas and Electricity Production Costs? *Chem. Eng. Res. Des.* **2017**, *119*, 209–220. [CrossRef]
30. Albarelli, J.Q.; Onorati, S.; Caliandro, P.; Peduzzi, E.; Meireles, M.A.A.; Marechal, F.; Ensinas, A.V. Multi-Objective Optimization of a Sugarcane Biorefinery for Integrated Ethanol and Methanol Production. *Energy* **2017**, *138*, 1281–1290. [CrossRef]
31. Dutta, A.; Sahir, A.; Tan, E.; Humbird, D.; Snowden-swan, L.J.; Meyer, P.; Ross, J.; Sexton, D.; Yap, R.; Lukas, J. Process Design and Economics for the Conversion of Lignocellulosic Biomass to Hydrocarbon Fuels. NREL/TP-5100-62455 and PNNL-23823. *Nrel* **2015**.
32. Khojasteh-Salkuyeh, Y.; Ashrafi, O.; Mostafavi, E.; Navarri, P. CO_2 utilization for Methanol Production; Part I: Process Design and Life Cycle GHG Assessment of Different Pathways. *J. CO2 Util.* **2021**, *50*, 101608. [CrossRef]
33. Lonis, F.; Tola, V.; Cau, G. Assessment of Integrated Energy Systems for the Production and Use of Renewable Methanol by Water Electrolysis and CO_2 Hydrogenation. *Fuel* **2021**, *285*, 119160. [CrossRef]
34. Shim, H.M.; Lee, J.; Yoo, Y.D.; Seung Yun, Y.; Kim, H.T. Simulation of DME Synthesis from Coal Syngas by Kinetics Model. *Korean J. Chem. Eng* **2009**, *26*, 641–648. [CrossRef]
35. Turton, R.; Shaeiwitz, J.A.; Bhattacharyya, D. *Analysis, Synthesis, and Design of Chemical Processes*; Prentice Hall: Hoboken, NJ, USA, 2018; ISBN 978-0-13-417740-3.

36. Li, L.; Manier, H.; Manier, M.A. Integrated Optimization Model for Hydrogen Supply Chain Network Design and Hydrogen Fueling Station Planning. *Comput. Chem. Eng.* **2020**, *134*, 106683. [CrossRef]
37. Bressanin, J.M.; Klein, B.C.; Chagas, M.F.; Watanabe, M.D.B.; de Mesquita Sampaio, I.L.; Bonomi, A.; de Morais, E.R.; Cavalett, O. Techno-Economic and Environmental Assessment of Biomass Gasification and Fischer–Tropsch Synthesis Integrated to Sugarcane Biorefineries. *Energies* **2020**, *13*, 4576. [CrossRef]

Article

Optimization and Tradeoff Analysis for Multiple Configurations of Bio-Energy with Carbon Capture and Storage Systems in Brazilian Sugarcane Ethanol Sector

Bruno Bunya [1], César A. R. Sotomonte [1,2], Alisson Aparecido Vitoriano Julio [1,3], João Luiz Junho Pereira [4], Túlio Augusto Zucareli de Souza [1,*], Matheus Brendon Francisco [1] and Christian J. R. Coronado [1]

[1] Mechanical Engineering Institute—IEM, Federal University of Itajubá—UNIFEI, Itajubá 37500-903, Brazil; cesar.sotomonte@unila.edu.br (C.A.R.S.); matheus_brendon@unifei.edu.br (M.B.F.); christian@unifei.edu.br (C.J.R.C.)

[2] Chemical Engineering Institute, Federal University of Latin American Integration—UNILA, Foz do Iguaçu 85870-650, Brazil

[3] Department of Planning, Aalborg University, Rendsburggade 14, 9000 Aalborg, Denmark

[4] Computer Science Division, Aeronautics Institute of Technology—ITA, São José dos Campos 12228-900, Brazil; joaoluizjp@ita.br

* Correspondence: tulio_zucareli@unifei.edu.br

Abstract: Bio-energy systems with carbon capture and storage (BECCS) will be essential if countries are to meet the gas emission reduction targets established in the 2015 Paris Agreement. This study seeks to carry out a thermodynamic optimization and analysis of a BECCS technology for a typical Brazilian cogeneration plant. To maximize generated net electrical energy (MWe) and carbon dioxide CO_2 capture (Mt/year), this study evaluated six cogeneration systems integrated with a chemical absorption process using MEA. A key performance indicator (gCO_2/kWh) was also evaluated. The set of optimal solutions shows that the single regenerator configuration (REG1) resulted in more CO_2 capture (51.9% of all CO_2 emissions generated by the plant), penalized by 14.9% in the electrical plant's efficiency. On the other hand, the reheated configuration with three regenerators (Reheat3) was less power-penalized (7.41%) but had a lower CO_2 capture rate (36.3%). Results showed that if the CO_2 capture rates would be higher than 51.9%, the cogeneration system would reach a higher specific emission (gCO_2/kWh) than the cogeneration base plant without a carbon capture system, which implies that low capture rates (<51%) in the CCS system guarantee an overall net reduction in greenhouse gas emissions in sugarcane plants for power and ethanol production.

Keywords: bio-energy; BECCS; multi-objective optimization; sugarcane bagasse

Citation: Bunya, B.; Sotomonte, C.A.R.; Vitoriano Julio, A.A.; Pereira, J.L.J.; de Souza, T.A.Z.; Francisco, M.B.; Coronado, C.J.R. Optimization and Tradeoff Analysis for Multiple Configurations of Bio-Energy with Carbon Capture and Storage Systems in Brazilian Sugarcane Ethanol Sector. *Entropy* **2024**, *26*, 698. https://doi.org/10.3390/e26080698

Academic Editors: Daniel Flórez-Orrego, Meire Ellen Ribeiro Domingos and Rafael Nogueira Nakashima

Received: 22 July 2024
Revised: 12 August 2024
Accepted: 14 August 2024
Published: 17 August 2024

1. Introduction

According to the report from the Intergovernmental Panel on Climate Change in 2018 [1], humans must reduce anthropogenic CO_2 emission levels by 45% from 2010 to 2030 and reach zero emissions by 2050 to limit global warming to 1.5 °C. The Paris Agreement from 2015 has set a goal for preventing global temperature increases by 2 °C, relative to pre-industrial levels, and seeks to limit temperature increases to 1.5 °C. In this agreement, Brazil pledged to reduce its GHG (greenhouse gas) emissions by 37% by 2030 and 43% by 2050, relative to 2005. Recently, Brazil reinforced its participation in reducing emissions to zero by 2050 at the 2021 Climate Summit.

Carbon capture and storage (CCS) systems and negative emission technologies (NETs) will be essential in meeting this target [2]. CCS systems are already available in the market; however, they are still expensive [3]. A complete CCS system can constitute 80% of the total cost of a plant, including capture, transportation, and storage [4]. A report released by the Global CCS Institute [5] presented different strategies for mitigating global warming and pointed out that bio-energy with carbon capture and storage (BECCS) technologies are crucial.

BECCS technologies refer to the integration of CCS systems with bioenergy-based systems, including biomass-fueled boilers and furnaces, biogas upgrading facilities, and ethanol plants. Biomass, as a renewable energy source, is considered carbon-neutral throughout its lifecycle [6]. Therefore, BECCS is viewed as the most viable approach for achieving negative emissions. This is especially true when compared to the application of CCS to fossil fuel-based systems, which can transform them into negative emission technologies at a cost of up to USD 1000 per tonne of CO_2 [7].

The main limitation, and what keeps the BECCS systems away from economic feasibility, is the energy penalty associated with its operation, as well as CO_2 compression, transportation, and storage processes [8]. Therefore, the tradeoff between energy efficiency and CO_2 capture is key to assessing the technical and economic feasibility of these systems. Fajardy et al. [9] emphasize that biomass residues are a more attractive option economically, since the energy allocated for planting can also be used for other purposes by diversifying the products' portfolio, like ethanol production from sugarcane. Sugarcane presents one of the highest efficiencies in converting solar energy into biochemical energy via photosynthesis [10], and it is the main biomass feedstock for energy in Brazil, accounting for 11.7 GW (406 thermoelectric plants) of installed capacity.

In fact, sugarcane represents one of the most important energy sources in the world, being widely used for bioethanol production and presenting a self-sustainable energy processing, often using sugarcane bagasse as a renewable solid fuel to simultaneously produce steam for process, bioethanol, and surplus electricity [11]. Moreover, the sugarcane processing sector is widely used for producing sugar and many other inputs for the food industry [12], and since sugarcane biomass has been also highlighted as a sustainable source of renewable hydrogen [13], its thermal cracking has proven to be a valuable way to obtain this energy vector [14].

Despite being a renewable resource, the sugarcane production chain has various environmental impacts, depending on the agricultural practices employed. These practices need to be properly managed to make sugarcane a more sustainable feedstock. A study focused on South Africa by Pryor et al. [15] showed that green cane harvesting could reduce energy inputs by 4% and greenhouse gas (GHG) emissions by 16%. However, mechanization leads to soil compaction and stool damage, resulting in lower yields and increased energy consumption and GHG emissions. Also, the proper use of sugarcane residues for energy production can increase the process efficiency even further [16].

Based on production records for 36 billion liters of ethanol in 2019, a potential capture of 44.77 tons of CO_2/year is estimated from the fermentation stage in the ethanol production process. For annual sugarcane production at 665.1 Mt, 246.1 Mt of CO_2/year can potentially be avoided via BECCS systems [17].

Among the available technologies for CCS systems, post-combustion is the most promising carbon capture method [6], given the relative ease of retrofitting existing thermal plants. In this process, CO_2 is removed from chemical absorption, which is the most widespread technique, given its technological maturity and potential for short-term applications [18], besides being applicable to sources of CO_2 between 3 and 20% in the gaseous mixture [19].

In the literature, Dubois and Thomas [20], analyzed three different post-combustion chemical absorption configurations and obtained specific energy consumption at 2.39 GJ/tCO_2 in the solvent regeneration for a mixture of MDEA 10% + PZ 30%. Bougie et al. [21] demonstrated that mixtures of MEA with other solvents like glycol monomethyl ether (DEGMEE) increased CO_2 absorption and reduced energy consumption by up to 78%, compared to traditional MEA at 30%. Li et al. [22] used aqueous ammonia to minimize energy consumption when capturing CO_2. The results indicated potential reductions of 3.3% in plant energy penalty efficiency compared to conventional MEA. Even though other solvents and mixtures may provide better results from an energy point of view and have high corrosion rates [23], MEA is the most used alternative for removing CO_2, mostly due to its costs [24].

Post-combustion technology based on MEA was evaluated for a BECCS system placed in the Brazilian sugarcane sector by [25], and although the energy penalty varied from 43% to 52%, investing in a BECCS system was placed as a better investment in comparison to a natural gas-based power plant. BECCS investments would be lower, and negative emissions might be achieved.

Therefore, several works on chemical absorption focus on the performance of pilot plants and models/simulations to find process improvements. In this work, the main objective is to investigate the technical feasibility of BECCS systems for use in the sugar energy sector using carbon capture technologies from chemical absorption under different Rankine cycle configurations. Multi-objective optimization will be performed using the metaheuristic Lichtenberg algorithm based on a thermodynamic cycle developed in the Aspen Plus® V11.

2. Bio-Energy with Carbon Capture and Storage—BECCS

2.1. Power Generation in the Sugar Energy Sector

The Brazilian sugar alcohol sector, with its varied production range, decades of technical knowhow, and appropriate use of industrial waste, is an essential model of sustainability [26]. Due to its advancements and practical knowledge of various methods, the sugarcane sector has incorporated modern cogeneration systems with reheating and regeneration [27], which provide heat and power for auxiliary equipment and plant utilities, besides surplus electricity, which is sold to the grid. Bagasse-fueled boilers operate at pressures ranging from 22 and 85 bar, with live steam at 320 °C, and they are superheated to 480–520 °C, which is the most common operation carried out with superheat steam at 480 °C and 65 bar. Table 1 summarizes the typical operating parameters for cogeneration plants in the sugar energy sector.

Table 1. Operational parameters of cogeneration plants in the Brazilian sugar energy sector.

	[28]	[27]	[29]
Boiler	22 bar/300 °C 65 bar/480 °C 100 bar/530 °C	85 bar/520 °C	67 bar/490 °C 100 bar/520 °C 100 bar/530 °C
Humid Bagasse	50%	50%	50%
Humid Chaff	15%	15%	15%
Humid Fibers	14%	14%	14%
Available Bagasse	280 kg/tc	280 kg/tc	-
Available Chaff	164 kg/tc	164 kg/tc	-
Operation	4464 h	4320 h	4300 h
Milling Capacity	500 tc/h	500 tc/h	465.12 tc/h
Process Steam Consumption	280 kg/tc 340 kg/tc 500 kg/tc	450 kg/tc	280 kg/tc 300 kg/tc 500 kg/tc

Based on the data compiled in Table 1, a plant was studies as a "base case" using typical characteristics for sugar and alcohol plants in Brazil. Table 2 summarizes the data for a sugar mill plant working with 2 Mt of sugarcane per year. The working operation regime was chosen to be 240 days or 5760 h per year, including the harvest and off-season periods, for the running time of the steam cycle, with no modulations to plant operation during the harvest and off-season periods. Figure 1 shows a simplified physical schematic of the proposed BECCS system, with the simulation being carried out using Aspen Plus® V11.

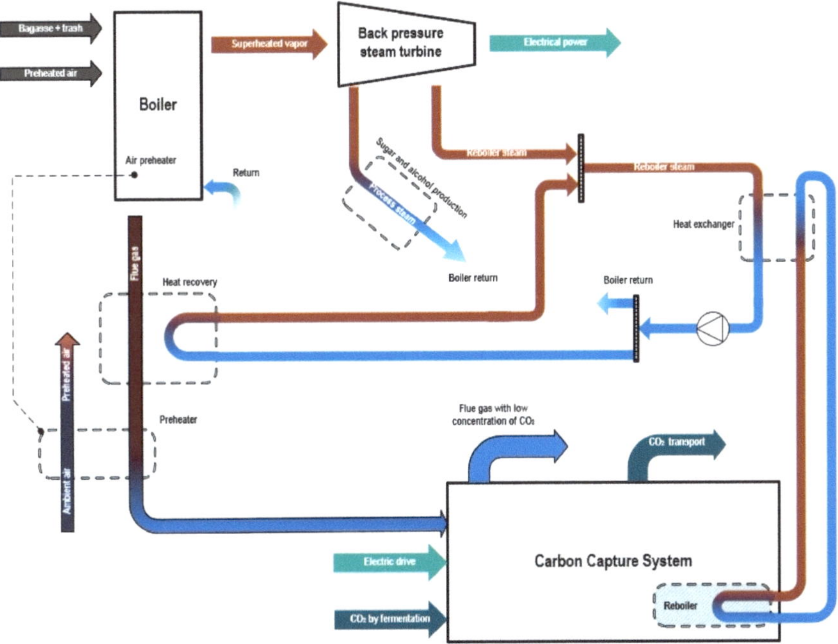

Figure 1. Integrated BECCS system.

Table 2. Parameters for sugarcane processing.

Annual milling capacity	2,000,000 tc/year
Vapor process consumption	430 kgv/tc
Annual operation time (harvest + between harvest)	5760 h
Cane processing/hour	347.2 tc/h
Bagasse/cane ratio	28%
Annual bagasse production	560,000 tb/year
Available bagasse for producing electricity (90%)	504,000 tb/year
Available bagasse	87.50 tc/h
Chaff/cane ratio	16.4 %
Annual chaff production	328,000 tp/year
Available chaff for producing electricity (10%)	32,800 tp/year
Chaff available	1.58 kg/s

2.2. Biomass Combustion

In first stage, bagasse and part of the straw (characterized in Table 3) produced in the field are used as fuel in the boiler. Bagasse with 50% moisture content and straw with 15% moisture content are fed into a yield reactor (RYield) to decompose the solid biomass into its main constituent elements before evaluating the combustion reaction in a Gibbs reactor (RGibbs), disregarding nitrogen oxide formation (Figure 2).

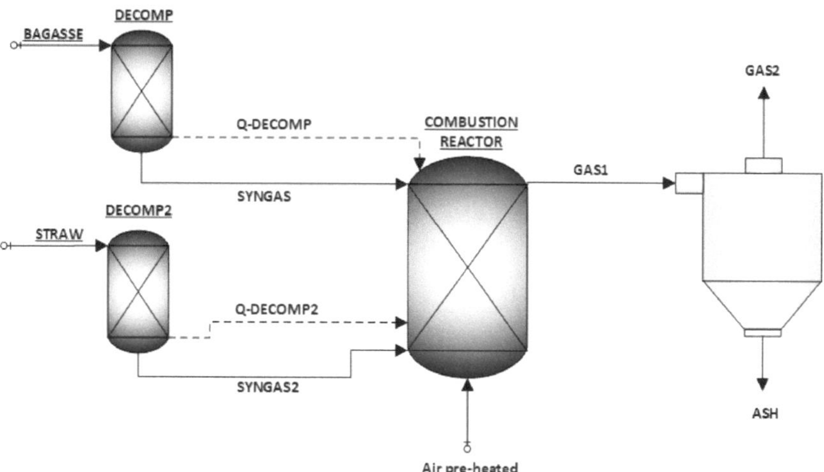

Figure 2. Biomass combustion simulation.

Table 3. Characterization of sugarcane bagasse and chaff.

Description	Bagasse [a] [b]	Chaff [c]
Immediate Analysis [% in mass] [a]		
Humidity	0.0	0.0
Fixed carbon	12.0	7.7
Volatile material	85.0	79.5
Ash	3.0	12.8
Chemical Analysis [% in mass] [b]		
Carbon	46.4	49.6
Hydrogen	6.1	6.4
Nitrogen	0.2	0.5
Chlorine	0.0	0.0
Sulfur	0.1	0.1
Oxygen	44.2	30.5
HHV [c]	19.30	20.04

a [30], b [31] and c [32].

In the Gibbs reactor, the simulation evaluated the biomass combustion reaction in the presence of preheated dry air. The excess air was based on other studies in the literature on bagasse plants from the sector. Carminati et al. [33] used 50% excess air. However, Rayaprolu [34] used a range from 30 to 50% excess air for burning bagasse in more modern boilers. We decided to use the average value of (40%) in the simulation. The combustion exhaust gases pass through a separator to remove ash, leaving only flue gas that is composed of O_2, N_2, CO_2, SO_2, and H_2O.

2.3. Cogeneration Cycle

The cogeneration cycle was based on modern cogeneration configurations known in the literature that are used by sugar and alcohol industries. Extractions of steam at 130 °C and 2.5 bar were used to meet plants themal demands, and the operation of the plant was carried out using backpressure turbines, allowing for more heat production downstream. Exhaust gases, which are the products of combustion in the boiler furnace, travel through four primary heat exchanger surfaces in the reheating cycles, namely a superheater, reheater, evaporator, and economizer (Figure 3). An internal heat recovery unit is used to increase the water flow temperature from 125 to 135 °C, which was used to provide heat to the reboiler in the carbon capture and storage (CCS) system. Exhaust gases were cooled to 80 °C by preheating the air used in the combustion simulation.

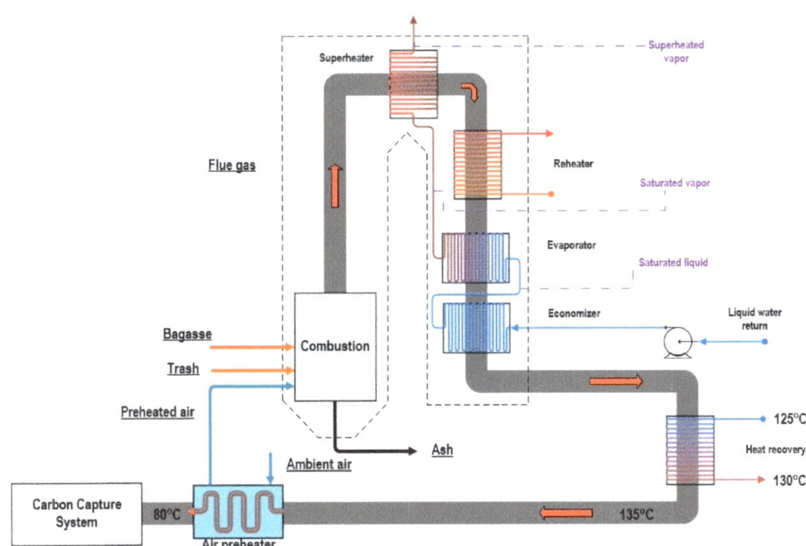

Figure 3. Biomass boiler configuration.

Six configurations were selected for power generation, as shown in Appendix A. For the simulations, a thermodynamic model based on the Peng Robinson Stryjek–Vera (PRSV) equation of state was used, considering all expansion and compression steps in the turbines and pumps using the isentropic efficiency model, while the heat exchangers were simulated using the on-design model.

The heat supplied in the process and in the reboiler was simulated using a simplified model for heat exchange (cooler). Electrical energy demand for driving motors, lighting, and other auxiliary equipment was taken as being 12 kWh per ton of processed cane. Table 4 summarizes the parameters used in the cogeneration cycle.

Table 4. Input parameters for all the evaluated configurations.

Isentropic efficiency of the turbine	85%
Isentropic efficiency of the pump	75%
Inlet temperature for ethanol production	130 °C
Input pressure for the process	2.5 bar
Output temperature for the process	90 °C
Output pressure for the process	1.3 bar
Inlet temperature for the reboiler	130 °C
Inlet pressure for the reboiler	2.5 bar
Output temperature for the reboiler	125 °C

2.4. Carbon Capture and Storage (CCS) System

Selecting the proper CO_2 capture technology is directly linked to the combustion or gas formation process. Once separated, CO_2 has to be compressed and transported in a supercritical state to its final destination. Although it is still considered an expensive technology, capturing CO_2 via chemical absorption using amine-based solutions is the most dominant technology on the market, and it has been labeled with a Technical Readiness Level of 9 [35]. Moreover, it has been widely studied with a specific focus on the performance of pilot plants and models/simulations for finding process improvements. Table 5 shows the typical carbon capture and storage system's operating parameters using chemical absorption and MEA as the solvent.

Table 5. Operational parameters of cogeneration plants in the Brazilian sugar energy sector

References	[36]	[37]	[38]	[39]	[40]	[41]	[42]
Plant/Simulation Absorber	Plant	Plant	Plant	Plant	Simulation	Plant/Simulation	Plant/Simulation
Combustion gases (Nm^3/h)	1550	30–110	350	293	368.8 (kg/s)	242–248 (kg/h)	72 (kg/h)
CO_2 (vol%)	14.2	3–14	15	13.5	13	5.5–9.9	5.4
CO_2 captured (%)	90	50–75	90	75–89	90	90	75–91
Solvent flow (m^3/h)	-	30–350	1300	800–1600	740 (kg/h)	-	200 (kg/h)
L/G rate	-	2.8	3.7	3.9–5.8	-	1.7–2.9	-
Temperature ($^\circ$C)	40	45–50	40	40–60	42	37–40	40
Stripper							
Reboiler specific heat duty (GJ/tCO$_2$)	3.5	3.98–5.01	3.92	3.77–4.36	3.76	5.2–7.4	5.01
Lean solvent (mol.CO_2/mol.MEA)	-	0.08–0.09	-	0.28–0.38	0.23	0.17–0.20	0.27
Rich solvent (mol.CO_2/mol.MEA)	-	0.11–0.14	-	0.46–0.53	0.49	0.38–0.44	0.38
Reboiler temperature ($^\circ$C)	98–113	120	113.8	105–110	103	108.7–110.4	112.85
Pressure (bar)	1.75–1.90	1–2.5	1.5	1	1.85	-	2

To simulate the separation of CO_2 via chemical absorption in Aspen Plus, the thermodynamic model ElecNRTL (Non-Random Two-Liquid) was used, which is widely used in the literature [43,44]. In Figure 4, a typical flow diagram of a CO_2 capture system via chemical absorption from the MEA solvent is presented.

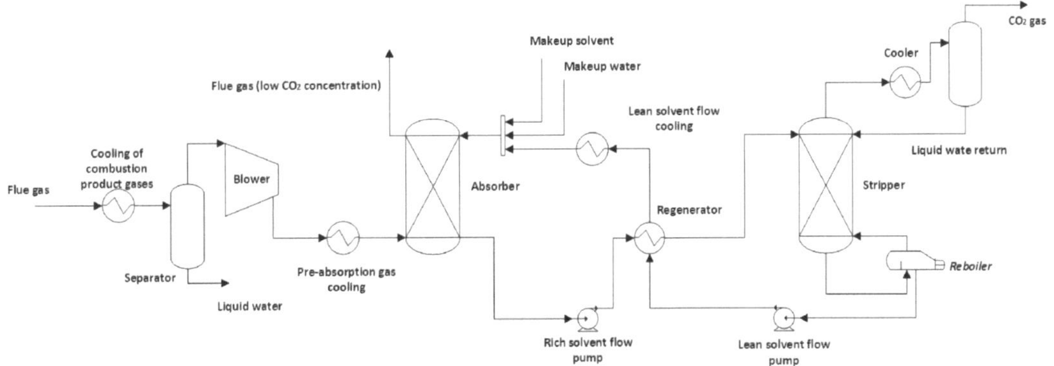

Figure 4. Typical schematic of a chemical absorption system.

At the beginning of the absorption process, exhaust gases leaving the boiler are cooled to 40 °C with the cooler, which is common in absorption columns (no more than 60 °C) to promote CO_2 absorption using MEA [35]. A separator is used to separate the condensate generated from cooling the gases to ensure there is no liquid in the gases at the blower's inlet. At the blower, gas pressure is increased (10 kPa) to overcome pressure drops in the absorption column, where gas is placed at the bottom of the column at approximately 50 °C. Both the lean CO_2 solvent and the makeup flow of water enter the top of the column at 37 °C/1.1 bar. The rich CO_2 solution is pumped to the stripper column at 2 bar, having passed through the regenerator to be heated to 105 °C; therefore, MEA degradation does not occur [40]. CO_2 is released from the solution at the top of the stripper, with 99% purity, and it is directed to the compression and transportation stages. At the bottom of the stripper column, the lean CO_2 solution exits the column at approximately 120 °C and returns to the absorption column, having passed through the regenerator and mixer, where MEA is replenished in the system. All operating parameters in the simulation are summarized in Table 6.

Table 6. Operational parameters for the chemical absorption CO_2 capture system simulated in Aspen Plus.

DATA	
Model Absorber	ELECNRTL
Calculation type	Equilibrium
N° of stages	12
Pressure (bar)	1
L/G ratio	3.8
Stripper	
Type	Kettle
Reflux rate	0.18
Boilup rate	0.14
N° of stages	20
Reboiler specific heat duty (GJ/tCO_2)	~3.9
Lean solvent (mol.CO_2/mol.MEA)	0.19
Rich solvent (mol.CO_2/mol.MEA)	0.47
Input flow temperature (°C)	105
Operational pressure (bar)	1.8

In addition to the CO_2 produced from the burning biomass, it also accounted the CO_2 generated from the plant's fermentation process. Unlike the CO_2 from exhaust gases, which need absorption systems for separation, the CO_2 from fermentation can be directly routed to the final transportation system, while considering the electrical power needed to compress it. Table 7 summarizes the correlations for producing CO_2 from the ethanol production process.

Table 7. Considerations for estimating the CO_2 captured form the fermentation process.

Parameter	Value	Reference
Ethanol production (L/tc)	86.3	[45]
CO_2 production per kg of ethanol [kg CO_2]	0.96	[17]

After capture, the CO_2 must be compressed at high pressures for transportation. The compression process was based on the configurations presented in [40]. Here, CO_2 was compressed up to 150 bar for transportation. The CO_2 flows produced by the plant were compressed from 2 to 128 bar using six compression stages, with a compression ratio equal to 2, and intermediate cooling down to 30 °C. After the last compression stage, the CO_2 was cooled again and compressed to 150 bar and then transported.

3. Parametrical Optimization Methodology

The technical and thermodynamic evaluation of the BECCS system involved four stages: (1) simulation of biomass combustion; (2) simulation of the Rankine cycle configurations; (3) simulation of the CCS system; and (4) parametric optimization of the BECCS system. The four steps are shown in the flowchart in Figure 5.

The thermodynamic problem in question can be statistically analyzed using variance analysis. An optimized matrix of experiments can be generated for the problem using the design of experiments. One must define the input variables—which are the variation intervals of each in the thermodynamic cycle—and the response variables.

After analyzing the cycle parameters, a multi-objective optimization can be performed to find the non-dominated solutions to the problem. All non-dominated solutions are optimal, as are those for which it is not possible to improve an objective without negatively affecting another objective. The set of these solutions is called the Pareto front. Meta-heuristics can better handle complex optimization problems where classical methods have limitations, as well as having the ability to handle optimization problems that do not have explicit objective functions. This approach is particularly useful for simultaneously assessing conflicting goals, such as the maximum cogeneration net power and minimum CCS energy penalty.

The Lichtenberg algorithm [46], will be applied for this. This meta-heuristic model was inspired by lightning and Lichtenberg figures, and examples of its application can be found in [47]. Also, the same metaheuristic model has already been validated for other renewable energy systems, such as steam reforming systems [48].

For optimization, one must define the search domain, i.e., the variation ranges for each variable, which are the same as in the design of experiments. So, the parameter optimizer must be adjusted. The following parameters were chosen based on the recommendations from Pereira et al. [46]: Pop = 100; Niter = 100; Rc = 200; Np = 106; S = 1; ref = 0.4; and M = 0.

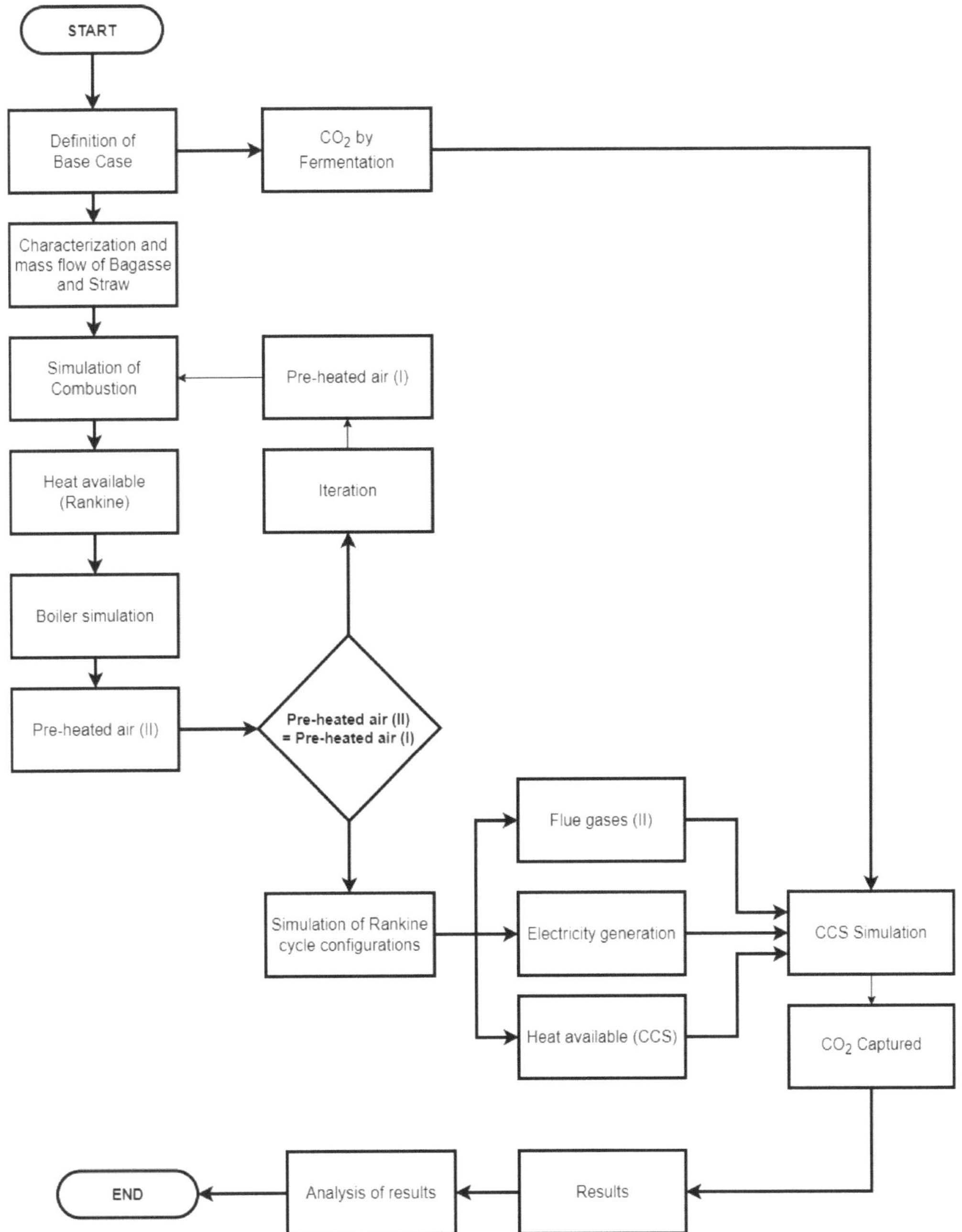

Figure 5. Flowchart of methodology.

A relevant indicator of the stripper is the specific consumption of thermal energy per mass of captured CO_2 (GJ/tCO_2 or $MJ/kgCO_2$), which varies between 3.5 and 7.4 GJ/tCO_2 (Table 5). It is of global interest that this indicator be as low as possible to reduce the plant's energy penalty. In this sense, two objective functions were considered in this optimization: maximizing CO_2 absorption in the CCS system and maximizing the net electrical power

(W_n) (or minimizing the energy penalty). These indicators are the most influential in determining the technical and economic feasibility of BECCS systems.

The following design variables were selected for this study: boiler outlet temperature, reheating temperature, turbine outlet pressure, and the pinch point in the economizer and regenerators. It is important to point out that in order to avoid the algorithm losing much of its efficiency, increasing the total number of simulations, and consuming more computational resources, only the design parameters for the cogeneration cycle were considered in the optimization process. Appendix A summarizes the input variables in the optimization cycle.

4. Results

As was mentioned before, the main technical barrier of BECCS systems is the energy penalty associated with CO_2 capture. For this reason, the reheating cycle with three regenerators was optimized using the net power of the cogeneration cycle as an objective function to evaluate the energy penalty associated with the CO_2 capture system. The results (Table 8) showed that net electrical power was 62.82 MWe, representing an energy efficiency of 31%, and emissions were equal to 1300 gCO_2/kWh.

The results show that all the configurations of the thermal system provided a perfect negative correlation between the objective functions for the operating range of the evaluated design variables. For the cogeneration cycles that discarded steam reheating (REG1, REG2, and REG3), the steam temperature at the boiler outlet had the greatest influence on the thermodynamic performance of the system. The higher the temperature of the steam at the turbine input, the greater the enthalpy variation during steam expansion in the equipment. Furthermore, high vaporization temperatures ensured that the steam exiting the last turbine stage remained saturated with pressure parameters close to the lower 250 kPa limit for meeting the plant's process steam quality conditions.

Raising temperatures close to 520 °C resulted in reduced steam mass flow in the cycle, generating less thermal energy for meeting CO_2 absorption. A lower temperature at the boiler's outlet increased steam availability for the processes. Under these operating conditions, higher pressures at the last turbine stage are needed to ensure that steam is saturated for alcohol and CCS production processes, leading to decreased power generation in the cogeneration cycle.

The vaporization pressure and pinch point are important design parameters. Similar to the evaporator temperature, the vaporization pressure is directly proportional to the enthalpy variation during steam expansion in the turbine, promoting power generation. The higher the pinch point, the higher the exhaust gas temperature at the inlet of the heat exchanger, favoring energy generation for the CCS process; however, this decreases the mass flow for the working fluid, decreasing power generation in the power cycle.

For configurations with steam reheating (Reheat1, Reheat2, and Reheat3), the results show that both the vaporization pressure and the low-pressure turbine discharge pressure influenced the thermal system the most, generating more electrical power or heat for downstream plant processes. On the other hand, the vaporization temperature had little influence on the evaluated objectives compared to the configuration without reheating. We also observed that there was a greater interaction between input parameters for the configurations with reheating, although they had little influence on the results when compared to the operating pressures in the cogeneration cycle.

Table 8. Optimal results for the cogeneration cycles.

	ReHeat3 (sem CCS)	REG1	REG2	REG3	ReHeat1	ReHeat2	ReHeat3
W_c [MWe]	-	5.30–4.81	4.81–4.63	4.63–4.62	4.62–4.45	4.45–4.34	4.34–4.31
W_n[MWe]	62.82	33.46–40.58	40.58–43.19	43.19–43.33	43.33–45.75	45.75–47.36	47.36–47.80
η [%]	31.01	16.52–20.03	20.03–21.32	21.32–21.39	21.39–22.58	22.58–23.38	23.38–23.60
ηcog [%]	88.02	77.53–77.05	77.05–78.33	78.33–78.40	78.40–79.60	79.60–80.39	80.39–80.61
CO_2 [Mt/year]	0.474	0.224–0.190	0.190–0.178	0.178–0.177	0.177–0.166	0.166–0.159	0.159–0.156
CO_2 capture	0	51.9–44.2	44.2–41.3	41.3–41.1	41.1–38.5	38.5–36.8	36.8–36.3
Emissions [g/kWh]	1300	1300–1215	1215–1190	1190	1190–1169	1169–1155	1155

Figure 6 shows the results of the multi-objective optimization for each evaluated configuration. As was mentioned above, we observed that, for each evaluated configuration, there was a negative linear correlation in which an increase in CO_2 capture capacity led to reduced electrical power generation in the cycle. The heat demand for stripping is critical for releasing CO_2 from the solvent, enabling its subsequent capture and separation. This demand encompasses sensible heat, which is needed to raise the solution's temperature; desorption heat, which is responsible for breaking the chemical bonds between CO_2 and the solution; and latent heat, which is essential for evaporating the solution's water content. Therefore, a higher CO_2 capture capacity requires an increased availability of heat in the Rankine cycle for CO_2 capture purposes. Consequently, a greater capture rate would lead to reduced water vapor available for power generation, resulting in a decrease in power output, commonly referred to as the energy penalty.

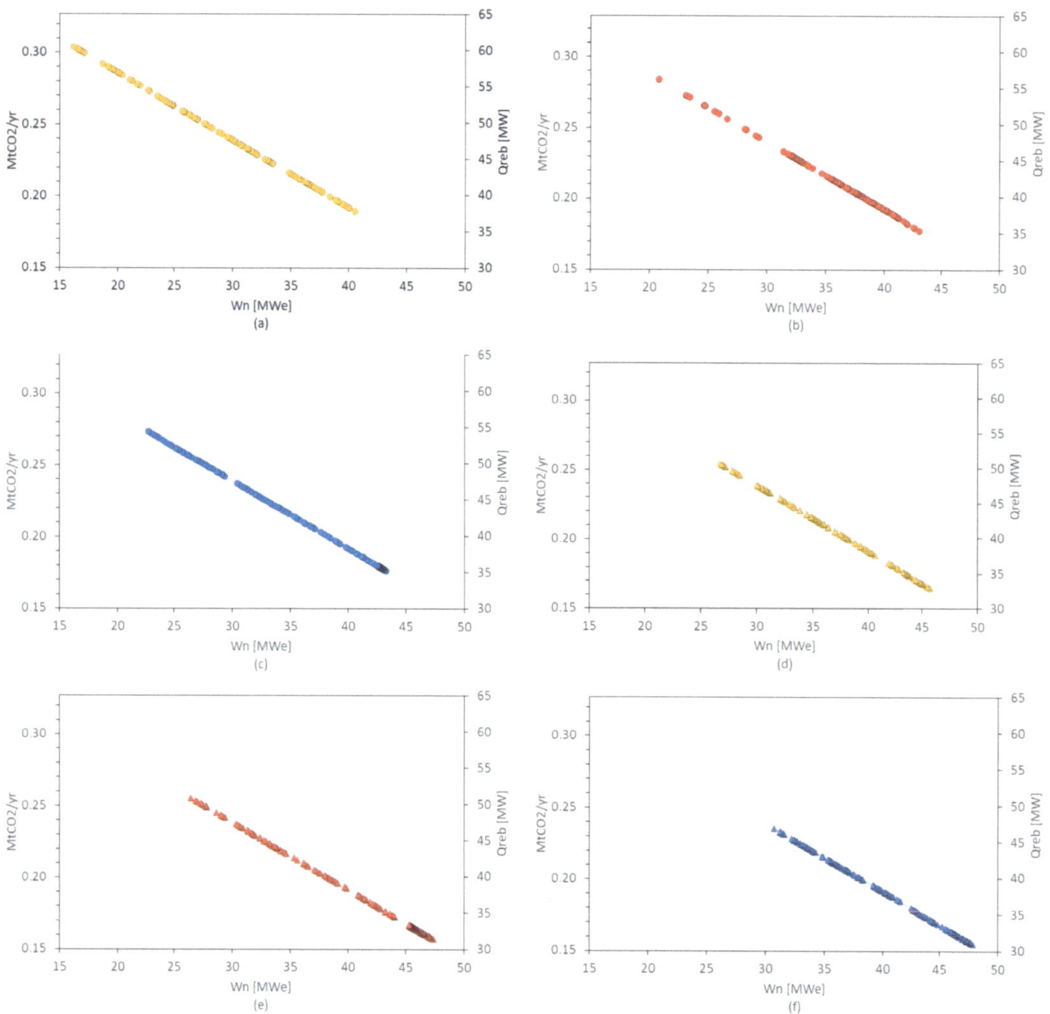

Figure 6. Power generated vs. CO_2 capture: (**a**) REG1, (**b**) REG2, (**c**) REG3, (**d**) Reheat1, (**e**) Reheat2, and (**f**) Reheat3.

Considering CO_2 absorption, the REG1 configuration was the thermodynamic cycle with the highest heat availability for the CCS system, at approximately 0.304 $MtCO_2$/yr, representing a 70.6% CO_2 capture percentage from nearby exhaust gases; however, it was limited to 16 MWe of net electric power generation (Figure 6a). The Reheat3 configuration was the technological option with the greatest capacity for generating electrical power (47.8 MWe) and for capturing CO_2, at 0.156 $MtCO_2$/yr (Figure 6f). The rest of the configurations showed electricity generation values and CO_2 capture levels within intermediate ranges between the two previously mentioned configurations.

The REG1 configuration is the least complex alternative (fewest devices), i.e., it is the least expensive thermodynamic cycle in terms of installation and maintenance. ReHeat3 is the opposite. Therefore, the more complex cogeneration cycle configurations that produced the same amount of electrical power and captured CO_2 were excluded from this analysis. Furthermore, the BECCS system showed lower specific emissions relative to the Reheat3 cycle without CCS (1300 gCO_2/kWh), discarding any set of optimal solutions above this restriction. Figure 7 shows the set of optimal solutions for all evaluated configurations.

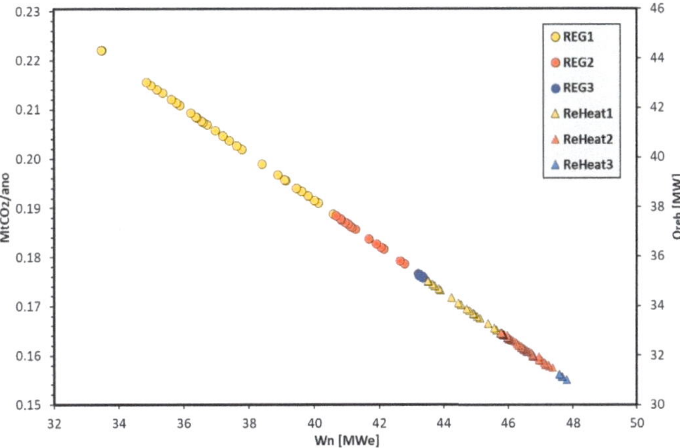

Figure 7. Pareto boundary of the objective functions.

The BECCS system was able to obtain a maximum capture of 0.224 Mt/year for REG1 and a CO_2 capture rate from exhaust gases close to 51.9%; however, it was limited to electrical power generation, which was 33.46 MWe. This was true up to 29.36 MWe, as less than 62.82 MWe was generated by the ReHeat3 configuration without CCS (14.49% penalty on the plant's electrical efficiency). On the other hand, Reheat3 (with CCS) resulted in more electrical power generation (47.80 MWe) and had a lesser penalty for electrical efficiency (7.41%); however, it had a minimum CO_2 capture rate of 36.3%, which was emitted by the plant.

Figure 8 shows three scenarios for percentages of CO_2 capture, as well as the respective electrical power required for compression at the plant. To compress the CO_2 generated from fermentation, approximately 2 MWe are needed. Considering the limiting scenario at 90% CO_2 capture from exhaust gases, 8.24 MWe would be needed to compress the CO_2 generated by the plant to its maximum capacity, which would be equivalent to 1.48 MWe of power for 0.1 Mt/year of CO_2 captured at the plant.

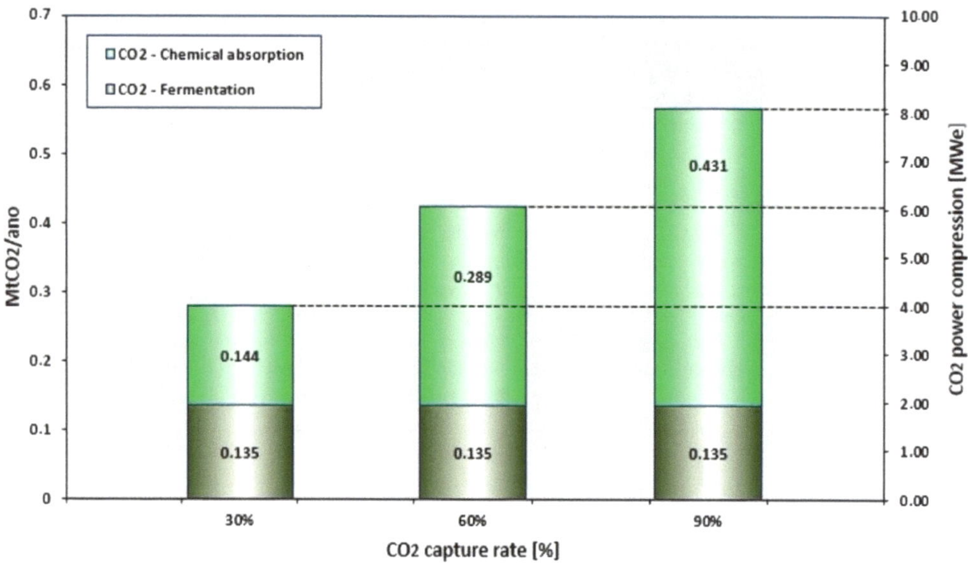

Figure 8. Results for the CO_2 compression system simulation.

Figure 9 shows a reduction in the net electric power generated by the BECCS system from the CO_2 compression system in the plant. The minimum capture point for CO_2 showed a reduction of 8.27% in the net electric power generated (52.11 to 47.8 MWe), while the maximum capture penalty was 13.67% (38 .76 to 33.46 MWe). For a theoretical scenario for a configuration with a greater CO_2 capture capacity, one could capture up to 88.73% of all the CO_2 generated at the plant; however, one would need to consume all the electrical power generated to meet the power demands of the compression system.

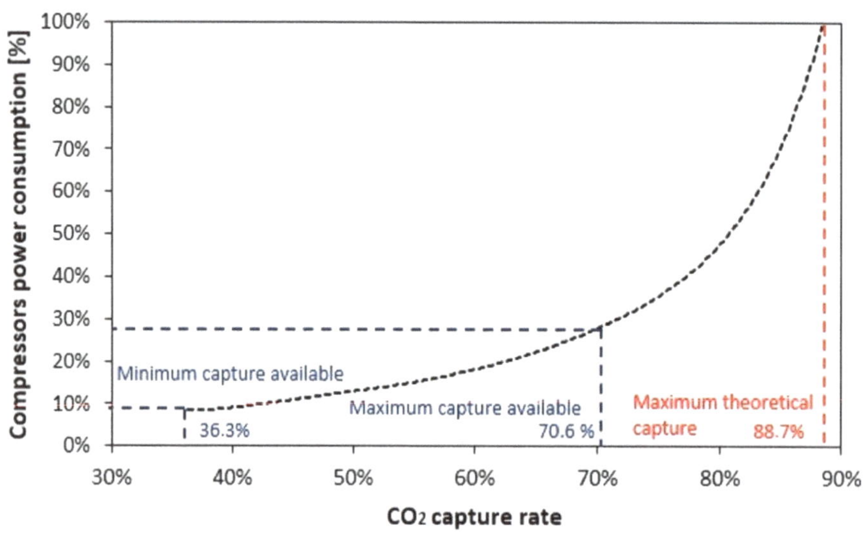

Figure 9. Curve for percentage power consumption of the compressors vs. percentage of CO_2 capture at the plant.

In this sense, the proposed carbon capture and storage (CCS) approaches have shown the potential to enhance the sustainability of sugarcane-derived bioethanol by further reducing its carbon footprint. Additionally, since CO_2 can serve as an input for various industrial processes and biofuel production, such as in Fischer–Tropsch synthesis, carbon capture could foster a greater integration between the sugarcane industry and other market sectors, advancing a circular and renewable economy.

5. Conclusions

This paper evaluated different bio-energy system configurations integrated with post-combustion chemical absorption (MEA) CO_2 capture technology. This work differs due to its approach of capturing carbon not only from the CO_2 of the fermentation process but also from the combustion of bagasse and sugarcane straw, in addition to considering the heat required to supply the ethanol production process in the plant, which globally implies a high thermal demand to be managed from the extractions of steam turbines.

The parametric analyses showed that it is challenging to define the best combination of pressure and temperature parameters, given that the objectives were conflicting (electrical power generation and CO_2 capture). Furthermore, of the evaluated configurations, different parameters with stronger influences were found for each configuration. Thus, we must use multi-objective and stochastic optimization methods to define the correct operational parameters and tradeoffs between generated electrical power and CO_2 capture.

From a power generation and carbon capture perspective, the results showed a trade-off for all the evaluated configurations of the BECCS system. The REG1 configuration resulted in the highest (51.9%) carbon capture with a 14.49% penalty on electrical efficiency (10.49% on the plant's cogeneration efficiency); therefore, it cannot capture all the CO_2 generated by the plant (theoretical limitation of 88.7% where all generated electricity would be used to compress the captured CO_2). The second analysis using (gCO_2/kWh) indicators showed that CO_2 capture is more expensive as more power must be used to capture the same amount of CO_2 in terms of mass, since less electrical power is generated and larger tons of CO_2 need to be compressed. CO_2 capture from 51.9% (0.224 Mt/year) would result in emission rates above 1300 g/kWh, which are higher than the plant's operating emissions with reheating and three regenerators without CCS. On the other hand, the Reheat3 configuration showed the best ratio at 1155 g/kWh (145 g less per kWh generated), with an even smaller penalty on the plant's electrical efficiency (7.41%); however, it was limited to a minimum capture level of 36.6% for all the CO_2 emitted at the plant.

The scenarios allowed us to reach reasonable results, where the BECCS system technically partially resulted in negative CO_2 emissions. This is a plant typical to the Brazilian sugarcane industry, with large demands for suppressed steam from ethanol and sugar production processes (115.5 MW). To capture 90% of all generated CO_2 from the bagasse and chaff combustion process, 198.4 MW would be needed, and 72% more heat would be destined to a secondary plant process. Future studies are needed to validate the operating ranges and configurations studied in this paper from an economic standpoint.

Author Contributions: B.B.: conceptualization, methodology, and writing—original draft preparation; C.A.R.S.: data curation, supervision, software, and validation; A.A.V.J.: methodology and writing—review and editing; J.L.J.P.: methodology and writing—review and editing; T.A.Z.d.S.: writing—review and editing; M.B.F.: writing—review and editing; C.J.R.C.: funding acquisition and supervision. All authors have read and agreed to the published version of the manuscript.

Funding: The authors would like to acknowledge the aid and financial support provided by Coordenação de Aperfeiçoamento de Pessoal de Nível Superior (CAPES) (Grant Number 88887.318226/2019-00); Conselho Nacional de Desenvolvimento Científico e Tecnológico (CNPq) (Proc. No. 305741/2019-5); and FAPEMIG (Fundação de Amparo à Pesquisa do Estado de Minas Gerais).

Data Availability Statement: Data are contained within the article.

Conflicts of Interest: The authors declare no conflicts of interest.

Appendix A. Cogeneration Cycle Simulation

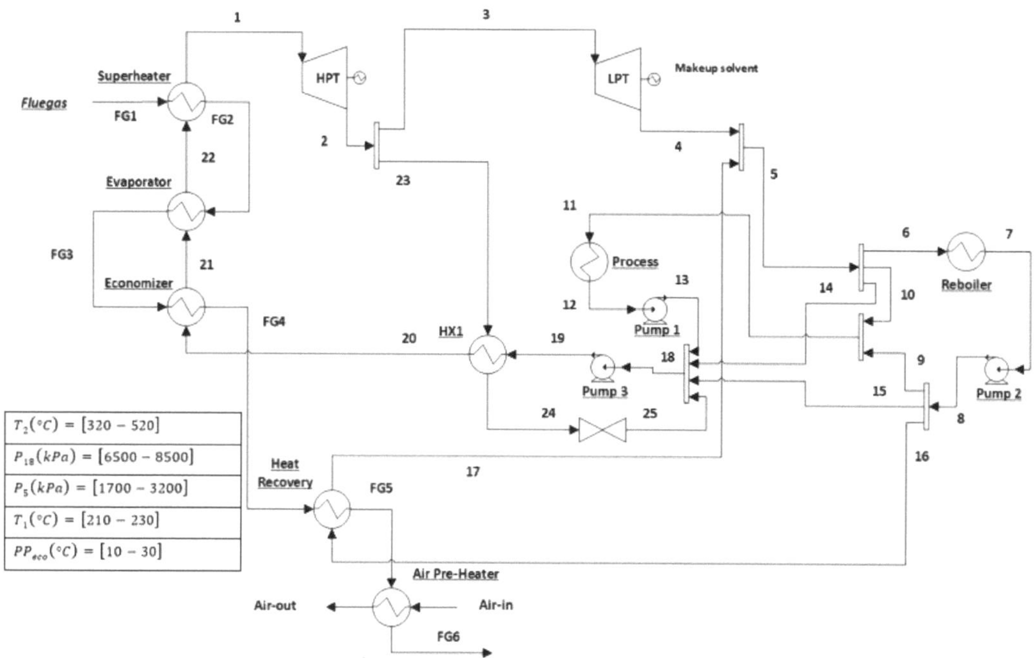

Figure A1. Cogeneration system with regeneration—REG1.

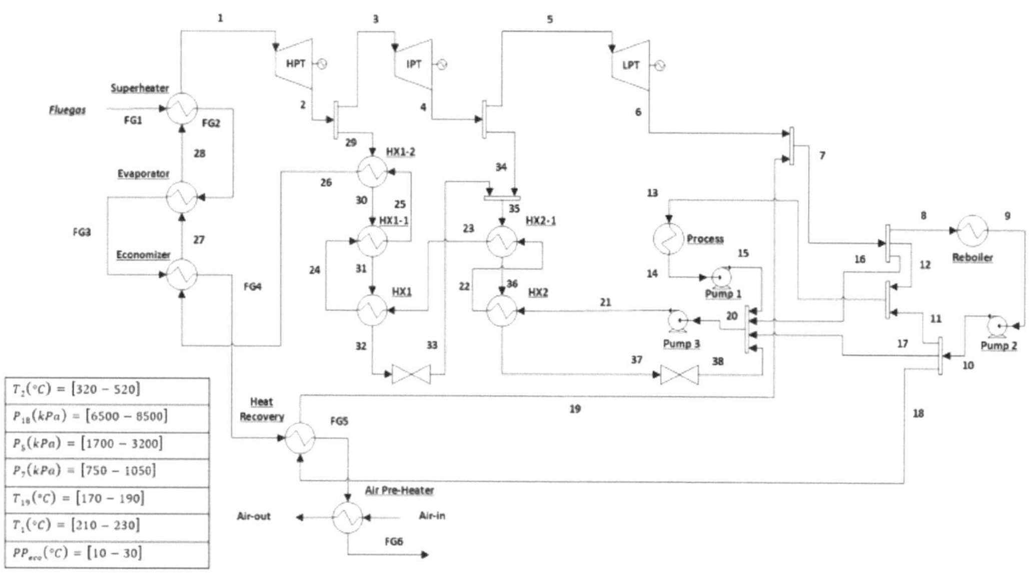

Figure A2. Cogeneration system with regeneration—REG2.

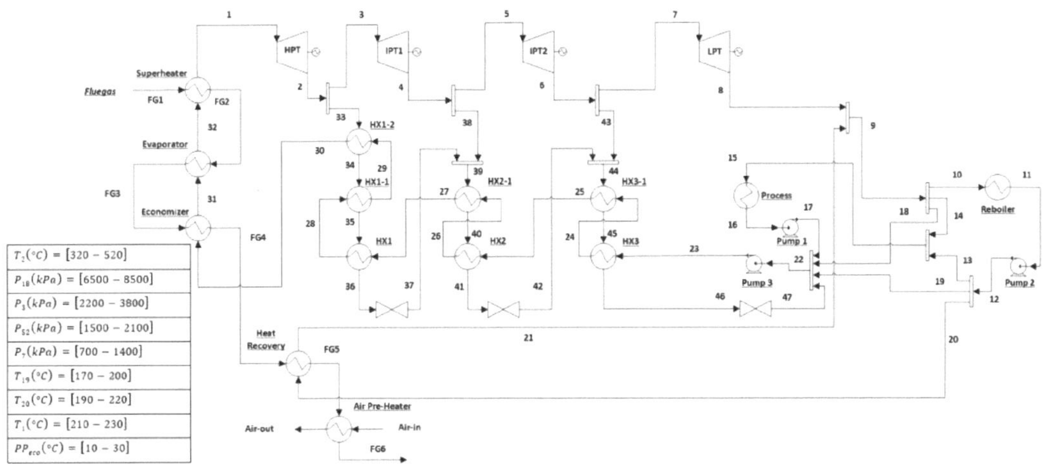

Figure A3. Cogeneration system with regeneration—REG3.

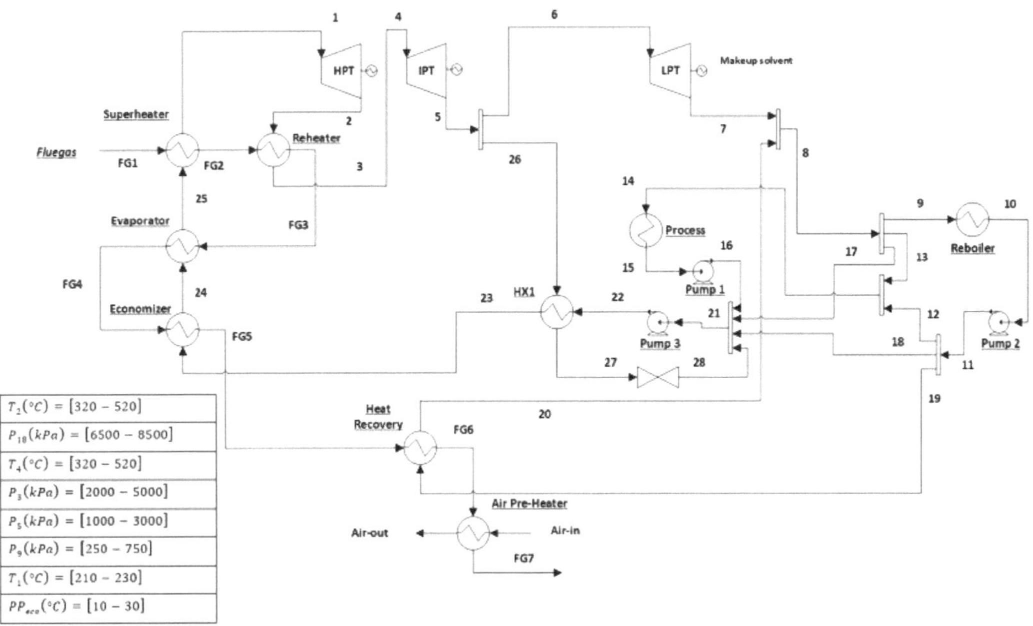

Figure A4. Cogeneration system with reheat and regeneration—Reheat1.

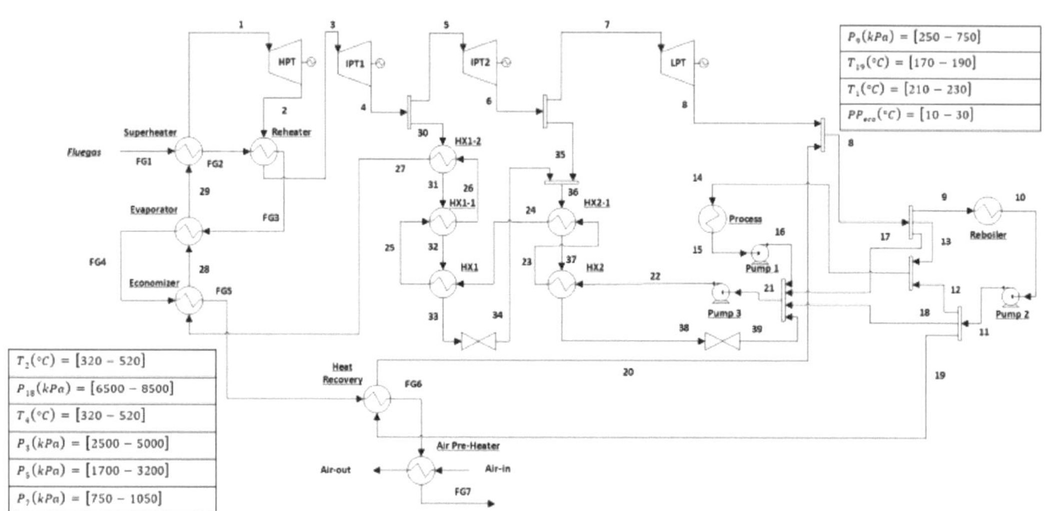

Figure A5. Cogeneration system with reheat and regeneration—Reheat2.

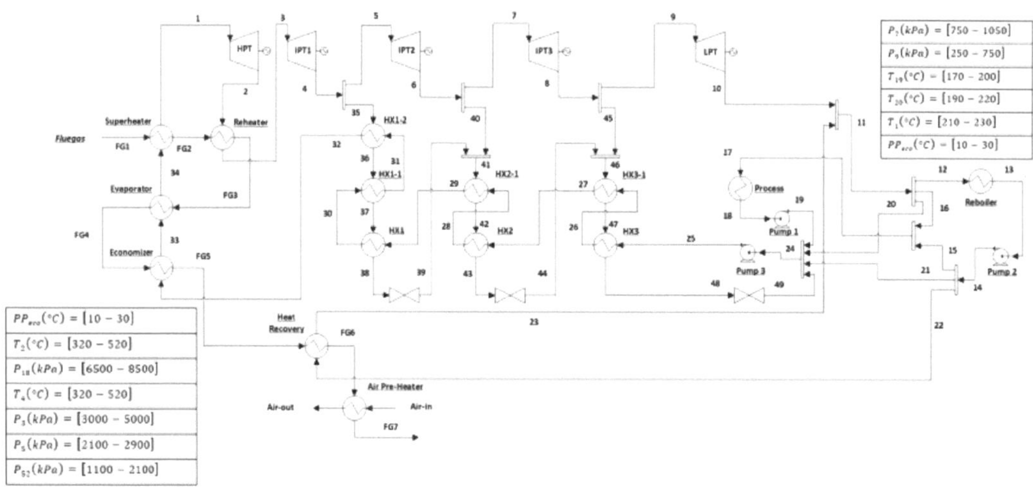

Figure A6. Cogeneration system with reheat and regeneration—Reheat3.

References

1. IPCC. *Global Warming of 1.5 °C. An IPCC Special Report on the Impacts of Global Warming of 1.5 °C above Pre-Industrial Levels and Related Global Greenhouse Gas Emission Pathways, in the Context of Strengthening the Global Response to the Threat of Climate Change*; Technical Report; Cambridge University Press: Cambridge, UK; New York, NY, USA, 2018. [CrossRef]
2. Intergovernmental Panel on Climate Change. *Climate Change 2014 Mitigation of Climate Change*; Cambridge University Press: Cambridge, UK, 2014. [CrossRef]
3. Wilberforce, T.; Baroutaji, A.; Soudan, B.; Al-Alami, A.H.; Olabi, A.G. Outlook of carbon capture technology and challenges. *Sci. Total Environ.* **2019**, *657*, 56–72. [CrossRef] [PubMed]
4. Blomen, E.; Hendriks, C.; Neele, F. Capture technologies: Improvements and promising developments. *Energy Procedia* **2009**, *1*, 1505–1512. [CrossRef]
5. Global CCS Institute. *Bioenergy and Carbon Capture: Perspective*; Technical Report; Global CCS Institute: Melbourne, Australia, 2019.
6. Emenike, O.; Michailos, S.; Finney, K.N.; Hughes, K.J.; Ingham, D.; Pourkashanian, M. Initial techno-economic screening of BECCS technologies in power generation for a range of biomass feedstock. *Sustain. Energy Technol. Assessments* **2020**, *40*, 100743. [CrossRef]

7. Du, Y.; Gao, T.; Rochelle, G.T.; Bhown, A.S. Zero- and negative-emissions fossil-fired power plants using CO_2 capture by conventional aqueous amines. *Int. J. Greenh. Gas Control* **2021**, *111*, 103473. [CrossRef]
8. Julio, A.A.V.; Escobar Palacio, J.C.; Rúa Orozco, D.J. Techno-economic and environmental comparison of carbon capture for standalone retrofitting and CO2 hubs in a coal-fueled power complex *Energy Convers. Manag.* **2024**, *315*, 118773. [CrossRef]
9. Fajardy, M.; Mac Dowell, N. The energy return on investment of BECCS: Is BECCS a threat to energy security? *Energy Environ. Sci.* **2018**, *11*, 1581–1594. [CrossRef]
10. Neves, M.F.; Kalaki, R.B. *Bioenergy from Sugarcane*, 1st ed.; Ourofino Agrociência: Guariba, SP, Brasil, 2020; p. 151.
11. Lozano, M.A.; dos Santos, R.; Santos, J.J.; Serra, L.M. Optimal modes of operation and product cost allocation in sugarcane steam cogeneration plants. *Therm. Sci. Eng. Prog.* **2024**, *52*, 102686. [CrossRef]
12. Minnu, S.N.; Bahurudeen, A.; Athira, G. Comparison of sugarcane bagasse ash with fly ash and slag: An approach towards industrial acceptance of sugar industry waste in cleaner production of cement. *J. Clean. Prod.* **2021**, *285*, 124836. [CrossRef]
13. Julio, A.A.V.; de Souza, T.A.Z.; Rocha, D.H.D.; Rodriguez, C.J.C.; Palacio, J.C.E.; Silveira, J.L. Environmental Footprints of Hydrogen from Crops. In *Environmental Footprints of Crops*; Muthu, S.S., Ed.; Springer Nature: Singapore, 2022; pp. 85–110. [CrossRef]
14. de Souza, T.; Rocha, D.; Julio, A.; Coronado, C.; Silveira, J.; Silva, R.; Palacio, J. Exergoenvironmental assessment of hydrogen water footprint via steam reforming in Brazil. *J. Clean. Prod.* **2021**, *311*, 127577. [CrossRef]
15. Pryor, S.W.; Smithers, J.; Lyne, P.; van Antwerpen, R. Impact of agricultural practices on energy use and greenhouse gas emissions for South African sugarcane production. *J. Clean. Prod.* **2017**, *141*, 137–145. [CrossRef]
16. Chipfupa, U.; Tagwi, A. Greenhouse gas emission implications of small-scale sugarcane farmers' trash management practices: A case for bioenergy production in South Africa. *Energy Nexus* **2024**, *15*, 100308. [CrossRef]
17. Moreira, J.R.; Romeiro, V.; Fuss, S.; Kraxner, F.; Pacca, S.A. BECCS potential in Brazil: Achieving negative emissions in ethanol and electricity production based on sugar cane bagasse and other residues. *Appl. Energy* **2016**, *179*, 55–63. [CrossRef]
18. Bhave, A.; Taylor, R.H.; Fennell, P.; Livingston, W.R.; Shah, N.; Dowell, N.M.; Dennis, J.; Kraft, M.; Pourkashanian, M.; Insa, M.; et al. Screening and techno-economic assessment of biomass-based power generation with CCS technologies to meet 2050 CO_2 targets. *Appl. Energy* **2017**, *190*, 481–489. [CrossRef]
19. Julio, A.A.V.; Castro-Amoedo, R.; Maréchal, F.; Martínez-González, A.; Escobar Palacio, J.C. Exergy and economic analysis of the trade-off for design of post-combustion CO_2 capture plant by chemical absorption with MEA. *Energy* **2023**, *128004*. [CrossRef]
20. Dubois, L.; Thomas, D. Comparison of various configurations of the absorption-regeneration process using different solvents for the post-combustion CO_2 capture applied to cement plant flue gases. *Int. J. Greenh. Gas Control* **2018**, *69*, 20–35. [CrossRef]
21. Bougie, F.; Pokras, D.; Fan, X. Novel non-aqueous MEA solutions for CO_2 capture. *Int. J. Greenh. Gas Control* **2019**, *86*, 34–42. [CrossRef]
22. Li, K.; Yu, H.; Feron, P.; Tade, M.; Wardhaugh, L. Technical and Energy Performance of an Advanced, Aqueous Ammonia-Based CO_2 Capture Technology for a 500 MW Coal-Fired Power Station. *Environ. Sci. Technol.* **2015**, *49*, 10243–10252. [CrossRef] [PubMed]
23. Chao, C.; Deng, Y.; Dewil, R.; Baeyens, J.; Fan, X. Post-combustion carbon capture. *Renew. Sustain. Energy Rev.* **2021**, *138*, 110490. [CrossRef]
24. Otitoju, O.; Oko, E.; Wang, M. Technical and economic performance assessment of post-combustion carbon capture using piperazine for large scale natural gas combined cycle power plants through process simulation. *Appl. Energy* **2021**, *292*, 116893. [CrossRef]
25. Restrepo-Valencia, S.; Walter, A. Techno-economic assessment of bio-energy with carbon capture and storage systems in a typical sugarcane mill in Brazil. *Energies* **2019**, *12*, 1129. [CrossRef]
26. Batlle, E.A.O.; Julio, A.A.V.; Santiago, Y.C.; Palácio, J.C.E.; Bortoni, E.D.C.; Nogueira, L.A.H.; Dias, M.V.X.; González, A.M. Brazilian integrated oilpalm-sugarcane biorefinery: An energetic, exergetic, economic, and environmental (4E) assessment. *Energy Convers. Manag.* **2022**, *268*, 116066. [CrossRef]
27. Díaz Pérez, Á.A.; Escobar Palacio, J.C.; Venturini, O.J.; Martínez Reyes, A.M.; Rúa Orozco, D.J.; Silva Lora, E.E.; Almazán del Olmo, O.A. Thermodynamic and economic evaluation of reheat and regeneration alternatives in cogeneration systems of the Brazilian sugarcane and alcohol sector. *Energy* **2018**, *152*, 247–262. [CrossRef]
28. Alves, M.; Ponce, G.H.; Silva, M.A.; Ensinas, A.V. Surplus electricity production in sugarcane mills using residual bagasse and straw as fuel. *Energy* **2015**, *91*, 751–757. [CrossRef]
29. Maluf, A.B. Avaliação Termoeconômica da Cogeração no Setor Sucroenergético com o Emprego de Bagaço, Palha, Biogás de Vinhaça Concentrada e Geração na Entressafra. Doctoral Thesis, University of Campinas, Campinas, Brazil, 2015.
30. Wienese, A. Boilers, Boiler Fuel and Boiller Efficiency. In Proceedings of the 75th Annual Congress South African Sugar Technologists' Association, Durban, South Africa, 31 July–3 August 2021; pp. 275–281.
31. Turn, S.Q.; Jenkins, B.M.; Jakeway, L.A.; Blevins, L.G.; Williams, R.B.; Rubenstein, G.; Kinoshita, C.M. Test results from sugar cane bagasse and high fiber cane co-fired with fossil fuels. *Biomass Bioenergy* **2006**, *30*, 565–574. [CrossRef]
32. Evaluation of cyclone gasifier performance for gasification of sugar cane residue—Part 1: Gasification of bagasse. *Biomass Bioenergy* **2001**, *21*, 351–369. [CrossRef]
33. Carminati, H.B.; Milão, R.d.F.D.; de Medeiros, J.L.; Araújo, O.d.Q.F. Bioenergy and full carbon dioxide sinking in sugarcane-biorefinery with post-combustion capture and storage: Techno-economic feasibility. *Appl. Energy* **2019**, *254*, 113633. [CrossRef]

34. Rayaprolu, K. *Boilers for Power and Process*; CRC Press: Boca Raton, FL, USA, 2009.
35. IEA. *Energy Technology Perspectives 2020—Special Report on Carbon Capture Utilisation and Storage*; International Energy Agency: Paris, France, 2020. [CrossRef]
36. Moser, P.; Schmidt, S.; Sieder, G.; Garcia, H.; Stoffregen, T. Performance of MEA in a long-term test at the post-combustion capture pilot plant in Niederaussem. *Int. J. Greenh. Gas Control* **2011**, *5*, 620–627. [CrossRef]
37. Mangalapally, H.P.; Hasse, H. Pilot plant study of two new solvents for post combustion carbon dioxide capture by reactive absorption and comparison to monoethanolamine. *Chem. Eng. Sci.* **2011**, *66*, 5512–5522. [CrossRef]
38. Kwak, N.S.; Lee, J.H.; Lee, I.Y.; Jang, K.R.; Shim, J.G. A study of the CO_2 capture pilot plant by amine absorption. *Energy* **2012**, *47*, 41–46. [CrossRef]
39. Stec, M.; Tatarczuk, A.; Więcław-Solny, L.; Krótki, A.; ązko, M.; Tokarski, S. Pilot plant results for advanced CO_2 capture process using amine scrubbing at the Jaworzno II Power Plant in Poland. *Fuel* **2015**, *151*, 50–56. [CrossRef]
40. Farajollahi, H.; Hossainpour, S. Application of organic Rankine cycle in integration of thermal power plant with post-combustion CO_2 capture and compression. *Energy* **2017**, *118*, 927–936. [CrossRef]
41. Akram, M.; Ali, U.; Best, T.; Blakey, S.; Finney, K.N.; Pourkashanian, M. Performance evaluation of PACT Pilot-plant for CO_2 capture from gas turbines with Exhaust Gas Recycle. *Int. J. Greenh. Gas Control* **2016**, *47*, 137–150. [CrossRef]
42. Notz, R.; Mangalapally, H.P.; Hasse, H. Post combustion CO_2 capture by reactive absorption: Pilot plant description and results of systematic studies with MEA. *Int. J. Greenh. Gas Control* **2012**, *6*, 84–112. [CrossRef]
43. Morgan, J.C.; Chinen, A.S.; Omell, B.; Bhattacharyya, D.; Tong, C.; Miller, D.C. Thermodynamic modeling and uncertainty quantification of CO_2-loaded aqueous MEA solutions. *Chem. Eng. Sci.* **2017**, *168*, 309–324. [CrossRef]
44. Chinen, A.S.; Morgan, J.C.; Omell, B.P.; Bhattacharyya, D.; Miller, D.C. Dynamic Data Reconciliation and Model Validation of a MEA-Based CO_2 Capture System using Pilot Plant Data. *IFAC-PapersOnLine* **2016**, *49*, 639–644. [CrossRef]
45. Macedo, I.C.; Seabra, J.E.; Silva, J.E. Green house gases emissions in the production and use of ethanol from sugarcane in Brazil: The 2005/2006 averages and a prediction for 2020. *Biomass Bioenergy* **2008**, *32*, 582–595. [CrossRef]
46. Pereira, J.L.J.; Francisco, M.B.; Diniz, C.A.; Antônio Oliver, G.; Cunha, S.S.; Gomes, G.F. Lichtenberg algorithm: A novel hybrid physics-based meta-heuristic for global optimization. *Expert Syst. Appl.* **2021**, *170*, 114522. [CrossRef]
47. Pereira, J.L.J.; Francisco, M.B.; da Cunha, S.S.; Gomes, G.F. A powerful Lichtenberg Optimization Algorithm: A damage identification case study. *Eng. Appl. Artif. Intell.* **2021**, *97*, 104055. [CrossRef]
48. de Souza, T.; Pereira, J.; Francisco, M.; Sotomonte, C.; Ma, B.; Gomes, G.; Coronado, C. Multi-objective optimization for methane, glycerol, and ethanol steam reforming using lichtenberg algorithm. *Int. J. Green Energy* **2023**, *20*, 390–407. [CrossRef]

Article

Combining Exergy and Pinch Analysis for the Operating Mode Optimization of a Steam Turbine Cogeneration Plant in Wonji-Shoa, Ethiopia

Shumet Sendek Sharew [1,2], Alessandro Di Pretoro [2,*], Abubeker Yimam [1], Stéphane Negny [2] and Ludovic Montastruc [2]

[1] School of Chemical and Bioengineering, Addis Ababa Institute of Technology (AAiT), Addis Ababa University, Addis Ababa P.O. Box 385, Ethiopia; shumet.sendek@aait.edu.et (S.S.S.); abubeker.yimam@aau.edu.et (A.Y.)
[2] Laboratoire de Génie Chimique, Université de Toulouse, CNRS/INP/UPS, Allée E. Monso 4, 31432 Toulouse, France; stephane.negny@ensiacet.fr (S.N.); ludovic.montastruc@ensiacet.fr (L.M.)
* Correspondence: alessandro.dipretoro@ensiacet.fr

Abstract: In this research, the simulation of an existing 31.5 MW steam power plant, providing both electricity for the national grid and hot utility for the related sugar factory, was performed by means of ProSimPlus® v. 3.7.6. The purpose of this study is to analyze the steam turbine operating parameters by means of the exergy concept with a pinch-based technique in order to assess the overall energy performance and losses that occur in the power plant. The combined pinch and exergy analysis (CPEA) initially focuses on the depiction of the hot and cold composite curves (HCCCs) of the steam cycle to evaluate the energy and exergy requirements. Based on the minimal approach temperature difference (ΔT_{lm}) required for effective heat transfer, the exergy loss that raises the heat demand (heat duty) for power generation can be quantitatively assessed. The exergy composite curves focus on the potential for fuel saving throughout the cycle with respect to three possible operating modes and evaluates opportunities for heat pumping in the process. Well-established tools, such as balanced exergy composite curves, are used to visualize exergy losses in each process unit and utility heat exchangers. The outcome of the combined exergy–pinch analysis reveals that energy savings of up to 83.44 MW may be realized by lowering exergy destruction in the cogeneration plant according to the operating scenario.

Keywords: pinch analysis; exergy; cogeneration plant; operating scenarios; ProSimPlus®

check for updates

Citation: Sharew, S.S.; Di Pretoro, A.; Yimam, A.; Negny, S.; Montastruc, L. Combining Exergy and Pinch Analysis for the Operating Mode Optimization of a Steam Turbine Cogeneration Plant in Wonji-Shoa, Ethiopia. *Entropy* **2024**, *26*, 453. https://doi.org/10.3390/e26060453

Academic Editors: Daniel Flórez-Orrego, Meire Ellen Ribeiro Domingos and Rafael Nogueira Nakashima

Received: 10 April 2024
Revised: 21 May 2024
Accepted: 22 May 2024
Published: 27 May 2024

1. Introduction

During the last decades, the efficient use of energy has become one of the aspects of major interest in a wide range of engineering fields. This increasing attention is due to both the particular concerns toward the environmental impact of industrial processes and to the relevant potential in terms of energy consumption reduction, and thus operating costs. In the process engineering domain, the best-established solutions for the mitigation of heating and cooling utilities are represented by process intensification and energy integration. The former involves the use of intensified equipment, able to perform multiple operations inside the same unit [1], while the latter consists of the rearrangement of the utility network with the purpose of using the process streams to be cooled as hot duty for the process streams to be heated and vice versa [2].

Regarding the second method, to increase the efficiency of energy usage in the heat exchanger network (HEN), a systematic approach called pinch analysis is conventionally employed. However, as Zhao et al. (2022) [3] point out, the primary drawback of pinch analysis is that it can only address processes involving heat transfer while it cannot address processes from the perspective of pressure and composition variations. All the stream parameters such as pressure, temperature, and composition that could potentially mitigate the drawback of pinch analysis are instead included in the so-called exergy analysis. The combination of exergy and pinch analysis applied to industrial processes is essential to determine which operations cause the greatest number of exergy casualties in heating and power facilities. Exergy is defined as the maximum amount of work that may be obtained from a certain thermal system as it moves towards a specific ultimate state when it is in balance with its environment. As a result of internal or external irreversibilities, exergy is not conserved as energy, and it is destroyed inside the system. To solve the internal or external irreversibilities in a process system other than energetic performance, which is based on the first law of thermodynamics, and exergetic performance, which is based on the second law, pinch technology, which is based on both, can be exploited. This work seeks to provide a novel combined pinch and exergetic-based systematic method for evaluating and enhancing current industrial processes from an energy perspective.

In order to integrate these two fragmented methodologies, a new technique called combined pinch and exergy analysis (CPEA) is applied to a cogeneration plant case study in this research. In particular, the plant simulation is based on the actual operating data of the Wonji-Shoa (Ethiopia) combined heat and power plant that provides the steam utility for the related sugar factory and, in addition, generates power for the national electricity grid. For a regular operating flow, the typical steam turbine is designed to ensure the turbo-alternator inlet steam flow of up to 164 t/h at 64 bar pressure and 505 °C temperature with uncontrolled extraction at 9 bar. At the desuperheater's output, the steam flowrate is equal to 20 t/h with regulated extraction at a pressure of 2.6 bar, resulting in a maximum steam flow of 117 t/h when the unit is operating at full capacity. Once the simulation to obtain all the other operating parameters is performed by means of ProSimPlus®, the CPEA can be then carried out. For the illustration of process streams on a temperature versus enthalpy (T–H) diagram, Linnhoff et al. (1982) [4] developed the composite curves (CCs), which are a common pinch analysis graphical tool. The hot or cold CCs are built from a combination of hot or cold streams that operate within a predetermined temperature range. To achieve the greatest amount of heat recovery potential and the least amount of hot and cold utility needs, the composite hot and cold streams can be horizontally shifted to approach one another along the enthalpy (ΔH) axis, until they are nipped off. The visual features of the cold and hot composite curves (CCs) and the grand composite curve (GCC) make it easier to find chances for heat integration, and they are highly helpful for gaining an in-depth knowledge of the issue [5]. The minimal energy objective for the operation may be then determined with the constraint of heat transfer going from higher to lower temperatures by keeping a minimum temperature difference as the driving force. Furthermore, the minimal permitted temperature differential ΔT_{lm} also represents a financial indicator of a nearly ideal trade-off between the cost of the initial investment (heat exchanger unit(s)) and the cost of running the process (energy) [6].

In this research, the exergy–pinch analysis method addresses the three possible functioning modes of the Wonji-Shoa CHP plant, mostly focusing on the heat integration part for the enhancement of the processing systems' energy efficiency. As better discussed in the dedicated section, each operation mode refers to a combination of the grid connection and sugar factory operating state (ON/ON, ON/OFF, OFF/ON). The variation in the electrical power provided by the plant for each scenario is managed by the regulation of the steam flowrate of up to 31.5 MW of useful power. The application of the exergy–pinch analysis could then highlight significant potential in terms of heat recovery when the process splits into a heat surplus zone below the pinch and a heat deficit region above the pinch. The main added value of combining pinch and exergy analysis with respect to standalone approaches, which are conventionally proposed, is represented by the possibility of simultaneously addressing both equipment technology and operation management improvements in the same analysis. Furthermore, since both the methodologies are suitable for energy and process systems that either already exist or need to be designed, when combining the two there is no need for methodological adaptations.

Better details about the specific application of these methodologies on the selected case study, along with the operating parameters resulting from the process simulation, are then provided in the following section.

2. The Wonji-Shoa Case Study

As mentioned in the introduction, the selected system for this research is an existing 31.5 MW cogeneration plant located in Ethiopia whose purpose is both to provide steam to the related sugar factory and to produce electricity for the national grid. The main advantages of this choice are the availability of the actual operating parameters [7] and the fact that the improvements obtained from the CPEA study could be effectively implemented in the real system. For a detailed analysis and discussion regarding the process side optimization, the reader could refer to the previous work of the authors [8] where the aspects concerning the plant operation are thoroughly presented. A general overview of the system layout in terms of material and energy fluxes is nevertheless provided in Figure 1 to outline the capacity and utility distribution of the entire plant.

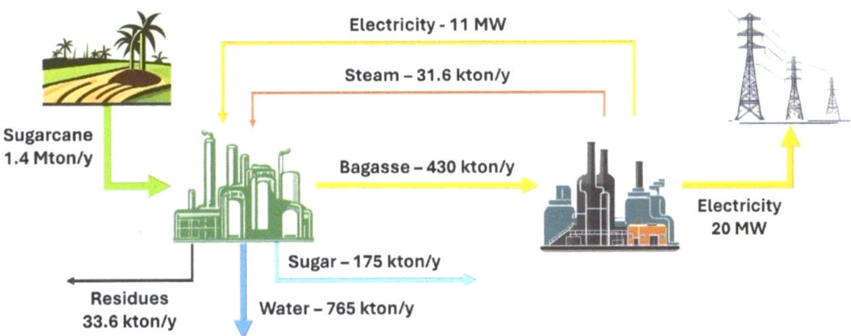

Figure 1. Simplified flow diagram of the Wonji-Shoa facility.

The present study focuses on the potential energy savings of the CHP section (grey block) and on the identification of possible operational upgrades to enhance energy efficiency as well as to further increase profitability. The Wonji-Shoa factory cogeneration plant was conceived to provide a 9 bar stream thermal duty at the desuperheater, 3 bar steam at the extraction, and for the supply of the electricity surplus to the grid for the maximization of the facility income. According to the required electricity demand and process operating conditions, the plant switches from one functioning mode to another throughout its yearly operation. In order to have a clear overview of the final results of this study, a brief description of the system units, simulated by means of ProSimPlus®, and the explanation of the three possible functioning modes are presented in this section.

The first operating mode, named Scenario I, refers to both power plant and sugar factory operation (Grid: ON/Factory: ON), and the corresponding system layout is reported in Figure 2. This is the default mode, and the turbo-generator capacity at full load corresponds to 31.5 MW. After the two steam extractions at 9 and 3 bar in correspondence of the turbines T-1 and T-2, the remaining steam is condensed and recycled with make-up water to reintegrate the water losses equal to 22% of the total flowrate. Once left the deaerator unit, whose purpose is to remove the dissolved gases, water feeds the boiler units whose heat is provided by the direct combustion of the bagasse by-product recovered from the sugar factory. The Scenario I functioning mode can be considered as the reference case since the other two result from layout modifications with respect to it.

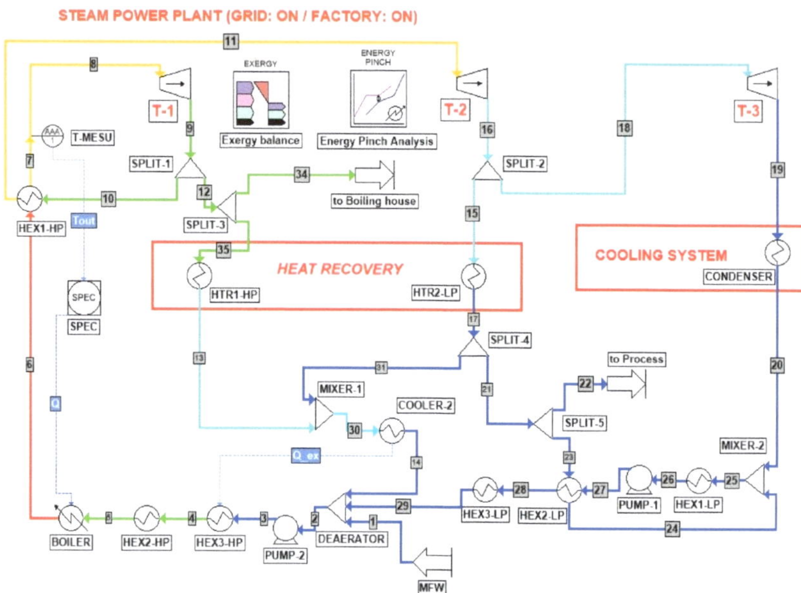

Figure 2. Steam turbine cogeneration plant for Scenario I (Grid: ON and Factory: ON).

Figure 3 shows the equivalent simulation layout for Scenario II (Grid: ON/Factory: OFF). Since the sugar factory is not operating, the absence of the stream splitting toward the process can be noticed. When running this mode, the cogeneration plant produces 130 ton/h of stream, with a reduced electricity production, and a single heat recovery solution was already implemented. Since there is no steam withdrawal, the make-up water valve can be closed, and the related mixing units are not active.

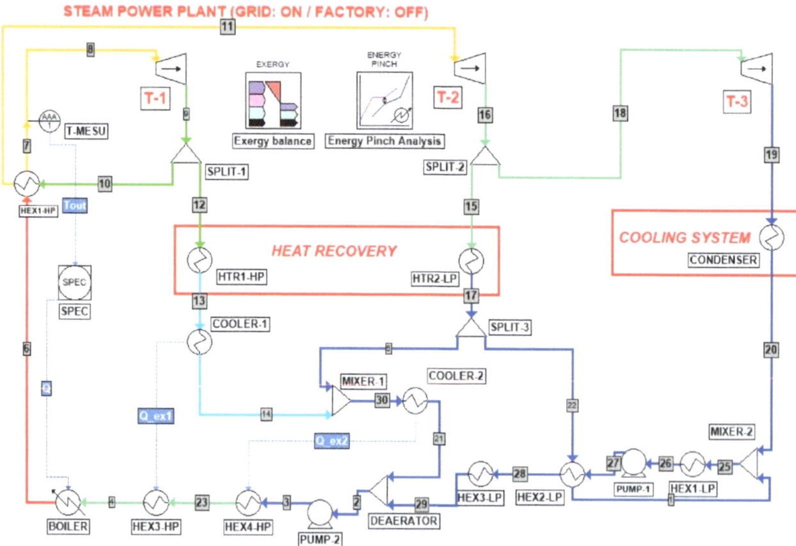

Figure 3. Steam turbine cogeneration plant for Scenario II (Grid: 'ON' and Factory: OFF).

Finally, in case the grid connection is down, Scenario III (Grid: OFF/Factory: ON) is represented by the process diagram in Figure 4. Since the sugar plant is the primary energy receiver, the turbine load can be decreased, and the boiler production is approximately equal to 59 ton/h of steam. As a consequence of the lower need for electricity production, there is no high-pressure extraction stream.

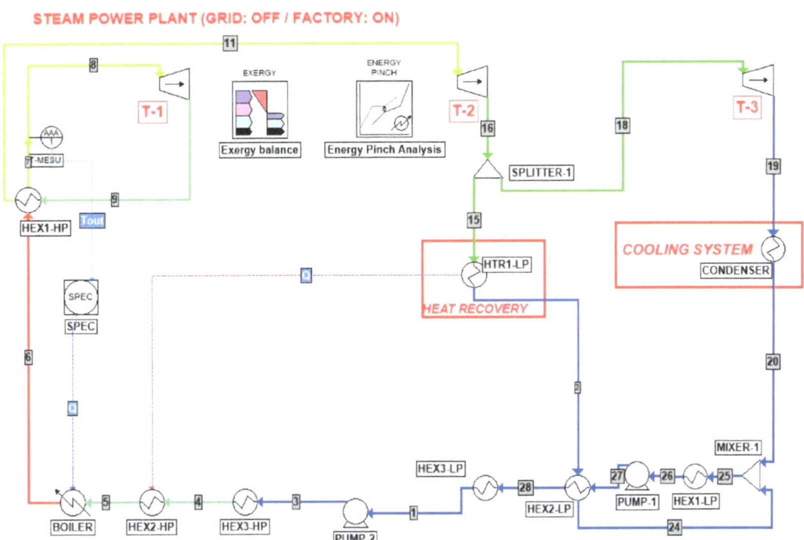

Figure 4. Steam turbine cogeneration plant for Scenario III (Grid: 'OFF' and Factory: ON).

Based on this process layout description, the exergy and pinch analysis methodologies are therefore presented in the next sections by referring to the unit and stream names provided in Figures 2–4.

3. Methodologies

Based on the process flow diagrams and the operating parameters available in the Wonji-Shoa factory manual [7], presented in the previous section, the research started with the implementation of the process simulation by means of ProSimPlus®. The reason lying behind the choice of this software is the availability of dedicated built-in tools for both pinch analysis and exergy balance, as can be pointed out in Figures 2–4. However, in order to obtain the desired outcome from these two modules, they need to be properly set up according to the analysis constraints and the operating conditions. Therefore, methodological details concerning exergy analysis, pinch analysis, and CPEA are discussed in the respective following sections in order to facilitate the physical interpretation of the obtained results and to enable the correlation of the outcome with the preliminary hypotheses.

3.1. Exergy Analysis

In this study, the exergy analysis tool is used to detect the units in the system that exhibit higher irreversibility, i.e., exergy destruction. Exergy is defined as the highest potential useful work achievable from an energy carrier under the circumstances imposed by an environment at a certain pressure P_0, temperature T_0, and amount of chemical elements. The goal of the exergy analysis is to locate, source, and quantify actual thermodynamic inefficiencies in process plants, such as power plants (Chao and Yan, 2006) [9]. The advantage of using exergy for the evaluation of cogeneration system performance is based on the fact that instead of treating heat and electricity equally, as is the case with more traditional energy approaches, exergy methods enable the "value" of cogeneration products to be evaluated on an equivalent basis [10].

In this study, condensation-extraction steam turbines can exploit the entire amount of produced bagasse to increase the amount of excess power generated by the cogeneration system which is sold to the national grid [11]. In particular, in this simulation, the energy provided by the bagasse combustion is represented by the boiler's heat duty term. The energy efficiency of a cogeneration facility producing both electricity and heat may be then defined as the ratio of useful energy output to the energy input given by Equation (1):

$$\eta_{cogen} = \frac{\dot{W}_{net} + \dot{Q}_{heat}}{\dot{E}_{in}} \tag{1}$$

where $\dot{W}_{net}$ is the net power output, $\dot{Q}_{heat}$ is the heating rate supplied by the plant, and $\dot{E}_{in}$ is the rate of the total energy input to the plant.

Regarding the general form of the implementation of the exergy balance used by the ProSimPlus® built-in module, it can be described by the following equation:

$$\frac{dB_{cv}}{dt} = \sum_j \left(1 - \frac{T_0}{T_i}\right) \cdot \dot{Q}_j - \left(\dot{W}_{cv} - P_0 \cdot \frac{dV_{cv}}{dt}\right) + \sum_i \dot{m}_i \cdot B_i - \sum_0 \dot{m}_0 \cdot B_0 - \dot{I} \tag{2}$$

The specific physical exergy of a stream (b_{ph}) is calculated by Equation (3):

$$B_{ph} = \dot{m} \cdot \left[(h - h_0) - T_0 \cdot (S - S_0)\right] + B_{ch} \tag{3}$$

where $\dot{m}$ is the stream flowrate, h and S are, respectively, the specific enthalpy and entropy, and B_{ch} is the term related to chemical exergy.

The specific physical exergy variation caused by temperature change is given by Equation (4):

$$B^T = C_p \cdot \left((T - T_0) - T_0 \ln \frac{T}{T_0}\right) \tag{4}$$

while the specific physical exergy due to pressure change is given by Equation (5):

$$B^P = R \cdot T \cdot \ln \frac{P}{P_0} \tag{5}$$

Finally, the specific chemical exergy, which can be calculated using conventional chemical exergy tables [12] in relation to environmental specifications, is given by

$$B_{ch} = \sum_{i=1}^k X_k \cdot B_k^{CH} + R \cdot T_0 \sum_{i=1}^k X_k \cdot \ln X_k \tag{6}$$

where X_k and R are the molar fraction of the component k-th and the universal gas constant. In all equations, the subscript "0" indicates the reference conditions of the analysis that were set equal to the ambient temperature and pressure, i.e., 25 °C and 101.3 kPa, respectively.

The exergy destruction rate $\dot{I}$ of a steady-state system is obtained from the equation

$$\dot{I} = \sum_j B_{qj} \cdot \dot{Q}_j - \left(\dot{W}_{cv}\right) + \sum_i B_i - \sum_0 B_{out} \tag{7}$$

where B_{in}, B_{out}, and B_D indicate the input exergy, output exergy, and exergy destruction of each unit, respectively. For the plantwide assessment of the exergy destruction, the unit-wise exergy destruction must be computed first. The bleed heat exchangers, condenser, expander, boiler, and auxiliary units lose some of the exergy from the fuel, while the remaining part is used to generate electricity with turbines. Finally, some exergy is wasted while turbines and pumps are working, according to their efficiency. The general energy and exergy efficiency is then equal to

$$\eta_I = \frac{B^{out}}{B_{in}} = 1 - \frac{\dot{I}}{B_{in}} \tag{8}$$

According to Figures 2–4, the exergy loss and the exergy efficiency for each of the steam turbine cogeneration cycle components may be computed as follows [13]:

$$B_{I,turbine} = \sum(\dot{m} \cdot b)_{in} - \sum(\dot{m} \cdot b)_{out} - \dot{W}_{out} \tag{9}$$

$$\eta_{Turbine} = \frac{\dot{W}_{out}}{\sum(\dot{m} \cdot b)_{in} - \sum(\dot{m} \cdot b)_{out}} \tag{10}$$

where $\dot{W}_{out}$ is the actual generated expansion work.

The pump's exergy loss and the related efficiency may be obtained as follows [14]:

$$B_{I,pump} = B_{D,pump} = \sum(\dot{m} \cdot b)_{in} - \sum(\dot{m} \cdot b)_{out} + \dot{W}_{in} \tag{11}$$

$$\eta_{pump} = \frac{\sum(\dot{m} \cdot B)_{out} - \sum(\dot{m} \cdot B)_{in}}{\dot{W}_{in}} \tag{12}$$

where $\dot{W}_{in}$ is the actual consumed power.

As concerns the heat exchangers, the exergy balance is stated as

$$B_{I,HE} = B_{in} - B_{out} = \sum(\dot{m} \cdot B)_{in} - \sum(\dot{m} \cdot B)_{out} \tag{13}$$

while the exergetic efficiency (η_{HEX}) is given by the ratio of the rise in the cold fluid's exergy to the reduction in the hot fluid's exergy:

$$\eta_{HEX} = \frac{\sum[\dot{m} \cdot B_{out} - \dot{m} \cdot B_{in}]_{Cold}}{\sum[\dot{m} \cdot B_{out} - \dot{m} \cdot B_{in}]_{Hot}} \tag{14}$$

Finally, the overall exergy loss in the cycle is given by the sum of all exergy losses in each involved unit operation, and the overall exergetic efficiency of the cycle can be finally calculated as

$$\eta_{cycle} = \frac{\dot{W}_{net}}{B_{fuel=heat\ duty}} \tag{15}$$

where $\dot{W}_{net}$ is the difference between $\dot{W}_{out}$ and $\dot{W}_{in}$:

$$\dot{W}_{net} = \dot{W}_{out} - \dot{W}_{in} \tag{16}$$

3.2. Pinch Analysis

Pinch analysis targets energy-saving strategies by means of modifications in the heat exchanger network design based on heat balances and operating temperatures. In industrial settings, the calculation of the lowest heating and cooling requirements usually reveals considerable energy savings. Process and energy integration, particularly pinch technology, is a very effective analytical tool for the selection of technological solutions aiming at increasing efficiency and optimizing production. When integrating energy conversion technologies, one must take into account the combined production of heat and power and the integration of steam networks, heat pumps, and refrigeration systems, as well as how the minimum energy requirement will be met by converting energy resources into process-useful energy. Thus, the exergy concept is integrated with pinch analysis in the context of process integration analysis to reduce the energy requirement of the process [14], optimize energy conversion system integration, and introduce polygeneration.

The fundamental concept behind pinch is the possibility to independently depict process heating and cooling requirements by using composite curve (CC) diagrams [6]. CCs are graphical representations of temperature–enthalpy profiles for the hot (HCC) and cold (CCC) streams, representing the process heat availability and demand, respectively. When evaluating the energy efficiency of a process, pinch-based methodologies identify potential energy recovery via heat transfer and determine the process's minimum energy requirement (MER). The heat exergy (B_q) provided by a stream which delivers a heat load (Q) from T_{in} to T_{out} is estimated for each linear segment in an enthalpy–temperature curve computed by Equation (17):

$$B_q = \dot{Q} \cdot \left(1 - \frac{T_0}{T_{lm}} \right) \tag{17}$$

where T_{lm} is the logarithmic mean temperature, and T_0 is the ambient temperature.

The heat provided by the HCC is directly reported in the T–H diagram while the delivered exergy corresponds to the area between the composite curve and the enthalpy axis by replacing the temperature axis with the Carnot factor ($1 - T_0/T$). The energy targets established by the composite curve (CC) and grand composite curve (GCC) in pinch analysis are exclusively expressed in terms of heat loads. However, in order to deal with systems including heat and power, the principles of both the CC and the GCC need to be expanded. A dedicated discussion about the shape and the interpretation of these graphical tools for the cogeneration plant case study are provided in Section 4. In addition, the GCC also depicts the difference between the available heat and the required amount.

Once the pinch has been discovered, the process may be divided into two distinct systems: one below the pinch and one above the pinch, as presented in Section 4 for this specific case study. In particular, the system behaves as a heat sink above the pinch and as a heat source below the pinch. Hence, in order to meet the process minimal energy targets, heat must not pass through the pinch, and there should not be external cooling and heating above and below the pinch, respectively. Thus, in case of insufficient heat in the hot streams above the pinch or insufficient cooling of the cold streams below the pinch, external utilities are required. According to Chen et al. (2016) [15], the overall goal of targeting multiple utilities is to maximize the usage of lower-cost utility levels while minimizing the use of higher-cost utility levels.

3.3. Combined Exergy and Pinch Analysis

In this study, a typical combined heat and power cycle steam turbine power plant was investigated in three different operation modes. The main technological benefit of cogeneration systems is their potential to enhance fuel efficiency in the production of electrical and thermal energy. In this context, the exergy concept is integrated with that of pinch analysis in process integration analysis to reduce fuel requirements (heat load) and to optimize the cogeneration cycle in steam turbine plants. Exergy analysis is carried out on all bottoming cycles to assess the exergy losses of the various components of the system. According to Bendig et al. (2012) [16], there is one holistic rule of exergy analysis aiming to minimize the area between the hot and cold composite curves of the integrated systems, including the energy recovery system. In the heat exchanger networking, the effect of temperature difference on the distance between the cold flow and heat flow is magnified by revealing that the greater the distance between the flows is, the greater the energy consumption and energy loss, and the lower the efficacy [17]. Hence, for the composite curve generation, an investigation of the fluctuation of the heating demand for a ΔT_{min} of 10 °C was performed. Also, the GCC simplifies the identification of heat integration and energy recovery possibilities [5], and it is extremely valuable for a deeper understanding of the situation. Since utility prices are affected by temperature, higher-temperature hot utilities are often more expensive than lower-temperature hot utilities in terms of exergetic costs [18]. Cold utilities at lower temperatures, on the other hand, are more expensive than those closer to the ambient temperature.

The entire energy demand indicated by the composite curve might be provided at many levels, capable of computing the total energy objective, while it cannot specify the quantity of energy that should be given to the process at different temperatures, whereas the grand composite curve (GCC) defines the quantity of each, as discussed in Section 4.

3.4. Process Modification Solutions

Since the final purpose of this study is the detection of eventual process improvement solutions based on the results of the CPEA, some observations concerning possible modifications are discussed in this last subsection. The primary energy-saving computation was articulated by subsequent approaches for calculating electricity generation from the installed cogeneration plant in various scenarios. In particular, a cogeneration unit that is functioning with the greatest theoretically achievable heat recovery from itself is said to be in a full cogeneration mode (Grid: ON/ Factory: ON), and the process is considered combined heat and power (CHP). Sometimes, the overall efficiency of the cogeneration plant may fall below the threshold value (75–80%) [19]; the cogeneration plant is then said to be a non-CHP system. Therefore, the process modification principles must be implemented for maximum heat recovery in both CHP and non-CHP scenarios.

3.4.1. Principles of Plus–Minus for Process Modification

In this principle, mostly the heat and material balance change, by shifting the position of composite curves with a subsequent impact on process energy targets. The $+/-$ process modification decisions are made on the amount of electricity generated on site in order to sell any surplus from the three operational scenarios. The basic concept here is to modify the way energy is generated and how production is carried out in response to external factors, such as power costs related to exergetic losses [20]. Moreover, the design of an appropriate heat recovery network can help in meeting the minimum theoretical energy requirement and the reduction in exergy losses due to irreversibilities. However, by employing thermodynamic criteria based on pinch analysis, it is feasible to find modifications in the relevant process parameters which will reduce energy requirements that could be governed by the plus–minus principle.

Useful guidelines in order to carry out this task are, firstly, that any increase in hot stream duty above the pinch point and decrease in cold stream duty above the pinch point results in a reduced hot utility target. Secondly, any increase in hot stream duty above the pinch region and any decrease in cold stream duty above the pinch region results in a lower hot utility target. A lowered cold utility is also the outcome of a drop in hot stream duties below the pinch region and an increase in cold stream duties below the pinch area [21].

3.4.2. Heat Pump Integration

Process utility systems rely heavily on some critical units such as heat engines and heat pumps. As Gundersen (2013) [6] stated, the heat pumps should be integrated over the pinch in such a way that it takes heat from the surplus zone below the pinch and transfers it to the deficit region above the pinch. Such an integration mechanism helps to reduce hot and cold fluid consumption. In general, heat transfer over a temperature difference results in exergy destruction owing to friction and material degradation. It is recommended that a correctly constructed heat integration network along with strategically located steam extraction (heat engine) stations can improve the system's heat integration [22]. In addition, Tiwari et al. (2012) [23] pointed out that the temperature differential between the cold and hot streams is also greater at higher pinch points, resulting in larger irreversibilities. As a result, the optimal placement of heat pumps (heat engines) in a particular system might be in two distinct locations according to the best compromise between process heat demand minimization and maximum power generation. One of the most energy efficient combinations of process heat demand and the generation of power is heat integration.

Since it is desirable to reduce the hot utility demand, the heat engine must be placed above the pinch temperature to reject heat into the process and the heat transferred to the heat sink. On the other hand, when the heat engine is placed below the pinch temperature, it brings energy from the process of an overall heat source. On the contrary, the integration of a heat engine across the pinch does not furnish any benefit.

The results of the cogeneration plant CPEA analysis and the subsequent detection of process improvements are therefore discussed in the following section.

4. Results and Discussion

In this section, the results obtained for the three alternative scenarios are discussed according to the pinch, exergy, and CPE analysis, based on the ProSimPlus® simulation results. The outcome will be analyzed and commented upon in conformity with the final purpose of this study, i.e., the assessment of the efficiency for retrofitting and the opportunity to improve the system performance by applying reasonable modifications. However, before addressing the energy optimization problem, some simulation results need to be discussed in order to have a better understanding of the phenomenological behavior of the system. The obtained flowrate values for the most relevant process streams and the power generated for each turbine section are then reported in Table 1 according to the system functioning mode.

Table 1. Flow summary and generated power for the CHP section.

Stream	Scenario I	Scenario II	Scenario III
Circulating flowrate [t/h]	165	130	59
HP steam extraction [t/h]	20	14	0
LP steam extraction [t/h]	115	72	43
T-1 steam inlet [t/h]	165	130	59
T-2 steam inlet [t/h]	145	116	59
T-3 steam inlet [t/h]	30	44	16
T-1 power [MW]	22.53	16.9	6.4
T-2 power [MW]	11.98	9.58	3.8
T-3 power [MW]	4.83	7.07	2.1

As already mentioned in the case study section, it can be noticed that the circulating flowrate is considerably affected by the amount of power that needs to be generated. However, the plant working at full capacity requires less steam than the exact proportional amount with respect to the other functioning modes. This aspect can already be interpreted as a symptom of higher efficiency, which will be later confirmed by the CPEA study. Moreover, although the total electricity production decreases in the case of a non-operational sugar factory (cf Scenario II), the power distribution between the three turbine sections is more homogeneous. In fact, since no steam should be sent to the factory, no constraints related to its pressure are applied at the second expansion.

Furthermore, it can be pointed out that the total electricity production at full capacity is higher than the declared 31.5 MW value. The first reason for overproduction is the need for producing the electricity that is consumed by pumps and other units within the CHP plant. In addition, the mismatch can be explained by the fact that the simulation results provide the ideal value of generated power without considering dispersions, which should be compensated by a corresponding production surplus. In fact, in the case that the grid connection is off (cf Scenario III), the 12 MW of total generated power is only slightly higher than the 11 MW served to the sugar factory.

With regard to the exergy and the thermal energy balances, exergy losses are provided in Section 4.2 while the heat capacity and enthalpic flows related to each stream are provided in Appendix A and discussed in the following section.

4.1. Targeting by Pinch

As a first result, the composite curves for each scenario are built according to the temperature and enthalpy levels of the process streams, as reported in Figures 5–7 (cf Appendix A for numerical values and stream properties). The specification of the minimum permissible temperature difference being equal to $\Delta T_{lm} = 10\ °C$, which is an economic parameter for the trade-off between investment costs (heat exchangers) and running costs (energy), determines the targets for the heat recovery system. Based on this hypothesis, the minimal external heating ($\dot{Q}_{H,min}$) and minimum external cooling ($\dot{Q}_{C,min}$) requirements can be assessed either directly from the CCs' graphics or by calculating the energy balances for each heat transfer section. These two values indeed are represented by the two parts where there is no curve overlapping, and they are equal to

- $\dot{Q}_{H,min}$ = 135.76 MW and $\dot{Q}_{C,min}$ = 63.28 MW with a pinch point at 185 °C for Scenario I (cf Figure 5);

- $\dot{Q}_{H,min}$ = 108.73 MW and $\dot{Q}_{C,min}$ = 79.7 MW with a pinch point at 171 °C for Scenario II (cf Figure 6);

- $\dot{Q}_{H,min}$ = 108.73 MW and $\dot{Q}_{C,min}$ = 49.10 MW with a pinch point at 131 °C For Scenario III (cf Figure 7).

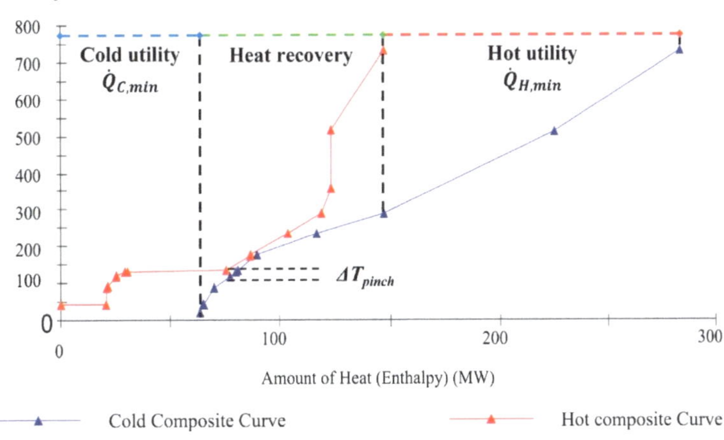

Figure 5. Hot and cold composite curves for Scenario I (Grid: ON and Factory: ON).

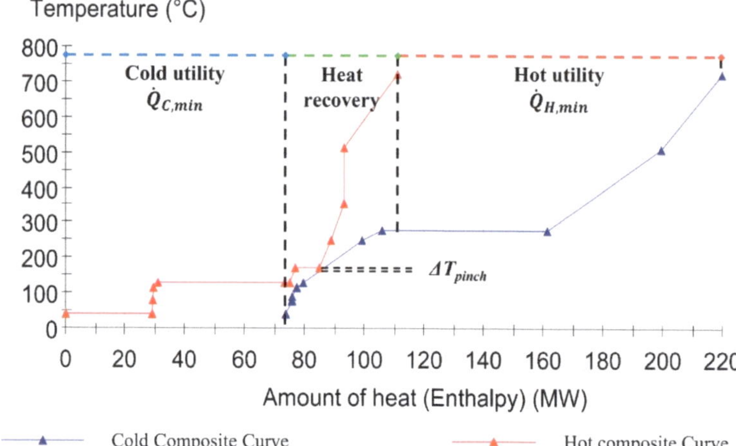

Figure 6. Hot and cold composite curves for Scenario II (Grid: ON and Factory: OFF).

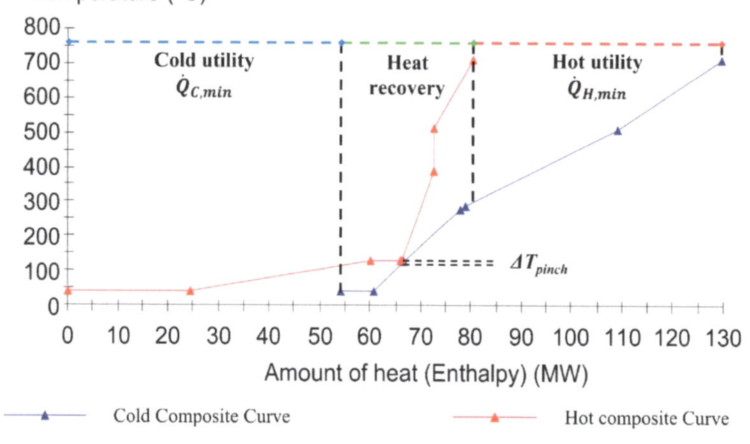

Figure 7. Hot and cold composite curves for Scenario III (Grid: OFF and Factory: ON).

In this instance, the satisfaction ratio in Scenario I achieved 18.24% and 12.20% for cold and hot fluids, respectively. Hence, with the assistance of the pinch analysis design and targeting skills, combined exergy and pinch analysis may provide an enhanced use of the exergy idea rather than pure thermal analysis, as accurately discussed in the next sections.

4.2. Exergy Analysis

Besides the pinch analysis, concerning the energy and temperature distribution of the process streams, the exergy analysis was carried out to better understand the unit-wise impact of equipment inefficiencies for each functioning mode. In particular, while the former was performed to optimize the heat transfer mechanism between cold and utilities for heat recovery, the latter aims at the identification of exergy destruction in the components of the cycle in order to identify the equipment that should be eventually improved.

As previously mentioned, the exergy analysis was performed based on the simulation results via the "Exergy balance" built-in tool of the ProSimPlus® process simulator for each scenario. The obtained results are reported in Table 2 only for units with non-zero exergy destruction values.

Table 2. Exergy loss values in process equipment [MW].

Unit	Scenario I	Scenario II	Scenario III
Deaerator	0.5309	0.0550	/
Pump-1	0.0029	0.0039	0.0049
Pump-2	0.1920	0.1217	0.0573
T-1	3.0204	2.2313	1.6772
T-2	1.4339	2.2313	1.0440
T-3	0.8384	1.2296	0.9454
Mixer-1	0.5669	0.0269	4.5162
Mixer-2	0.0454	0.0075	/
HEX1-HP	3.1515	2.2257	0.8308
HEX2-LP	0.3249	0.2924	0.5585

As a first remark, it can be immediately noticed that the three scenarios exhibit analogous exergy destruction distribution over the different modules (with the exception of the inactive ones). In general, it can be observed that pumps have a low impact on exergy losses and that this value is mainly proportional to the circulating flowrate, as for the deaerator and the mixers in the first two scenarios. On the contrary, in Scenario III, the mixer inefficacy increases due to the relevant thermal difference between the streams to be mixed. In addition, it can be observed that there are some specific units which play a role of major impact in terms of exergy destruction. In particular, the heat exchanger HEX1-HP accounts for around 31% of total exergy destruction, followed by HEX2-LP. Since both these exchangers are dedicated to heat recovery, it can be concluded that the energy integration already present in the cycle is not optimal for the energy efficiency of the process. Furthermore, a considerable contribution can be pointed out for the turbine sections T-1, T-2, and T-3, which are responsible for 30%, 14%, and 8% of the plant's exergy destruction, respectively. However, the irreversibility due to the turbines can be mitigated only by means of an equipment efficiency improvement, e.g., unit replacement.

The analysis coupling both the exergy and pinch results is therefore discussed in the next section to comment on additional aspects aimed at the improvement of the process layout.

4.3. Combined Exergy and Pinch Analysis

The general purpose of this approach is the estimation of the entire avoidable and unavoidable exergy losses for the global process and the specific process units, revealing potential improvements for heat recovery by means of an exchanger network (HEN).

As discussed in the previous sections, for this specific case study, Scenario I exhibits particular potential in terms of energy savings and integration. The impact of the ΔT_{lm} affects the quantity of energy and exergy level in the hot and cold streams, particularly for steam turbine extraction steam, low-pressure steam (LP), intermediate pressure (IP), and high-pressure (HP) steam. In fact, this is the only aspect playing a critical role in the heat transfer and heat recovery capacity reduction, resulting in exergy losses. Thus, the heat duty and maximum recovery along with its integration satisfaction ratio from the extraction and condensate streams have been collected for the three operational scenarios.

Moreover, for the cogeneration plant, it can be observed that pinch analysis is able to minimize energy usage in electricity production by enhancing energy recovery. Before moving to the quantitative analysis of the outcome, it is worth noting that Scenario I already includes a heat recovery loop for the exchangers HEX1-HP and HEX2-LP, with energy savings corresponding to 23.7 MW and 3.06 MW, respectively.

However, based on the results concerning units' irreversibility, it can be also noted that some unit operations do not need to be included in the process system integration, while it would be better to redesign or directly exclude them from the economic opportunities point of view. For instance, Mixer-1 and the deaerator in Scenario I; Mixer-1, Mixer-2, the condenser, and the deaerator in Scenario II; and finally, Mixer-1 in Scenario III must be either redesigned or excluded from the system to maximize the cogeneration system economic advantage over the equipment purchase cost.

Furthermore, the pressure (or saturated temperature) and the extraction mass flowrates were fixed at the optimal position of the turbine blades (at 9 and 2.6 bar) in order to utilize the minimum amount of fuel (heat duty) for the boiler feed water heating. Tables 3 and 4 show the extraction quantities, saturation temperatures, and the quantity of heat (heat duty) consumed by the cogeneration plant in each scenario.

Table 3. Maximum heat energy recovery for the hot and cold sides.

Scenario	Fluid	Heat Duty [MW]			Satis. Ratio [%]
		Minimum	Actual	Maximum	
I	Cold	63.28	119.96	146.71	18.24
	Hot	135.76	192.44	219.19	12.21
II	Cold	79.66	/	118.57	/
	Hot	108.73	/	147.64	/
III	Cold	53.97	/	80.12	/
	Hot	49.10	/	75.25	/

Table 4. Maximum heat energy recovery and integration potential.

Property	Scenario I	Scenario II	Scenario III
Max energy recovery [MW]	83.44	36.88	26.15
Pinch temperature [°C]	185	171	131
Actual integration ratio [%]	32.07	/	/
IPI [1] #1 [%]	45.6	/	/
IPI [1] #2 [%]	29.5	/	/

[1] IPI stands for integration potential indicator.

In case the cost of energy was higher than the selected value, the targeted investment would overcome the maximum threshold even in case of 32% integration ratio (Table 4, Scenario I) at a pinch point temperature of 185 °C. In this situation, Scenario I is the most feasible integration configuration, with a maximum energy recovery equal to 83.44 MW and integration potential indicators #1 and #2 up to 45.61% and 29.54%, respectively.

The possibility of heat pump integration is then discussed in the next section.

4.4. Heat Pump Placement

For both the heat supply of the process and power generation, the GCC can help locate the heat pump position in a process system. The most energy efficient combination for this purpose is achieved by integrating heat pumps to allow waste heat to be used for process heating. However, before exploiting the waste heat (represented in this research by extracts and condensate recycling), economic drawbacks must be considered. The economics of heat pump placement are indeed determined by the balance of process heat savings versus the electricity consumption for heat pumping. To make the heat pump alternative cost-effective, a high process heat duty and a small temperature difference across the heat pump cycle are required.

Figures 8–10 depict the background process's grand composite curves (GCCs) for the three scenarios, respectively. They were used to determine whether heat pumps could enhance the cogeneration system economic benefit via energy saving. This graphic contains the same basic data of interest as the composite curves (CCs) (i.e., the position of the pinch and the minimal external heating and cooling), but it additionally conceals information about process-to-process heat transfer. For each of the obtained GCC graphics, a net heat surplus can be derived by transferring the surplus heat from one interval to another with a heat deficit at lower temperature, by forming a special feature heat pocket below the pinch region.

In this study then, the GCCs not only display how much external heating and cooling is necessary, but they also illustrate at what temperatures such external heating and cooling are needed. In Scenarios II and III, above and below the pinch, the GCC exhibits an area of little temperature change and substantial enthalpy change. However, in Scenario I, the area of high temperature change does not correspond to a significant enthalpy change compared with the other two operational scenarios. In Figures 8–10, the pointed "nose" at the pinch suggests that a heat pump may be placed for reasonably considerable savings in heating ($\dot{Q}_{H,min}$) and cooling ($\dot{Q}_{C,min}$) demand throughout the modest temperature shift. As a result, the energy savings will be significant for a small power consumption, leading to high performance. Therefore, the integration of a heat pump in Scenarios II and III is quite beneficial in terms of heat recovery. However, in Scenario I, the heat pump alternative appears to be uneconomical since the temperature differential across the heat pump is fairly wide, resulting in a high power consumption of the additional unit.

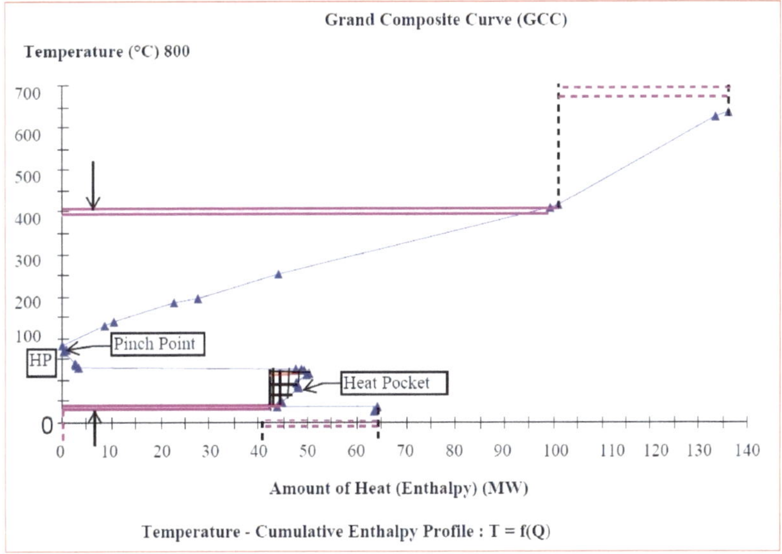

Figure 8. GCC with heat pocket for heat pump integration (Scenario I).

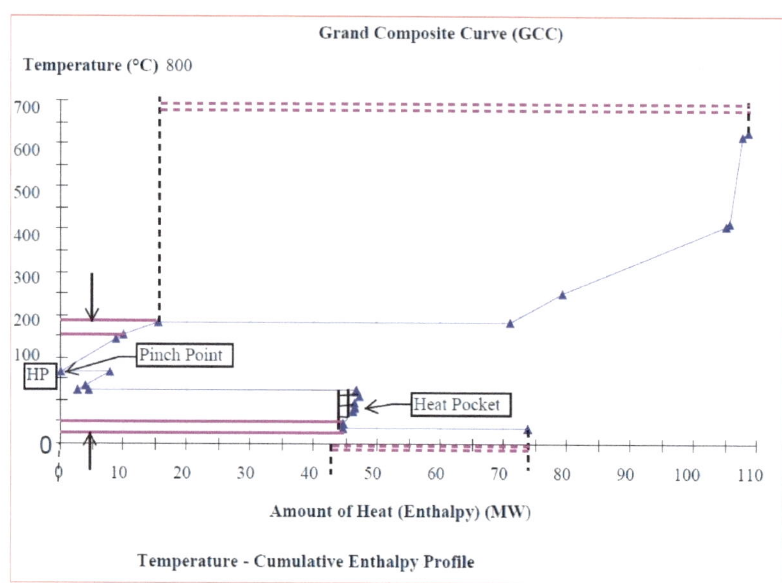

Figure 9. GCC with heat pocket for heat pump integration (Scenario II).

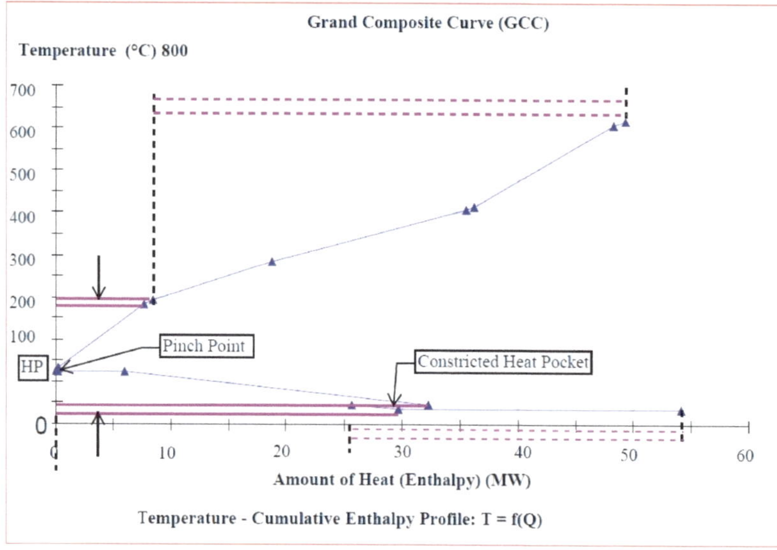

Figure 10. GCC with heat pocket for heat pump integration (Scenario III).

5. Conclusions

In this research, the implementation of the exergy concept aligned with pinch-based methodologies for analyzing the optimal integration of energy conversion systems in steam turbine cogeneration plants was investigated. The methodology proved to be effective for the desired purpose and, based on the obtained outcome, some important conclusions of general validity can be drawn:

- The application of the CPEA method successfully allows the detection of steam usage reduction opportunities and effective waste heat recycling requirements by means of dedicated heat integration;
- For each specific operating mode, the maximum energy recovery, and thus steam consumption, boundary can be quantitatively estimated. In particular, for the Wonji-Shoa cogeneration plant, up to 83.44 MW, 36.88 MW, and 26.15 MW can be recovered for Grid/Factory scenarios ON/ON, ON/OFF, and OFF/ON, respectively. These values represent a highly significant increase in cogeneration system efficiency and potential electricity surplus from the already installed condensing and extracting turbines;
- In addition, the CPEA method allows the quantification of the energy-saving potential indicators in case of a plant revamping decision. In fact, although Scenarios II and III already include waste heat recycling and reuse, the analysis highlighted further heat savings in the case a better energy integration is implemented. For instance, Scenario I exhibits integration potential indicators #1 and #2 equal to 45.61% and 26.54%, respectively;
- The highest exergy destruction cannot always be correlated with the highest energy recovery potential. As for this specific case study, even if Scenario III exhibits the highest irreversibility, the operating mode with the largest potential for irreversibility reduction is Scenario II;
- In terms of methodology, each tool can be correlated to a specific purpose. For pinch analysis, the composite curves (CCs) are used to calculate the process's lowest energy and exergy demand. First, part of the exergy requirement is calculated by accounting for an exergy loss caused by the differential temperature (ΔT_{ml}). Then, the remaining part is given by the sum of three contributions: the exergy created as an energy surplus between the pinch point and the ambient temperature, the exergy required above the pinch point, and the exergy required for minimal cooling and heating. In this case, the GCC diagrams are used to calculate the process's net cumulative heat surplus and heat deficit, serving as a useful interface between the process and the utility sections.

Based on these observations, the CPEA approach is worth further studies for more detailed and complex applications. In particular, it would be interesting to assess how the results in terms of optimal energy savings and exergy recovery could be exploited as a decisional tool for the selection of process and utility system operating modes and related optimal scheduling. This analysis, coupled with economic and environmental indicators, could exhibit great potential in terms of simultaneous profitability and sustainability optimization.

Author Contributions: Conceptualization, S.S.S., L.M. and S.N.; methodology, S.S.S., L.M. and S.N.; software, S.S.S. and A.D.P.; writing—original draft preparation, S.S.S. and A.D.P.; writing—review and editing, L.M. and S.N.; supervision, A.Y., L.M. and S.N. All authors have read and agreed to the published version of the manuscript.

Funding: This research received no external funding.

Data Availability Statement: Data are contained within the article.

Acknowledgments: The authors gratefully acknowledge the personnel of the "Wonji-Shoa Sugar Factory" and the "Wonji Research Center", particularly Girum Asfaw, the former Head of the "Wonji Research Center", for enabling the use of factory operational data for this study.

Conflicts of Interest: The authors declare no conflicts of interest.

Glossary

Symbol	Definition	Unit
B_{ch}	Chemical exergy	MW
$B_{(D,pump)}$	Exergy destruction rate of pump	MW
B_{fuel}	Fuel exergy (heat duty)	MW
B_{heat}	Exergy of heat stream	MW
$B_{(I,HE)}$	Exergy of heat exchanger	MW
$B_{material}$	Exergy of material stream	MW
B_{ph}	Physical exergy	MW
$B_{(P,k)}$	Component's exergy	MW
B_{system}	Exergy of the system	MW
B^T	Physical exergy due to temperature change	MW
B_{work}	Work rate	MW
b_i	Specific exergy	MW
CC	Composite curve	acronym
CCC	Cold composite curve	acronym
C_p	Specific heat capacity	(kJ/kg. K)
CPEA	Combined pinch and exergy analysis	acronym
$\dot{E}_{in}$	The rate of energy input to the plant	MW
GCC	Grand composite curve	acronym
h	Specific enthalpy at a temperature T	(KJ/kg)
h_0	Specific enthalpy at a temperature T_0	(KJ/kg)
HCC	Hot composite curve	acronym
HEN	Heat exchanger network	acronym
$\dot{Q}$	Exergy destruction rate (irreversibility)	MW
$\dot{m}$	Mass flow rate	(t/h)
$\dot{Q}$	Heating power	MW
S	Entropy	(kJ/kg. K)
S_0	Entropy at a dead state temperature	(kJ/kg. K)
T_0	Temperature at dead state	°C
T_i $(i = 1, 2, 3)$	Turbine units (1, 2, and 3)	-
$\dot{W}_{cv}$	Rate of work performed in control volume	MW
$\dot{W}_{net}$	Net power output	MW
$\dot{W}_{out}$	The actual generated expansion work	MW
$\dot{W}_{in}$	The actual power consumed in the pump	MW
X_k	Mass fraction of component k-th	kg/kg
η_{cogen}	Energy efficiency of the cogeneration facility	-
η_{cycle}	Overall exergetic efficiency of the cycle	-
η_{pump}	Pump's exergetic efficiency	-
ΔT_{lm}	Logarithmic mean temperature difference	°C

Appendix A

This appendix reports thermal data for each heat exchanger network stream according to the cogeneration plant functioning mode.

Table A1. Stream result data for the HEN (Scenario I, cf Figure 2).

#	Type	State	$\dot{m} \cdot C_p$ [MW/°C]	T_{in} [°C]	T_{out} [°C]	$\dot{Q}$ [MW]
31	Cold	V	0.0386	134.105	138.053	0.1526
27	Cold	L	0.0419	46.156	121.091	3.0620
20	Cold	L	39.0877	46.065	46.075	0.3909
1	Cold	L	0.0628	25.000	90.336	4.1029
3	Cold	L	0.2081	91.803	291.477	41.5614
10	Cold	V	0.0863	237.251	511.982	23.6963
5	Cold	L	0.2656	180.000	730.537	146.2280
4	Hot	L	0.2181	291.477	180.000	24.3143
13	Hot	V	0.2250	175.949	138.053	8.5279
23	Hot	V	0.3058	134.105	124.091	3.0620
19	Hot	LV	1983.750	46.085	46.075	19.8375
24	Hot	L	0.0058	124.091	46.075	0.4557
28	Hot	L	0.0411	121.091	95.000	1.0726
30	Hot	V	11.3234	138.053	134.005	45.8372
14	Hot	L	0.0895	134.005	90.336	3.9057
29	Hot	L	0.04090	95.000	90.336	0.1908
25	Hot	LV	45.6022	46.085	46.075	0.4560
35	Hot	V	0.0091	237.251	175.949	0.5561
6	Hot	V	0.1099	730.537	514.982	23.6963
15	Hot	V	0.0659	358.743	134.005	14.8014

Table A2. Stream result data for the HEN (Scenario II, cf Figure 3).

#	Type	State	$\dot{m} \cdot C_p$ [MW/°C]	T_{in} [°C]	T_{out} [°C]	$\dot{Q}$ [MW]
27	Cold	L	0.0546	46.15	82.28	1.9738
20	Cold	L	12.8248	46.06	46.08	0.1283
29	Cold	L	0.0552	95.00	119.99	1.3791
3	Cold	L	0.1646	121.36	281.77	26.4079
4	Cold	L	0.2209	281.77	723.08	97.4858
10	Cold	V	0.0689	253.27	511.98	17.8217
13	Hot	V	788.9600	175.96	175.95	7.8896
22	Hot	V	0.0405	134.01	85.28	1.9738
5	Hot	V	640.2730	134.02	134.00	6.4027
14	Hot	L	0.0168	175.95	134.00	0.7056
1	Hot	L	0.0035	85.28	46.08	0.1368
30	Hot	LV	4215.2100	134.02	134.00	42.1521
21	Hot	L	0.0983	134.00	119.99	1.3773
25	Hot	LV	13.6991	46.09	46.08	0.1370
12	Hot	V	0.0090	253.27	175.95	0.6931

Table A2. *Cont.*

#	Type	State	$\dot{m} \cdot C_p$ [MW/°C]	T_{in} [°C]	T_{out} [°C]	$\dot{Q}$ [MW]
19	Hot	LV	2909.5000	46.09	46.08	29.0950
6	Hot	V	0.0857	723.08	514.99	17.8226
15	Hot	V	0.0412	358.74	134.00	9.2667

Table A3. Stream result data for the HEN (Scenario III, cf Figure 4).

#	Type	State	$\dot{m}\cdot C_p$ [MW/°C]	T_{in} [°C]	T_{out} [°C]	$\dot{Q}$ [MW]
27	Cold	L	0.0690	46.16	131.01	5.8536
3	Cold	L	0.0751	132.39	281.77	11.2178
5	Cold	L	0.1017	281.77	713.18	43.8573
20	Cold	L	661.7250	46.07	46.08	6.6173
9	Cold	V	0.0350	291.83	511.99	7.7020
2	Hot	V	585.3600	134.02	134.01	5.8536
24	Hot	LV	0.2775	134.01	46.08	24.3982
25	Hot	LV	2440.1300	46.09	46.08	24.4013
6	Hot	V	0.0389	713.18	514.99	7.7020
19	Hot	V	0.1267	135.97	46.08	11.3895
15	Hot	V	0.0246	392.69	134.01	6.3742

References

1. Di Pretoro, A.; Fedeli, M.; Ciranna, F.; Joulia, X.; Montastruc, L.; Manenti, F. Flexibility and environmental assessment of process-intensified design solutions: A DWC case study. *Comput. Chem. Eng.* **2022**, *159*, 107663. [CrossRef]
2. Di Pretoro, A.; Manenti, F. Pinch Technology. In *Non-Conventional Unit Operations. SpringerBriefs in Applied Sciences and Technology*; Springer: Cham, Switzerland, 2020. [CrossRef]
3. Zhao, Y.; Zhang, Y.; Cui, Y.; Duan, Y.; Huang, Y.; Wei, G.; Mohamed, U.; Shi, L.; Yi, Q.; Nimmo, W. Pinch combined with exergy analysis for heat exchange network and techno-economic evaluation of coal chemical looping combustion power plant with CO_2 capture. *Energy* **2022**, *238*, 121720. [CrossRef]
4. Linnhoff, B.; Townsend, D.W.; Boland, D.; Hewitt, G.F.; Thomas, B.E.A.; Guy, A.R.; Marsland, R.H. *User Guide on Process Integration for the Efficient Use of Energy*, 1st ed.; IChemE: Rugby, UK, 1982.
5. Pina, E.A.; Palacios-Bereche, R.; Chavez-Rodriguez, M.F.; Ensinas, A.V.; Modesto, M.; Nebra, S.A. Reduction of process steam demand and water-usage through heat integration in sugar and ethanol production from sugarcane—Evaluation of different plant configurations. *Energy* **2017**, *138*, 1263–1280. [CrossRef]
6. Gundersen, T. 4—Heat Integration—Targets and Heat Exchanger Network Design. In *Woodhead Publishing Series in Energy, Handbook of Process Integration*; Klemes, J., Ed.; Woodhead publishing: Trondheim, Norway, 2013; pp. 129–167.
7. Wonji-Shoa factory, Operational manual of Wonji-Shoa sugar factory. 2013.
8. Sharew, S.; Montastruc, L.; Yimam, A.; Negny, S.; Ferrasse, J.-H. Optimal efficiency of biomass conversion from bio-based byproducts to biofuel production in the ethiopian sugar industry: A case study in Wonji-Shoa sugar factory, Ethiopia. In *31st European Symposium on Computer Aided Process Engineering of Computer Aided Chemical Engineering*; Turkay, M., Gani, R., Eds.; Elsevier: Amsterdam, The Netherlands, 2021; pp. 2009–2017. [CrossRef]
9. Chao, Z.; Yan, W. Exergy cost analysis of a coal-fired power plant based on structural theory of thermoeconomics. *Energy Convers. Manag.* **2006**, *47*, 817–843.
10. Ensinas, A.V.; Modesto, M.; Nebra, S.A.; Serra, L. Reduction of irreversibility generation in sugar and ethanol production from sugarcane. *Energy* **2009**, *34*, 680–688. [CrossRef]
11. Kanoglu, M.; Dincer, I. Performance assessment of cogeneration plants. *Energy Convers. Manag.* **2008**, *50*, 76–81. [CrossRef]
12. Kotas, T.J. *The Exergy Method of Thermal Plant Analysis*; Butterworth Heinemann Ltd.: Oxford, UK, 1985.
13. Sanjay, Y.; Singh, O.; Prasad, B.N. Energy and exergy analysis of steam cooled reheat gas–steam combined cycle. *Appl. Therm. Eng.* **2007**, *27*, 2779–2790. [CrossRef]
14. Manassaldi, J.; Mussati, S.; Scenna, N. Optimal synthesis and design of Heat Recovery Steam Generation (HRSG) via mathematical programming. *Energy* **2010**, *36*, 475–485. [CrossRef]
15. Chen, X.; Chang, C.; Wang, Y.; Feng, X. An energy hub approach for multiple-plants heat integration. *Chem. Eng. Trans.* **2016**, *52*, 571–576.
16. Bendig, M.; Marechal, F.; Favrat, D. Defining the Potential of Usable Waste Heat in Industrial Processes with the Help of Pinch and Exergy Analysis. *Chem. Eng. Trans.* **2012**, *29*, 103–108.
17. Rudiyanto, B.; Raga, T.; Prasetyo, T.; Eko Rahmanto, D.; Nuruddin, M.; Pambudi, N.; Wibowo, K. Analysis Heat Exchanger Network Steam Power Plant in Using Pinch (case study in PT POMI Unit 3 power Plant Paiton). *Int. J. Heat Technol.* **2020**, *38*, 439–446. [CrossRef]
18. Wan Alwi, S.R.; Zainuddin, A.M. Simultaneous energy targeting, placement of utilities with flue gas, and design of heat recovery networks. *Appl. Energy* **2016**, *161*, 605–610. [CrossRef]
19. Gambini, M.; Vellini, M. High Efficiency Cogeneration: Performance Assessment of Industrial Cogeneration Power Plants. *Energy Procedia* **2014**, *45*, 1255–1264. [CrossRef]
20. Pablos, C.; Merino, A.; Acebes, L.F. Modeling On-Site Combined Heat and Power Systems Coupled to Main Process Operation. *Processes* **2019**, *7*, 218. [CrossRef]

21. Lythcke-Jørgensen, C.E.; Haglind, F.; Clausen, L.R. Exergy analysis of a combined heat and power plant with integrated lignocellulosic ethanol production. *Energy Convers. Manag.* **2014**, *85*, 817–827. [CrossRef]
22. Hamssin, F.; Universiti Teknologi PETRONAS, Tronoh, Perak, Malaysia. Analysis of Pinch and Approach point of Heat Recovery Steam Generator (HRSG) of cogeneration plant. 2017, *Unpublished work*.
23. Tiwari, A.K.; Hasan, M.M.; Islam, M. Effect of Operating Parameters on the Performance of Combined Cycle Power Plant. *Sci. Rep.* **2012**, *1*, 351.

Article

Exergoeconomic Analysis of a Mechanical Compression Refrigeration Unit Run by an ORC

Daniel Taban [1], Valentin Apostol [1], Lavinia Grosu [2], Mugur C. Balan [3], Horatiu Pop [1], Catalina Dobre [1,4,*] and Alexandru Dobrovicescu [1,*]

1. Department of Engineering Thermodynamics, National University of Science and Technology Politehnica Bucharest, 060042 Bucharest, Romania; tabandaniel@yahoo.com (D.T.); valentin.apostol@magr.ro (V.A.); horatiu.pop@upb.ro (H.P.)
2. Lab Energet Mech & Electromagnetism (LEME), University of Paris Nanterre, 50 Rue Sevres, F-92410 Ville d'Avray, France; mgrosu@parisnanterre.fr
3. Department of Thermodynamics, Technical University of Cluj-Napoca, 400114 Cluj-Napoca, Romania; mugur.balan@termo.utcluj.ro
4. Academy of Romanian Scientists, Ilfov 3, 050044 Bucharest, Romania
* Correspondence: catalina.dobre@upb.ro (C.D.); adobrovicescu@yahoo.com (A.D.)

Abstract: To improve the efficiency of a diesel internal combustion engine (ICE), the waste heat carried out by the combustion gases can be recovered with an organic Rankine cycle (ORC) that further drives a vapor compression refrigeration cycle (VCRC). This work offers an exergoeconomic optimization methodology of the VCRC-ORC group. The exergetic analysis highlights the changes that can be made to the system structure to reduce the exergy destruction associated with internal irreversibilities. Thus, the preheating of the ORC fluid with the help of an internal heat exchanger leads to a decrease in the share of exergy destruction in the ORC boiler by 4.19% and, finally, to an increase in the global exergetic yield by 2.03% and, implicitly, in the COP of the ORC-VCRC installation. Exergoeconomic correlations are built for each individual piece of equipment. The mathematical model for calculating the monetary costs for each flow of substance and energy in the system is presented. Following the evolution of the exergoeconomic performance parameters, the optimization strategy is developed to reduce the exergy consumption in the system by choosing larger or higher-performance equipment. When reducing the temperature differences in the system heat exchangers (ORC boiler, condenser, and VCRC evaporator), the unitary cost of the refrigeration drops by 44%. The increase in the isentropic efficiency of the ORC expander and VCRC compressor further reduces the unitary cost of refrigeration by another 15%. Following the optimization procedure, the cost of the cooling unit drops by half. The cost of diesel fuel has a major influence on the unit cost of cooling. A doubling of the cost of diesel fuel leads to an 80% increase in the cost of the cold unit. The original merit of the work is to present a detailed and comprehensive model of optimization based on exergoeconomic principles that can serve as an example for any thermal system optimization.

Keywords: exergy analysis; diesel engine; heat exchanger; exergy destruction; exergoeconomic correlation; exergoeconomic cost assessment; exergoeconomic factor

Citation: Taban, D.; Apostol, V.; Grosu, L.; Balan, M.C.; Pop, H.; Dobre, C.; Dobrovicescu, A. Exergoeconomic Analysis of a Mechanical Compression Refrigeration Unit Run by an ORC. *Entropy* **2023**, *25*, 1531. https://doi.org/10.3390/e25111531

Academic Editors: Daniel Flórez-Orrego, Meire Ellen Ribeiro Domingos and Rafael Nogueira Nakashima

Received: 12 October 2023
Revised: 29 October 2023
Accepted: 30 October 2023
Published: 10 November 2023

1. Introduction

The increase in the cost of fossil fuels, demand for energy, and environmental concerns have led to the analysis, design, and development of thermal systems that can also convert low- and medium-temperature-level heat sources into mechanical work. One such thermal system, used on a large scale to recover waste heat, is the Rankine cycle with organic fluids (ORC) [1].

Daniarta, S. et al. [2] analyzed the benefits for using an ORC for recovering the heat released by the baking ovens in an automotive paint shop. The use of the ORC succeeds in reducing the heat loss that accompanies the manufacturing process, contributing to ameliorating the employees' comfort while also producing electrical energy.

The mechanical power produced by an ORC is often used to drive a VCRC.

Aphornratana, S. and Sriveerakul, T. [3] presented a heat-fed refrigeration cycle concept that combines an ORC and a mechanical vapor compression (VCRC) refrigeration plant using a unitary assembly consisting of a free piston compressor–expander. The two systems use the same working fluid and share the same condenser. Also, the authors of paper [4] presented a thermally activated cooling cycle consisting of an ORC and a VCRC. The system can be powered by solar thermal energy, geothermal energy, or various waste heat flows. The ORC expander shaft was directly coupled to the VCRC compressor shaft to reduce power conversion losses.

Hu, K. et al. [5] focused their attention on the analysis of a refrigerating system based on an ORC. The refrigeration was obtained in a two-stage transcritical CO_2 mechanical compression cycle. The analysis was based on economic, energetic, and exergetic criteria. The influences of the intermediary and high-pressure CO_2 were shown. The analysis pointed out that the boiler of the ORC had the largest exergy destruction, while the evaporator of the refrigerator had the highest investment cost.

The advantage of a combined ORC-VCRC compared with absorption refrigeration systems [6] is that when the refrigerating effect is not required, all the thermal energy can be converted into power and used for other applications.

Li, T. et al. [7] analyzed the use of different working fluids in a combined cooling, heating, and power cogeneration system composed of an ORC supplied with geothermal heat and a two-stage mechanical compression refrigerator that offers two levels of cold temperatures. The influence of the geothermal fluid temperature on the turbine net power output was discussed. The values of the mechanical-power-generated cooling and heating when using different working fluids were shown.

Yue, C. et al. [8] proposed that the energy source supplying the secondary systems of a vehicle should be constituted by a combined ORC-VCRC. They used R134a, R245fa, n-propane, and cyclopentane and found R134a to be the most suitable working fluid. The proposed combined system was from 9.2% to 9.8% more efficient than the conventional power system.

Other studies have focused on applying ORC technology to recover waste heat from diesel engines. The authors of paper [9] presented an alternative for using the waste heat from a diesel engine in an ORC, which, in turn, produced the energy needed to drive the compressor of a VCRC that achieved the cooling effect at different temperature levels.

Ochoa, G. et al. [10] analyzed the use of solar energy as a heat source in ORC cycles.

To increase the overall performance of the combined ORC-VCRC, numerous studies were carried out based on the principles of thermodynamics and economic analysis to determine the operating characteristics of the system, such as the right working fluid, the temperature of the heat source, and the evaporation temperature of the working fluid.

Bett, A.K. and Jalilinasrabady, S. [11] combined exergy with pinch analysis to optimize an ORC system that recovers geothermal energy. Different ORC fluids were comparatively analyzed. The use of a regenerative internal heat exchanger was considered. The variation in the pinch temperature difference in the ORC evaporator imposed the inlet pressure of the ORC fluid in the turbine. The optimal boiling pressure for the net generated power was found for each ORC fluid.

Camero, A.B. et al. [12] used an advanced exergoenvironmental analysis to discover perspectives for improving an ORC that recovers waste heat from a natural gas engine. The work, based on the concepts of endogenous and exogenous exergy destruction, pointed out that most key parts of the system were weakly connected with the others and could be optimized practically and individually. The analysis revealed that the greatest endogenous exergy destruction happens in the heat transfer process between the engine exhaust gases and ORC fluid, where about 70% of the available energy of the exhaust gases is destroyed. A reduction in this exergy destruction will increase the overall system performance.

Kim and Perez-Blanco, H. [13], looking for the best exergetic efficiency for an ORC-VCRC, analyzed eight fluids, including R143a, R134a, R152a, ammonia, R22, R600, isobutane, and propane. They proposed isobutane as a suitable working fluid to obtain a high coefficient of performance (COP) and an increased exergetic efficiency.

Saleh, B. [14] studied the behavior of an ORC-VCRC for 14 pure working fluids, including hexane, butane, R152a, R245fa, isobutane, RC318, R236ea, R236fa, R1234ze (E), C5F12, RE245cb2, R245ca, R6010, and R6012. From the exergy analysis, the author found that by working with R6012, the ORC-VCRC reached the maximum coefficient of performance.

Nazari, N. et al. [15] conducted an exergoeconomic analysis and multiobjective optimization for a solar- or biomass-based trigeneration system. In this work, exergy analysis is used only to point out the thermodynamic performance of the system and not as an instrument to find ways for improvement. The economic correlations for calculating the purchase costs of each piece of equipment, except for the compressor, are mainly economic but not exergoeconomic and are based on the size and not on the coefficient of performance of the equipment. The optimization search is based on a multi-verse optimization algorithm and not on exergoeconomic principles.

Wang, Q. et al. [16] analyzed, from the technoeconomic point of view, the performance of an organic Rankine cycle with dual-level heat sources. An exergetic analysis was performed giving only the results of the exergy destructions in each piece of equipment but without any mathematical model to support the analysis.

Louay Elmorsy et al. [17] compared integrated solar combined-cycle configurations based on exergoeconomic evaluation. Five configurations were analyzed. Linear Fresnel and parabolic through collectors and a solar tower were discussed based on exergoeconomic principles. The paper offers detailed economic and exergoeconomic analyses for the best promising configurations.

Ibrahim et al. [18] analyzed a solar distillation system for cost optimization. The authors used the exergoeconomic method to evaluate the cost parameters. To investigate, for freshwater production, the effects of the key variables on the exergoeconomic cost, a sensitivity analysis was carried out. The analysis was performed on an existing system; and after the exergoeconomic optimization, the cost of the fresh water decreased by about 45%.

Rangel-Hernandez et al. [19] used exergoeconomics to compare the operation of a domestic refrigerator when R1234yf was used as a drop-in replacement for R134a. From the point of view of the unit exergy cost, R134a performed better than R1234yf under different operating conditions due to less exergy destruction in the system.

Fergani, Z. et al. [20] evaluated an organic Rankine cycle with a zeotropic mixture, using exergy-based methods. For parametric optimization, a multiobjective approach was applied. A comparative analysis of zeotropic mixtures and pure fluids was conducted as well. For each piece of equipment, economic and exergoeconomic correlations were given. The conclusion of the work is that the application of zeotropic mixtures as working fluids for ORCs improves the exergoeconomic performances compared to those obtained using the pure components.

Luo, I. et al. [21] evaluated, from thermodynamic and economic viewpoints, the performance of a vapor compression refrigeration machine with CO_2 as the working fluid driven by a Brayton cycle. The refrigeration was dedicated to hot climate conditions. The objective of the optimization was the maximization of the exegetic efficiency. The optimization search was based on a genetic algorithm and revealed a 4.7% exergetic efficiency when the system operated only as a refrigerator and 22% in cogeneration with heat capacities.

Tashtoush, B. et al. [22] conducted an exergoeconomic analysis of a solar-energy-driven organic Rankine cycle combined with an ejector refrigeration system to produce cooling and electrical power. Exergy and exergoeconomic analyses of the combined system were carried out to predict the cost of the inefficiencies present in the key components of the system.

To further improve the overall performance of ORC-based systems, researchers have used several optimization techniques. Salim, M.S. and Kim, M. [23] presented the application of a genetic algorithm (GA) for the thermoeconomic optimization of the combined ORC-VCRC system to determine the optimal component sizes and thermal performance of the system. The authors recommended R245fa as the working fluid for the optimal thermoeconomic performance of the system. In paper [24], the authors used the TOPSIS technique based on the entropy method to determine the optimal mass fraction of mixtures to be used as working fluids in a simple ORC and concluded that a mass fraction of 0.1/0.9 was optimal for mixtures, such as pentane/butane, hexane/pentane, and isohexane/pentane. Also, in paper [25], Bademlioglu, A.H. et al. presented a multiobjective optimization for the simple ORC system, using the Taguchi technique and gray relational analysis (GRA) to determine the order of importance of each input parameter in the system and their effects on the energetic and exergetic performances of the system. The authors found that by choosing the optimal condensing and evaporation temperatures, the turbine isentropic efficiency and heat-exchanger efficiencies were the most effective, followed by optimizing the superheating temperature, pinch temperature differences in the condenser and evaporator, and isentropic efficiency of the pump.

The literature review reveals the interest in the optimization of the ORC-VCRC compound system. In most approaches, exergy analysis is used for estimating the magnitude of inefficiencies; and for the optimization procedure, researchers appeal to mathematical methods.

In the present work the optimization of an ORC-VCRC will be conducted in an open view based on only exergoeconomic principles. The exergoeconomic optimization offers the possibility to follow the effect of any local structural or operational change on the monetary cost of the overall system and the desired product.

The concept proposed for analysis in this paper combines an ORC with a mechanical vapor compression refrigeration cycle (VCRC) to form a thermally activated cooling system by recovering heat from the combustion gases of a stationary internal combustion engine (ICE). The ORC-VCRC system uses the direct gear ratio of the mechanical work generated by the ORC expander. Thus, the shaft of the ORC expander directly engages the shaft of the VCRC compressor and uses R1224yd(Z) as single working fluid to drive both thermodynamic cycles. In this sense, a hydraulic route has been implemented with a single condenser to circulate the working fluid required for both cycles.

The present work aims to identify and optimize the constructive solutions of the ORC-VCRC-coupled system by means of an exergoeconomic analysis. The exergoeconomic optimization method aims to identify the exergy destruction of each functional area and assign its monetary cost. Areas with a high cost of exergy destruction will be the first targets of the optimization procedures, and the effect of each local cost reduction will be verified at the global level. To quantify the value of a zonal destruction, the fuel, product, loss, and exergetic performance coefficients were evaluated for each operational area, with the aim of obtaining the maximum amount of product from a limited resource at the minimum monetary cost.

The scheme of the ORC-VCRC combined installation and representation of the cycle in the p–h diagram are presented in Figure 1.

The refrigerating cycle (1Tv-6-7-8-1Tv) (Figure 1) is run directly by the ORC (1P-2-3-4-1P) (Figure 1). The refrigerating task of the VCRC is to chill water. When the VCRC works, the mechanical work supplied by the expander of the ORC is totally used to run the VCRC compressor (process 7–8, Figure 1). In the boiler, the ORC recovers the heat carried out by the combustion products of the diesel engine (process 2–3, Figure 1).

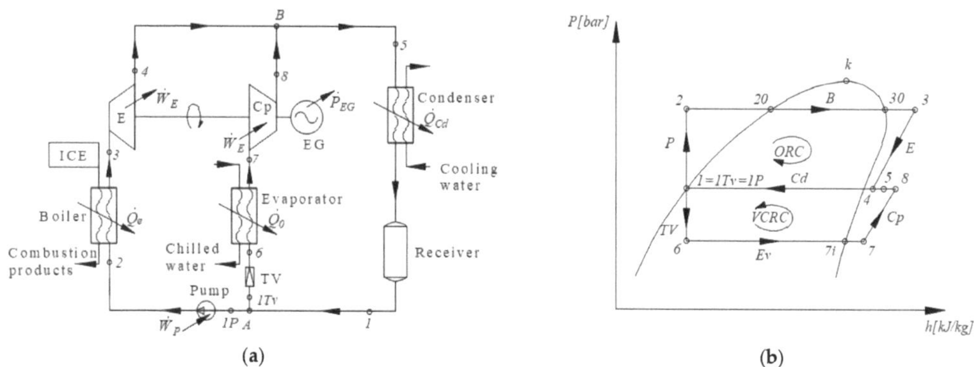

Figure 1. ORC-VCRC system: (**a**) flowchart; (**b**) ORC-VCRC cycles in the p–h diagram.

The characteristic measures (Table 1) for the analysis of the optimized coupling were taken from the internal combustion engine (ICE) that equips the ICE-ORC experimental stand located within the Department of Thermotechnics, Engines, Thermal Equipment, and Refrigeration at the National University of Science and Technology Politehnica Bucharest. The stand is equipped with a diesel-type engine, model 4TNV98TGGEHR manufactured by Yanmar, and was designed and built to experimentally investigate the possibilities for improving the performance coefficient of an internal combustion engine by coupling it with an ORC installation with the role of recovering the heat dissipated by the engine.

Table 1. The characteristics of the ICE-ORC-VCRC coupled system.

Main System Parameter	Value	Unit of Measure
Maximum mechanical power	40	kW
The temperature of the combustion gases in the 100% regime, equal to the temperature of the hot fluid at the entrance to the ORC boiler, $t_\alpha = t_{i,g}$	420	°C
Outlet temperature of the combustion gases from the boiler, $t_\beta = t_{o,g}$	140	°C
Measured mass-flow rate of combustion gases, m_g	0.0534	kg/s
Specific heat of combustion gases, c_g	1.14	kJ/kgK
The temperature of the cooling water at the entrance to the condenser, $t_\xi = t_{iw,Cd}$	15	°C
Condenser outlet water temperature, $t_\tau = t_{ow,Cd}$	22	°C
Condenser cooling water mass-flow rate, $\dot{m}_{w,c}$	0.5	kg/s
The temperature of the cooled water at the entrance to the evaporator, $t_\gamma = t_{iw,Ev}$	8	°C
The temperature of the cooled water at the exit from the evaporator, $t_\delta = t_{ow,Ev}$	3	°C
Ambient temperature, t_o	15	°C
ORC-VCRC working fluid	R1224yd(Z)	-

The operating conditions were established starting from the parameters of the heat source for the ORC (engine exhaust gases) and the condenser cooling water. The maximum evaporation temperature of the organic fluid was imposed considering the restriction that the temperature of the combustion gases at the exit from the boiler evaporator should not fall below 140 °C to avoid condensation and the formation of sulfuric acid (H_2SO_4) in and implicit corrosion of the ORC boiler.

Compared with absorption- or adsorption-based refrigeration systems, the ORC–VCRC combined system is flexible: During hot summer periods, all the available thermal energy can be converted into mechanical energy and then into a cooling effect; and during winter, when there is no need for cold, the system can produce only electrical energy.

2. Mathematical Modeling of the System

The mathematical model is based on mass, energy, and exergy balances for different constructive structures of the ORC-VCRC scheme.

The importance of the exergetic analysis is the possibility to identify and evaluate the magnitude of the dysfunctions in the system. Based on the exergetic analysis, a strategy can be established to improve the performance of a system by making constructive and operational changes.

The analysis of the functional scheme and installation cycle (Figure 1) highlights the following sequence of work processes: (2–20) heating the ORC liquid to the boiling temperature, (20–30) evaporating the ORC liquid in the ORC boiler, (30-3) superheating the vapors in the ORC boiler, (3–4) expanding the vapors to the condensing pressure in the ORC expander, (7–8) compressing the superheated vapors to the condensing pressure in the VCRC compressor, (5) mixing the streams in states (4) and (8), and (5-1) condensing the saturated liquid in the ORC-VCRC condenser. (1TV) is the part of the stream (1) that expands in the throttling valve (TV) of the refrigeration plant. (1P) is the part of the stream (1) that is compressed in the ORC liquid pump (P). (1P-2) is the liquid compression in the ORC pump. (1TV-6) is the throttling of the liquid agent in the VCRC throttling valve. (6–7) is the evaporation and superheating in the VCRC evaporator. (A) is the nodal point for separating the condensed liquid (1) from the streams (1TV) to the TV throttling valve of the VCRC and from (1P) to the ORC liquid pump. (B) is the nodal mixing point of currents (4) and (8).

The analysis was made based on the following assumptions:

- there is no heat loss in the heat exchangers;
- the pressure losses in all the components of the installation are neglected;
- the expansion in the expander, compression in the compressor, and pressure increase in the pump are irreversible adiabatic processes characterized by the isentropic efficiencies η_{sE}, η_{sCp}, and η_{sP};
- the expander is directly engaged with the compressor, and power is transferred without mechanical losses;
- the working fluid in the ORC-VCRC system is R1224yd(Z), a new type of refrigerant that is used to replace R245fa.

The mathematical model of the operation of the ORC-VCRC system is built based on the choice of some decision-making parameters.

2.1. Choice of Decision-Making Parameters

2.1.1. The Boiling Temperature in the ORC Boiler

The temperature difference, Δt_{minB}, is chosen between the outlet temperature of the combustion gases, $t_{o,g}$, and the boiling temperature of the working fluid in the boiler, t_B (Figure 2a). The result is the boiling temperature of the working fluid in the boiler, t_B.

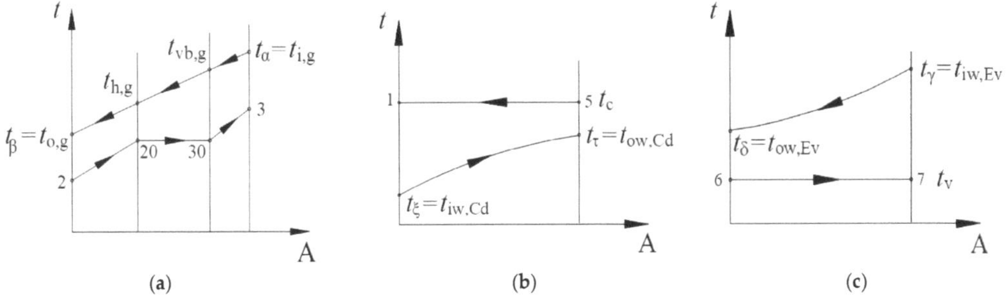

Figure 2. Temperature–Area diagrams: (**a**) ORC boiler; (**b**) ORC-VCRC condenser; (**c**) VCRC evaporator (for the ORC-VCRC, working fluid notations correspond to those in Figure 1).

2.1.2. Condensation Temperature

The temperature difference, Δt_{minCd}, is chosen between the outlet temperature of the water from the condenser, $t_{ow,Cd}$, and the condensation temperature of the working fluid, t_C (Figure 2b), as follows:

$$t_C = t_{ow,Cd} + \Delta t_{minCd} \tag{1}$$

2.1.3. Evaporation Temperature

The temperature difference, Δt_{minEv}, is chosen between the outlet temperature of the chilled water from the evaporator, $t_{ow,Ev}$, and the evaporation temperature of the working fluid, t_v (Figure 2c), as follows:

$$t_v = t_{ow,Ev} - \Delta t_{minEv} \tag{2}$$

2.1.4. The Temperature of the ORC Fluid at the Entrance to the Expander

The degree of superheating of the vapors in the boiler, $\Delta t_{oh,B}$, is chosen (Figures 1b and 2a), resulting in temperature t_3 at the exit from the boiler as follows:

$$t_3 = t_B + \Delta t_{oh,B} \tag{3}$$

2.2. Energetic Analysis

Based on the energy balances of the component devices, the flow rates of the working fluids, thermal loads of the heat exchangers, and mechanical powers are calculated (Figure 1).

The organic fluid mass-flow rate results from the energy balance of the ORC boiler as follows:

$$\dot{Q}_B = \left|\dot{Q}_g\right| = \dot{m}_g \cdot c_g \left(t_{i,g} - t_{o,g}\right) \tag{4}$$

$$\dot{m}_{ORC} = \dot{Q}_B / (h_3 - h_2) \tag{5}$$

The power produced in the expander is as follows:

$$\dot{W}_E = \dot{m}_{ORC} (h_3 - h_4) \tag{6}$$

The power input of the working fluid drive pump is as follows:

$$\left|\dot{W}_P\right| = \dot{m}_{ORC} (h_2 - h_1) \tag{7}$$

The specific mechanical work required to drive the compressor is as follows:

$$\left|w_{Cp}\right| = h_8 - h_7 \tag{8}$$

Considering that the VCRC compressor is directly driven by the expander without mechanical losses,

$$\left|\dot{W}_{Cp}\right| = \dot{W}_E - \left|\dot{W}_P\right| \tag{9}$$

The mass-flow rate of the fluid in the refrigeration cycle can be calculated as follows:

$$\dot{m}_{VCRC} = \left|\dot{W}_{Cp}\right| / \left|w_{Cp}\right| \tag{10}$$

Because the condenser is common to the two combined cycles, its thermal power is as follows:

$$\left|\dot{Q}_{Cd}\right| = (\dot{m}_{ORC} + \dot{m}_{VCRC})(h_5 - h_1) \tag{11}$$

The energy balance of the evaporator specifies the refrigeration power of the system as follows:

$$\dot{Q}_{Ev} = \dot{m}_{VCRC}(h_7 - h_6) \tag{12}$$

The energy performance coefficients are calculated for each individual subsystem and the overall system as follows:

$$COP_{ORC} = \frac{\dot{W}_E}{\left|\dot{Q}_B\right|} \tag{13}$$

$$COP_{VCRC} = \frac{\dot{Q}_{ev}}{\left|\dot{W}_{cp}\right|} \tag{14}$$

$$COP_{ORC-VCRC} = \frac{\dot{Q}_{ev}}{\left|\dot{Q}_B\right|} \tag{15}$$

2.3. Exergetic Analysis

Exergy represents the maximum mechanical work that a system can release or the minimum mechanical work it must consume to reach total equilibrium with its environment. In energy-transfer processes, a part of the exergy is consumed (destroyed) owing to irreversibility.

Minimizing the exergy destruction in the key components of an energy system provides a strategy to follow to optimize the structure and the way the system works.

The exergy destruction in each operating area of the ORC-VCRC system is calculated based on the Gouy–Stodola equation or the exergy balance equations.

The analysis of the heat transfer in the ORC boiler was carried out by dividing the boiler into distinct areas for liquid heating, evaporation, and vapor superheating as follows:

- Liquid-heating area (Figure 2a)

The amount of heat required to heat the working fluid in the liquid state in the boiler is as follows:

$$\dot{Q}_h = \dot{m}_{VCRC}(h_{20} - h_2) \tag{16}$$

From the energy balance in this area, the inlet temperature of the combustion gases in this area is as follows:

$$\dot{Q}_h = \dot{Q}_{h,g} = \dot{m}_g c_g \left(t_{h,g} - t_{o,g}\right) \tag{17}$$

The exergy destruction associated with heat transfer at the finite temperature difference in the liquid-heating zone is as follows:

$$\dot{I}_{\Delta T_{,h}} = \left|\dot{Ex}_Q^{T_{g,h}}\right| - \dot{Ex}_Q^{T_{ORC,h}} \tag{18}$$

$$\left|\dot{Ex}_Q^{T_{g,h}}\right| = \left|\dot{Q}_{g,h}\right|\left(1 - \frac{T_o}{T_{g,h}}\right) \tag{19}$$

$$\dot{Ex}_Q^{T_{ORC,h}} = \dot{m}_{ORC}(h_{20} - h_2 - T_o(s_{20} - s_2)) \tag{20}$$

where

$$T_{g,h} = \frac{t_{h,g} - t_{o,g}}{\ln\dfrac{T_{h,g}}{T_{o,g}}} \tag{21}$$

- ORC fluid evaporation zone (Figure 2a)

The amount of heat required to evaporate the working fluid in the boiler is as follows:

$$\dot{Q}_{vb} = \dot{m}_{ORC}(h_{30} - h_{20}) \tag{22}$$

The exergy of the boiling heat of the ORC fluid in the boiler is as follows:

$$\dot{Ex}_Q^{T_{ORC,vb}} = \dot{m}_{ORC}(h_{30} - h_{20} - T_0(s_{30} - s_{20})) \tag{23}$$

On the side of the combustion gases in the evaporation area of the ORC fluid, the energy-balance equation specifies the following:

$$\left.\dot{Q}\right|_{g,vb} = \dot{Q}_{ORC,vb} \tag{24}$$

$$\left.\dot{Q}\right|_{g,vb} = \dot{m}_g c_g \left(t_{vb,g} - t_{h,g}\right) \tag{25}$$

in which the variable is the temperature entering the area of the combustion gases, $t_{vb,g}$.

The average thermodynamic temperature of the combustion gases in the evaporation zone is calculated as follows:

$$T_{g,vb} = \left(t_{vb,g} - t_{h,g}\right)/\ln\left(T_{vb,g}/T_{h,g}\right) \tag{26}$$

The exergy of the combustion gases is as follows:

$$\left|\dot{Ex}_Q^{T_{g,vb}}\right| = \left.\dot{Q}\right|_{g,vb}\left(1 - T_0/T_{g,vb}\right) \tag{27}$$

The exergy destruction due to heat transfer at the finite temperature difference for the evaporation of the working fluid is calculated using the following relation:

$$\dot{I}_{\Delta T,vb} = \left|\dot{Ex}_Q^{T_{g,vb}}\right| - \dot{Ex}_Q^{T_{ORC,vb}} \tag{28}$$

- Overheating zone of organic fluid vapors (Figure 2a)

The exergy destruction due to heat transfer at the finite temperature difference for superheating the working fluid is as follows:

$$\dot{I}_{\Delta T,oh} = \left|\dot{Ex}_Q^{T_{g,oh}}\right| - \dot{Ex}_Q^{T_{ORC,oh}} \tag{29}$$

$$\dot{Ex}_Q^{T_{ORC,oh}} = \dot{m}_{ORC}(h_3 - h_{30} - T_0(s_3 - s_{30})) \tag{30}$$

$$\left|\dot{Ex}_Q^{T_{g,oh}}\right| = \left.\dot{Q}\right|_{g,oh}\left(1 - T_0/T_{g,oh}\right) \tag{31}$$

$$T_{g,oh} = \left(t_{i,g} - t_{vb,g}\right)/\ln\left(T_{i,g}/T_{vb,g}\right) \tag{32}$$

Considering the exergy consumption in the boiler's functional areas, the total exergy destruction due to heat transfer at the finite temperature difference in the boiler is as follows:

$$\dot{I}_{\Delta T,B} = \dot{I}_{\Delta T,h} + \dot{I}_{\Delta T,vb} + \dot{I}_{\Delta T,oh} \tag{33}$$

Considering Equations (20), (28), and (32), the total exergy consumption in the ORC boiler, which is also the exergetic fuel of the global system, is as follows:

$$\left|\dot{Ex}_{Q,g}\right| = \left|\dot{Ex}_Q^{T_{g,h}}\right| + \left|\dot{Ex}_Q^{T_{g,vb}}\right| + \left|\dot{Ex}_Q^{T_{g,oh}}\right| \tag{34}$$

For the other equipment components of the ORC-VCRC group (Figure 1), the following exergy consumptions or losses are identified:

- exergy destruction in the expander;

$$\dot{I}_E = \dot{m}_{ORC} T_0 (s_4 - s_3) \tag{35}$$

- exergy loss in the condenser;

$$\dot{L}_{Cd} = \dot{m}_{ORC-VCRC} [h_5 - h_1 - T_0 (s_5 - s_1)] \tag{36}$$

- exergy destruction due to the mixing of the working fluid at the entrance to the common condenser (state 5, Figure 1);

$$\dot{I}_m = T_0 (\dot{m}_{ORC-VCRC} \cdot s_5 - \dot{m}_{ORc} \cdot s_4 - \dot{m}_{VCRC} \cdot s_8) \tag{37}$$

- exergy destruction in the pump;

$$\dot{I}_P = T_0 \cdot \dot{m}_{ORC} (s_2 - s_1) \tag{38}$$

- exergy destruction in the throttling valve;

$$\dot{I}_{TV} = T_0 \cdot \dot{m}_{VCRC} (s_6 - s_1) \tag{39}$$

- exergy destruction in the compressor.

$$\dot{I}_{Cp} = T_0 \cdot \dot{m}_{VCRC} (s_8 - s_7) \tag{40}$$

The product of the combined ORC-VCRC system is the exergy of the refrigerating effect achieved in the evaporator of the refrigerating cycle, as follows:

$$\left| \dot{Ex}_Q^{T_{VCRC,v}} \right| = \left| \dot{Ex}_{Q_0} \right| = -\dot{m}_{VCRC} (h_7 - h_6 - T_0 (s_7 - s_6)) \tag{41}$$

The exergetic performance of the combined ORC-VCRC system is specified by the global exergetic efficiency and share of the exergy consumption and losses, associated with each equipment or process, in the exergy consumption of the system.

Noting that the product of the global system is defined by Equation (41) and that the fuel used upon entering the system is defined by Equation (34), the exergetic efficiency of the ORC-VCRC system is as follows:

$$\eta_{ex} = \frac{\left| \dot{Ex}_Q^{T_{VCRC,v}} \right|}{\left| \dot{Ex}_{Q,g} \right|} = \frac{\left| \dot{Ex}_{Q_0} \right|}{\left| \dot{Ex}_{Q,g} \right|} \tag{42}$$

The weight of the destruction or loss of exergy in the exergy consumption of the global system is defined as follows:

$$\psi_i = \frac{\dot{I}_i}{\left| \dot{Ex}_{Q,g} \right|} \tag{43}$$

2.4. Results and Discussion for the Basic ORC-VCRC Cycle

For the basic ORC-VCRC scheme (Figure 1), the energetic and exergetic studies were carried out for the fluid R1224yd(Z) under the following conditions: $t_0 = 15\ °C$; $\dot{m}_g = 0.0534\ kg/s$; $t_{i,g} = 480\ °C$; $t_{o,g} = 140\ °C$; $c_g = 1.14\ kJ/(kgK)$; $t_{ow,Cd} = 22\ °C$; $t_v = -5\ °C$; $\Delta T_{minB} = 10\ K$; $\Delta T_{oh,B} = 6\ K$; $\Delta T_{minCd} = 6\ K$; $\eta_{sE} = 0.85$; $\eta_{sCp} = 0.85$; $\eta_{sP} = 0.6$.

The results of the energetic and exergetic analyses of the ORC-VCRC-coupled cycle are presented in Table 2 and Figure 3.

Table 2. The results of the energetic and exergetic analyses for the basic ORC-VCRC scheme.

COP_{VCRC} (-)	COP_{ORC} (-)	$COP_{ORC-VCRC}$ (-)	η_{ex} (%)	$\dot{m}_{VCRC}$ (kg/s)	$\dot{m}_{ORC}$ (kg/s)	$\dot{m}_{ORC-VCRC}$ (kg/s)	ψ_{Cp} (%)	ψ_E (%)	$\psi_{\Delta T,B}$ (%)	$\psi_{\Delta T,h}$ (%)
6.108	0.1657	0.9468	14.91	0.1089	0.07605	0.1849	4.605	5.464	51.45	28.07

$\psi_{\Delta T,oh}$ (%)	$\psi_{\Delta T,vb}$ (%)	ψ_P (%)	$\psi_{L,Cd}$ (%)	ψ_{TV} (%)	ψ_m (%)	$\dot{Q}_B$ (kW)	$\dot{Q}_{Cd}$ (kW)	$\dot{Q}_{Ev}$ (kW)	$\dot{W}_E$ (kW)	$\left|\dot{W}\right|_P$ (kW)	$\left|\dot{W}\right|_{Cp}$ (kW)
2.425	20.96	0.8721	19.22	3.21	0.267	17.05	33.18	16.14	2.825	0.1823	2.6427

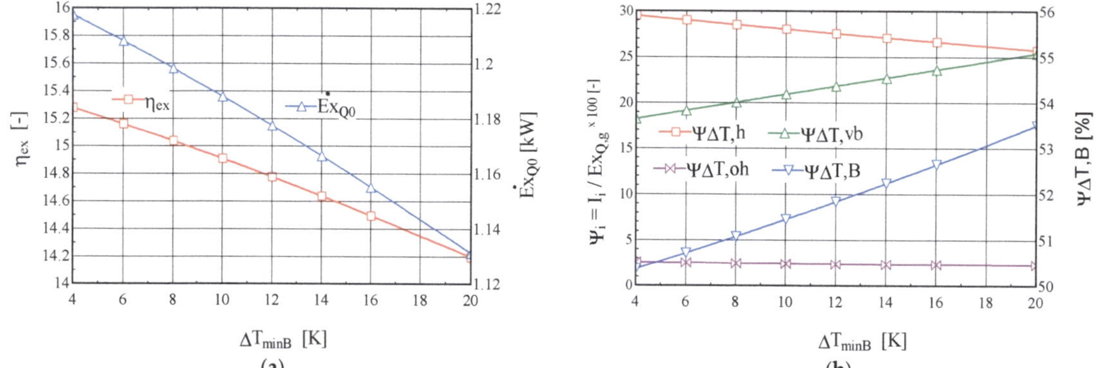

Figure 3. The ORC-VCRC system: (**a**) the exergetic performance coefficient and the exergy of the heat current extracted in the VCRC evaporator, depending on the minimum temperature difference, Δt_{minB}, in the ORC boiler. (**b**) The shares of the exergy consumption in the overall system, and the exergy destructions from the heating, boiling, and overheating areas of the boiler, depending on the minimum temperature difference, Δt_{minB}.

The exergy analysis highlights the high exergy consumption due to the heat transfer at the finite temperature difference from the ORC boiler. The largest temperature differences, which also determine the highest exergy consumption, are found in the ORC liquid heating zone, $\psi_{\Delta T,h} = 28.08\%$, and evaporation zone, $\psi_{\Delta T,vb} = 20.96\%$; a reduction in these destructions is the way forward in the optimization procedure. A decrease in the temperature difference in the ORC boiler would lead to a decrease in the exergy destruction in this area, leading to an increase in the cooling power of the combined system. The loss of exergy in the condenser is more related to the temperature difference between the ORC and VCRC fluids and the cooling water, which is influenced to some extent by the variation in the organic fluid mass-flow rates with respect to the variation in the other decision-making parameters.

To understand how the decision-making parameters functionally and constructively influence the evolution of the ORC-VCRC system, a sensitivity study of the energetic and exergetic performances and exergy destruction and losses of the combined system was conducted by varying the decision-making parameters (Δt_{minB}, Δt_{minCd}, and $\Delta t_{oh,B}$).

2.4.1. The Behavior of the ORC-VCRC Cycle Obtained by Varying the Temperature Difference, Δt_{minB}, in the ORC Boiler

The results of the sensitivity study of the change in the temperature difference, Δt_{minB}, in the ORC boiler are presented in Figure 3.

When the temperature difference, Δt_{minB}, increases, the exergy destruction due to the heating process, $\psi_{\Delta T,h}$, decreases and that due to the evaporation process, $\psi_{\Delta T,vb}$, increases (Figure 3b). Overall, the exergy destruction in the boiler increases with increasing temperature difference between the fluids, which leads to a decrease in the exergetic efficiency of the cycle and a decrease in the exergy of the refrigerating power (Figure 3a).

2.4.2. The Behavior of the ORC-VCRC Cycle Obtained by Varying the Overheating Degree, $\Delta t_{oh,B}$, in the ORC Boiler

Increasing the degree of overheating in the boiler has a positive effect, leading to an increase in the exergetic efficiency of the combined cycle and an increase in the exergy of the refrigerating power (Figure 4a).

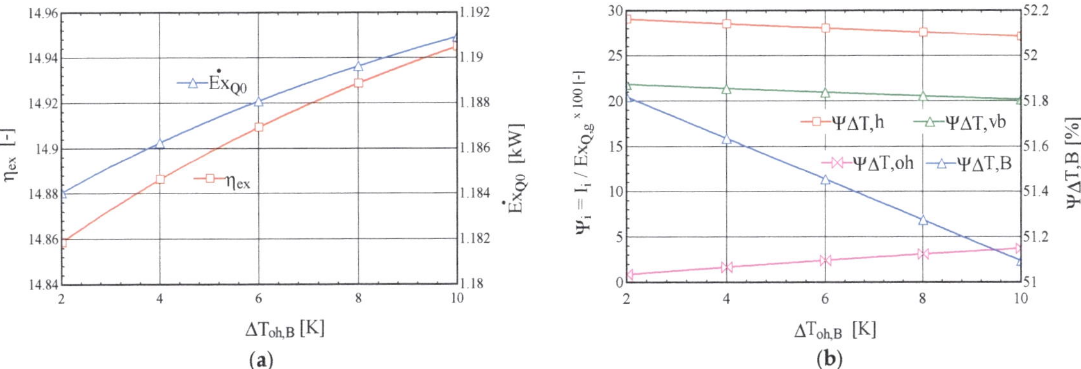

Figure 4. The ORC-VCRC system: (**a**) the exergetic performance coefficient and the exergy of the heat stream extracted in the VCRC evaporator, depending on the degree of overheating in the ORC boiler, $\Delta t_{oh,B}$. (**b**) The shares of the exergy consumption in the global system, and the exergy destructions from the heating, boiling, and overheating areas of the boiler, depending on the degree of overheating in the ORC boiler, $\Delta t_{oh,B}$.

Under the conditions of the constant thermal flow transferred to the ORC boiler from the combustion gases, the increase in the degree of overheating, $\Delta t_{oh,B}$, of the ORC fluid leads to the redistribution of thermal loads from the boiler in the liquid-heating, -evaporation, and -superheating phases. The relative exergy destructions from the phases of liquid heating, $\psi_{\Delta T,h}$, and evaporation, $\psi_{\Delta T,vb}$, which have the highest weights, decrease faster than the increase in the weight of the exergy destruction associated with overheating, $\psi_{\Delta T,oh}$ (Figure 4b); overall, the exergy destruction caused by the irreversibility of the heat transfer at the finite temperature difference in the boiler, $\psi_{\Delta T,B}$, decreases, leading to an increase in the exergy efficiency of the global system and an increase in the exergy of the refrigerating power (Figure 4a).

2.4.3. The Behavior of the ORC-VCRC Cycle Obtained by Varying the Minimum Temperature Difference, Δt_{minCD}, in the Condenser

When the minimum temperature difference, Δt_{minCD}, in the condenser increases, the exergetic efficiency of the ORC-VCRC system and the exergy of the refrigerating power decrease (Figure 5).

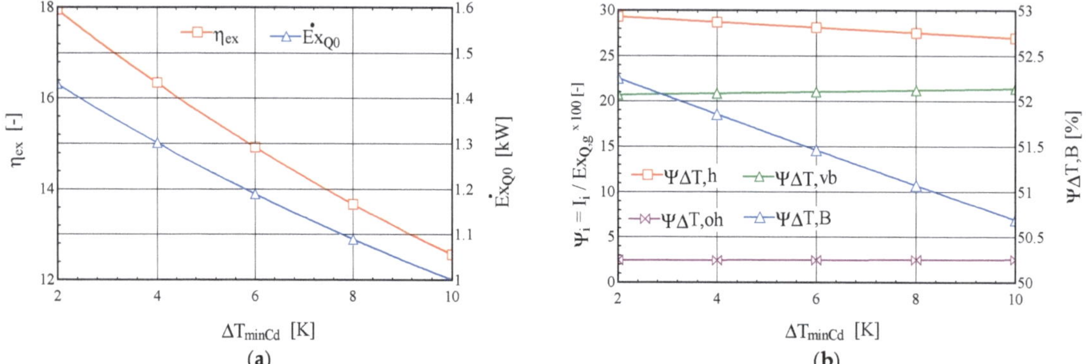

Figure 5. The ORC-VCRC system: (**a**) the exergetic performance coefficient and the exergy of the heat current extracted in the VCRC evaporator, depending on the temperature difference, Δt_{minCD}, in the condenser. (**b**) The shares of the exergy consumption in the overall system, and the exergy destructions in the heating, boiling, and overheating areas of the boiler, depending on the temperature difference, Δt_{minCD}, in the condenser.

The increase in the minimum temperature difference, Δt_{minCD}, in the condenser leads to an increase in the condensation temperature, which leads to a decrease in the refrigerating power and its exergy, $\dot{E}x_{Q_0}$ (Figure 5a). The decrease in the exergy of the refrigeration power is the factor that determines the decrease in the exergy efficiency of the global system despite the reduction in the relative exergy destruction, $\psi_{\Delta T,h}$, associated with the heating of the ORC fluid in the boiler and the consequent reduction in the total exergy destruction in the boiler, $\psi_{\Delta T,B}$ (Figure 5b).

Figure 6 presents, in $\left(\Theta = 1 - \frac{T_0}{T}\right)$ coordinates, the exergetic temperature factor—H—curve of the hot stream represented by the combustion gases and the composite curve of the cold stream represented by the ORC fluid from the ORC boiler.

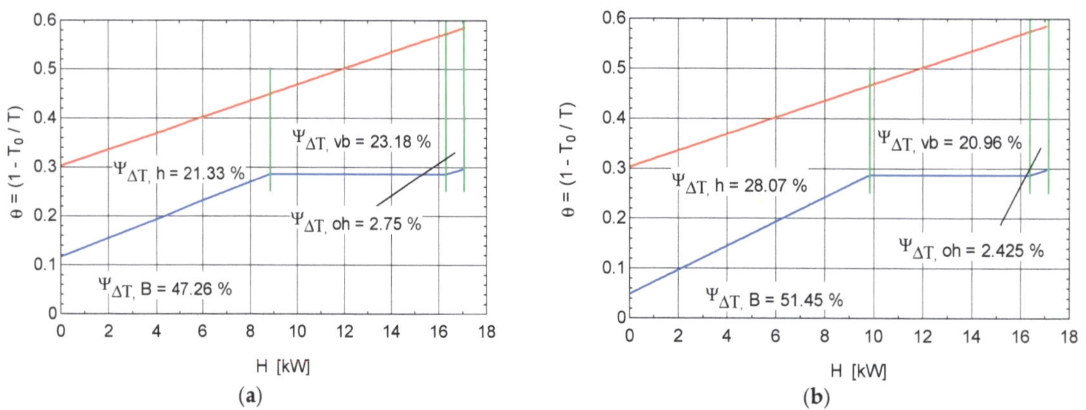

Figure 6. θ-H diagram of hot and cold currents from the ORC boiler: (**a**) simple ORC system (Figure 1); (**b**) ORC system with internal recuperative heat exchanger (Figure 7).

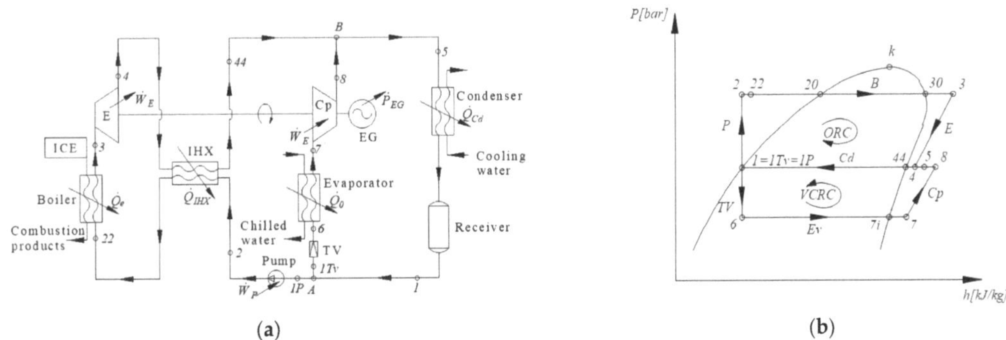

Figure 7. ORC-VCRC system containing ORC with internal regenerative heat exchanger; (**a**) scheme of the ORC-VCRC system; (**b**) the thermodynamic cycle of the ORC-VCRC system in the p–h diagram.

The area between the flue-gas curve and H axis gives the measure of the total exergy of the heat transferred from the flue gases, and the area between the composite curve of the working fluid (ORC), in the heating, evaporating, and superheating stages, and the H axis gives the measure of the exergy received by the working fluid. The areas between the two curves give the measure of the exergy destruction in each distinct system component, i.e., heater, evaporator, and superheater.

Figure 6a shows that for the simple ORC system (Figure 1), a large part of the amount of exergy introduced through the combustion gases is consumed in the boiler as follows: 28.07% in the heater, 20.96% in the evaporator, and only 2.425% in the superheater. The recorded values are due to the large temperature difference in the different areas of the boiler.

From Figure 6a, it becomes evident that the largest area between the two curves (and the most exergy destruction) is associated with heat transfer at a too-large temperature difference in the heating and evaporating stages of the working fluid. A reduction in these areas and, therefore, the amount of exergy consumed, can be achieved for the same evaporation temperature by increasing the inlet temperature of the ORC fluid in the heating zone; this can be accomplished, if the ORC fluid allows (i.e., if it is a dry type), by changing the structure of the ORC cycle by introducing an internal recuperative exchanger (Figure 7). The greatest effect for reducing the amount of exergy destruction in the boiler is the increase in the evaporation temperature to the limit allowed by the temperature difference at pinch in this device.

2.4.4. ORC-VCRC Scheme with an Internal Heat Exchanger on the ORC Circuit

Figure 7 shows the scheme of the ORC-VCRC with an internal heat exchanger on the ORC circuit.

Table 3 presents the results of the mathematical modeling based on the exergetic analysis of the operation of the ORC-VCRC cycle with an internal heat exchanger on the ORC circuit for a 100% engine load.

Table 3. Energy and exergy results obtained for the ORC-VCRC scheme with internal heat exchanger on the ORC circuit.

COP_{VCRC} (-)	COP_{ORC} (-)	$COP_{ORC-VCRC}$ (-)	η_{ex} (%)	$\dot{m}_{VCRC}$ (kg/s)	$\dot{m}_{ORC}$ (kg/s)	$\dot{m}_{ORC-VCRC}$ (kg/s)	ψ_{Cp} (%)	ψ_E (%)	$\psi_{\Delta T,B}$ (%)	$\psi_{\Delta T,h}$ (%)
6.108	0.1883	1.076	16.94	0.1237	0.08642	0.2101	5.233	6.21	47.26	21.33

$\psi_{\Delta T,oh}$ (%)	$\psi_{\Delta T,vb}$ (%)	ψ_P (%)	$\psi_{L,Cd}$ (%)	ψ_{TV} (%)	ψ_m (%)	$\dot{Q}_B$ (kW)	$\dot{Q}_{Cd}$ (kW)	$\dot{Q}_{Ev}$ (kW)	$\dot{W}_E$ (kW)	$\left.\dot{W}\right|_P$ (kW)	$\left.\dot{W}\right|_{Cp}$ (kW)
2.751	23.18	0.991	19.34	3.648	0.1194	17.05	33.38	18.34	3.21	0.2072	3.0028

By analyzing the results shown in Figure 6a compared with those shown in Figure 6b, the area between the two curves decreased when the internal heat exchanger was introduced to the ORC scheme, especially in the heating and evaporating stages of the working fluid. This decrease in the surface area leads to a decrease in the share of the exergy destruction in the ORC boiler by 4.19% and, finally, an increase in the overall exergetic efficiency by 2.03% and, implicitly, in the COP of the ORC-VCRC installation with the internal heat exchanger.

3. Exergoeconomic Optimization of the Basic Scheme of the ORC-VCRC Coupled System

3.1. Exergoeconomic Analysis: General Principles

Exergoeconomic analysis is the only investigation method and optimization procedure that takes into account the fact that any thermodynamic system interacts with the following two environments:

(a) a physical environment determined by a system of intensive parameters, such as pressure, temperature, and chemical potential;

(b) an economic environment characterized by prices of raw materials and equipment and sets of regulations to ensure sustainable development.

In defining the method for searching for optimal functional and constructive solutions, the exergoeconomic analysis is based on the union between the thermodynamics of irreversible processes and the economic analysis.

Given that energy is a conserved measure (principle I of thermodynamics) and that, in principle, it is not consumed and, therefore, cannot conceptually appear in economic balance sheets, another non-conserved measure, i.e., exergy, must be found to define the consumption of usable energy.

The exergoeconomic analysis lays the foundation for establishing a methodology for directly searching for the optimum system constructively and functionally, while offering users clear rules for improving the studied system, for which effects can be followed step by step.

The exergetic analysis method makes the connection between the system and its surrounding physical environment with which it interacts, using the concept of exergy, which quantifies the value of the use of each exergy stream, highlighting the place and size of the consumed (destroyed) exergy.

Written in an economic tone, the exergetic balance equation (Figure 8a)

$$\Sigma Ex_i = \Sigma Ex_o + \Sigma I \tag{44}$$

becomes

$$\dot{Ex}_F = \dot{Ex}_P + \Sigma \dot{Ex}_L + \Sigma \dot{I} \tag{45}$$

where

$\dot{Ex}_F$—the exergy current of the resources used (generically called fuel);

$\dot{Ex}_P$—the exergy current of the product;

$\dot{Ex}_L$—the exergy of the loss currents;

$\dot{I}$—consumption or destruction of exergy due to irreversibility (of work processes).

The exergoeconomic analysis balances the economics accounting for both the monetary flows related to the operating process and those of investments (Figure 8b).

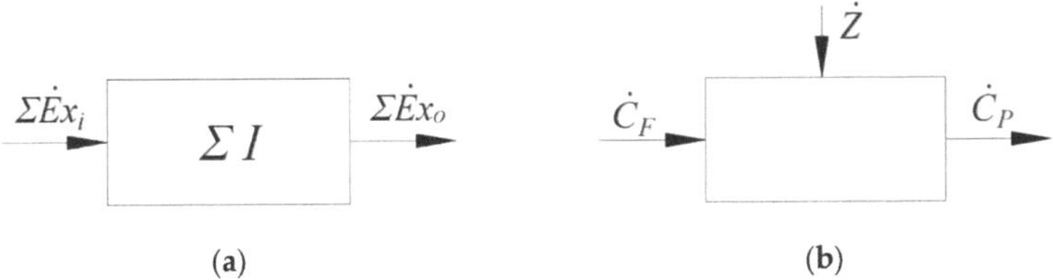

Figure 8. Balance sheets: (**a**) exergy balance for a control volume; (**b**) exergoeconomic balance.

The monetary-balance equation is of the following form:

$$\dot{C}_P = \dot{C}_F + \dot{Z} \tag{46}$$

where

$\dot{C}_P\left[\frac{EUR}{s}\right]$—the monetary cost of the product stream;

$\dot{C}_F\left[\frac{EUR}{s}\right]$—the monetary cost of the fuel flow;

$\dot{Z}\left[\frac{EUR}{s}\right]$—the amortization rate of the invested capital.

For the calculation of the amortization rate of the invested capital, a capital recovery factor, CRF, with the following formula is applied:

$$CRF = \left[1 + \frac{(1+i_{ef})^n}{(1+i_{ef})^{n-1}}\right] \tag{47}$$

where $i_{ef} = 0.05$ represents the interest (5% per year), and n = 10 represents the economic life period.

The amortization rate of the capital invested for equipment is calculated as follows:

$$\dot{Z}\left[\frac{EUR}{s}\right] = Z[EUR] \cdot \frac{CRF}{n_h \cdot 3600} \tag{48}$$

where $n_h = 7000$ h is the number of operating hours per year, and the amortization rate of the invested capital per hour is given as follows:

$$\dot{Z}_h\left[\frac{EUR}{hour}\right] = \dot{Z}\left[\frac{EUR}{s}\right] \cdot 3600[s] \tag{49}$$

If the amount of exergy losses can be reduced by sealing, isolation, or recovery, a reduction in the amount of exergy destruction, which is the essential source for increasing the performance coefficient, is achieved by increasing the investment expenses. The antagonistic evolution of the operating cost (caused by exergy destruction) and investment cost leads to the optimal solution given by the minimization of the total cost as the sum of the operating and investment costs.

The exergoeconomic optimization method aims to identify the exergy destruction of each functional area and assign its monetary cost.

Areas with high exergy-destruction costs will be the first targets of the optimization procedures, and the effect of each local cost reduction will be verified at the global level.

To quantify the value of a zonal destruction, the fuel, product, loss, and exergetic performance coefficients will be evaluated for each operational area.

We will proceed with the search for the functionally and constructively optimal model for the ORC-VCRC combined system in which the power Rankine cycle is simple and does not have an internal recuperative heat exchanger.

The scheme of the VCRC installation operated with the ORC system shown in Figure 9 was divided into the following seven operating areas: (1) ORC Evaporator (Boiler);

(2) Expander; (3) Compressor; (4) Condenser; (5) Throttling Valve; (6) VCRC Evaporator; (7) ORC Liquid Pump, to which node A for splitting and node B for mixing of substance currents are added.

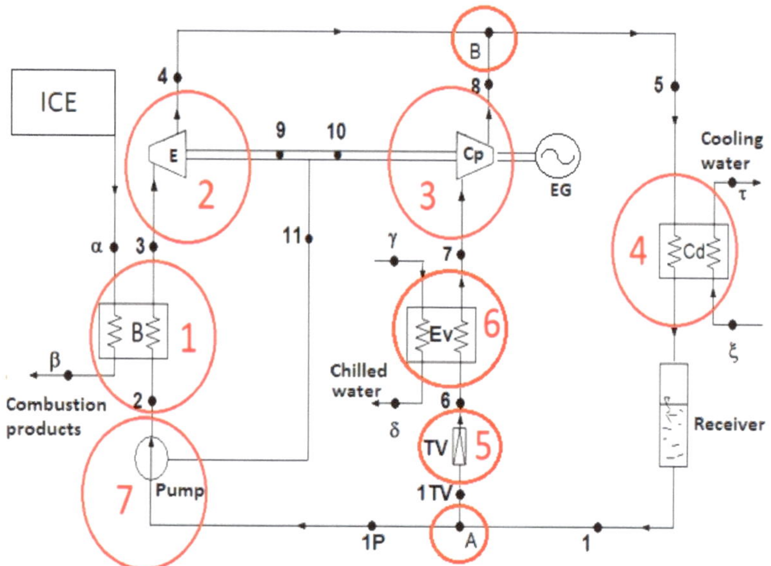

Figure 9. The ORC-VCRC system divided into functional areas.

The exergetic concepts of Fuel, Product, Exergy Loss, and Destruction for each functional area are presented in Table 4 [26].

Table 4. The concepts of Fuel, Product, Exergy Loss, and Destruction for functional areas.

Equipment	Fuel	Product	Exergy Loss	Destruction
Boiler	$\dot{E}x_\alpha$	$\dot{E}x_3 - \dot{E}x_2$	$\dot{E}x_\beta$	$\dot{E}x_\alpha - \dot{E}x_3 + \dot{E}x_2 - \dot{E}x_\beta$
Expander	$\dot{E}x_3 - \dot{E}x_4$	$\left|\dot{W}_D\right|$		$\dot{E}x_3 - \dot{E}x_4 - \dot{W}_D$
Compressor	$\left|\dot{W}_{Cp}\right|$	$\dot{E}x_8 - \dot{E}x_7$		$\left|\dot{W}_{Cp}\right| - \dot{E}x_8 + \dot{E}x_7$
Condenser	$\dot{E}x_5 - \dot{E}x_1$		$\dot{E}x_\tau - \dot{E}x_\xi$	$\dot{E}x_5 - \dot{E}x_1 - \dot{E}x_\tau + \dot{E}x_\xi$
Throttling Valve	$\dot{E}x_{1TV}$	$\dot{E}x_6$		$\dot{E}x_{1TV} - \dot{E}x_6$
Evaporator	$\dot{E}x_6 - \dot{E}x_7$	$\dot{E}x_\delta - \dot{E}x_\gamma$		$\dot{E}x_6 - \dot{E}x_7 - \dot{E}x_\delta + \dot{E}x_\gamma$
ORC Pump	$\left|\dot{W}_P\right|$	$\dot{E}x_2 - \dot{E}x_{1P}$		$\left|\dot{W}_P\right| - \dot{E}x_2 + \dot{E}x_{1P}$

Observing from relationship (46) that the destruction, I, of a zonal exergy is a part of the exergy of the fuel, it is logical to assign it as the unitary monetary cost, which is the unitary cost of the fuel in that specific area.

3.2. Exergoeconomic Monetary Cost Assessment

Each internal unitary monetary cost represents a part of the overall system's fuel cost and capital investment. These costs are determined by the system's interaction with the market.

A methodology must be found to allocate these costs imposed by the market to each internal current.

There are 19 currents: 1–11, 1P, 1TV, α, β, γ, δ, τ, and ξ, for which 19 relationships are needed between them.

Nine monetary-balance relationships of the Equation (46) type can be written for the seven established functional areas and for the two nodes A and B.

The rest of the equations, up to 19, will represent auxiliary relationships between the monetary costs of the various exergy currents.

The auxiliary equations are built based on some rules that can be stated based on the following principles of economic common sense [26]:

Rule 1 → For an unconsumed fuel stream exiting a subsystem, the monetary cost per unit of exergy is equal to the unit cost of the fuel stream feeding the subsystem;

Rule 2 → If a subsystem has several products, the monetary cost per unit of exergy of each product is the same;

Rule 3 → If a component of a product has several output currents, each is assigned the same unit monetary cost;

Rule 4 → In the absence of an external evaluation, the loss streams are assigned a zero unitary monetary cost;

Rule 5 → In the absence of an external assessment, the unitary exergy costs of the input currents in the global system are equal to their exergies.

A mathematical model based on the exergoeconomic analysis is written for each functional zone (Figure 9) as follows:

- Zone 1. The ORC evaporator (boiler)

The exergoeconomic monetary-balance cost (Equation (46)) of the ORC boiler is as follows:

$$c_\alpha \cdot \dot{E}x_\alpha + c_2 \cdot \dot{E}x_2 - c_\beta \cdot \dot{E}x_\beta - c_3 \cdot \dot{E}x_3 + \dot{Z}_B = 0 \tag{50}$$

The exergy of the combustion gases, $\dot{E}x_\alpha$, is calculated as the sum of their thermomechanical and chemical exergies, at a certain operating mode of the engine, compared to the standard composition of the ambient environment.

Because combustion gases are the unconsumed part of diesel fuel (internal combustion engine fuel), they will be assigned the diesel fuel unit exergetic monetary cost (*Rule 1*). It is considered that the mass exergy of diesel is $\dot{E}x_{CHF} = 41{,}800$ (kJ/kg) and has a cost of 1.7 (EUR/kg) as follows:

$$c_F = c_\alpha = 4.07 \cdot 10^{-5} \left[\frac{EUR}{kJ} \right] \tag{51}$$

Because the exergy of current β (Figure 9) is a loss (Table 4), according to *Rule 4*,

$$c_\beta = 0 \tag{52}$$

- Zone 2. The expander

The exergoeconomic monetary-balance cost (Equation (46)) of the expander is as follows:

$$c_3 \cdot \dot{E}x_3 - c_9 \cdot \dot{W}_E - c_4 \cdot \dot{E}x_4 + \dot{Z}_E = 0 \tag{53}$$

Because current 4 represents an unconsumed part of fuel 3, as per *Rule 1*,

$$c_3 = c_4 \tag{54}$$

The unit monetary costs of currents 9–11 are equal because they represent parts of the same product (*Rule 3*) as follows:

$$c_{10} = c_9 \tag{55}$$

$$c_{11} = c_9 \tag{56}$$

- Zone 3. The compressor

The exergoeconomic monetary-balance cost (Equation (46)) of the compressor is as follows:

$$c_7 \cdot \dot{Ex}_7 + c_{10} \cdot \left| \dot{W}_{Cp} \right| - c_8 \cdot \dot{Ex}_8 + \dot{Z}_{Cp} = 0 \tag{57}$$

- Zone B. The mixing point

At the mixing point, B, the following monetary-balance equation can be written:

$$c_4 \cdot \dot{Ex}_4 + c_8 \cdot \dot{Ex}_8 = c_5 \cdot \dot{Ex}_5 \tag{58}$$

- Zone 4. The condenser

The condenser of the ORC-VCRC system is a dissipative zone and has no exergetic product (Table 4).

Its monetary-balance equation, based on exergoeconomic criteria, is as follows:

$$c_5 \cdot \dot{Ex}_5 + c_\xi \cdot \dot{Ex}_\xi - c_1 \cdot \dot{Ex}_1 - c_\tau \cdot \dot{Ex}_\tau + \dot{Z}_{Cd} = 0 \tag{59}$$

Because the exergy of the cooling water of the condenser that is not being used represents a loss, according to *Rule 4*,

$$c_\xi = 0 \tag{60}$$

$$c_\tau = 0 \tag{61}$$

- Zone A. The splitting point

The following monetary-balance equation can be written for the splitting point, A:

$$c_1 \cdot \dot{Ex}_1 + c_{1TV} \cdot \dot{Ex}_{1TV} - c_{1P} \cdot \dot{Ex}_{1P} = 0 \tag{62}$$

where the specific monetary cost of currents 1TV and 1P is the same as those representing multiple products (*Rule 2*) as follows:

$$c_{1TV} = c_{1P} \tag{63}$$

- Zone 5. The throttling valve

The exergoeconomic monetary-balance cost (Equation (46)) of the throttling valve is as follows:

$$c_{1TV} \cdot \dot{Ex}_{1TV} - c_6 \cdot \dot{Ex}_6 + \dot{Z}_{TV} = 0 \tag{64}$$

- Zone 6. The VCRC evaporator

The exergoeconomic monetary-balance cost (Equation (46)) of the evaporator is as follows:

$$c_6 \cdot \dot{Ex}_6 + c_\gamma \cdot \dot{Ex}_\gamma - c_7 \cdot \dot{Ex}_7 - c_\delta \cdot \dot{Ex}_\delta + \dot{Z}_{Ev} = 0 \tag{65}$$

The exergy current, 7, is the unconsumed part of the fuel, 6, that feeds the evaporator. According to *Rule 1*, the two currents have the same monetary unit cost as follows:

$$c_7 = c_6 \tag{66}$$

According to *Rule 5*, a value is assigned to current γ as follows:

$$c_\gamma = 1 \tag{67}$$

It is observed that the value of the unitary monetary cost of the chilled water current, γ, is not important in the calculation because the state, γ, is taken as a reference for the calculation of the exergy of the chilled water ($\dot{Ex}_\gamma = 0$).

- Zone 7. The ORC liquid pump

  For the ORC liquid pump, the monetary-cost balance equation is as follows:

$$c_{1P}\cdot \dot{E}x_{1P} + c_{11}\cdot \left|\dot{W}_P\right| - c_2\cdot \dot{E}x_2 + \dot{Z}_P = 0 \tag{68}$$

Equations (50)–(68) represent the 19 equations of the mathematical model for calculating the 19 monetary-unit costs.

3.3. Exergoeconomic Cost Correlations of Component Equipment

Unlike thermoeconomic cost correlations, in which the purchase cost of equipment is estimated as a function of the amount of material resources used, exergoeconomic correlations specify the cost according to the exergetic performance coefficient (or defined based on the second law of thermodynamics) or the decision-making variables that define the exergetic performance (such as exergy destruction) and function of the size of the exergetic product of the equipment [27–32].

(a)   VCRC compressor

The exergoeconomic cost correlation is of the following form [27]:

$$Z_{Cp} = 71\frac{\dot{m}_{VCRC}}{0.9 - \eta_{sCp}}\cdot r_{Cp}\cdot \ln r_{Cp}\,[EUR] \tag{69}$$

where the performance coefficient of the equipment defined based on the second law of thermodynamics is the isentropic efficiency, η_{sCp}, of the compression process, and the exergetic product is as follows:

$$P_{ex})_{Cp} = \dot{m}_{VCRC}\cdot r_{Cp}\cdot \ln r_{Cp}\left[\frac{kg}{s}\right] \tag{70}$$

which indicates the antagonistic effect of the variation in the compressed-gas flow rate $\dot{m}_{VCRC}$ with the compression ratio, $r_{Cp} = \frac{P_c}{P_v} = \frac{P_8}{P_7}$ (Figure 9).

The investment cost, $Z_{Cp}[EUR]$, has been updated to the level of 2020.

(b)   ORC expander

The exergoeconomic correlation for the acquisition cost of the expander (Zone 2, Figure 9) takes into account the isentropic efficiency of the expansion, η_{sE}, as a measure of the perfection of the equipment through the prism of the second law of thermodynamics. The cost of the expander is proportional to the exergetic product as a measure of the mechanical work produced by expansion as follows [27]:

$$P_{ex})_D = \dot{m}_{ORC}\cdot \ln \frac{P_3}{P_4}\left[\frac{kg}{s}\right] \tag{71}$$

These notations correspond to those in Figure 9, which shows the technological scheme of the installation.

With these considerations and accounting for the cost update at the level of 2020, the exergoeconomic cost correlation for the expander is as follows [31]:

$$Z_E = 479\frac{\dot{m}_{ORC}}{0.92 - \eta_{sE}}\cdot \ln \frac{P_3}{P_4}\,[EUR] \tag{72}$$

(c)   ORC pump

For the pump, a form of exergoeconomic correlation like that used for the compressor is chosen as follows [28]:

$$Z_P = 10.38 \frac{\dot{m}_{ORC}}{0.9 - \eta_{sP}} r_P \cdot \ln r_P \, [\text{EUR}] \tag{73}$$

in which the performance of the equipment in relation to the second law of thermodynamics is described by the isentropic compression efficiency, η_{sP}, and the exergetic product correlates the pumped flow rate, $\dot{m}_{ORC}$, with the pressure increase ratio, $r_P = \frac{p_2}{p_1}$ (Figure 9).

(d) Heat exchangers

In the case of the heat exchangers, to specify the cost of the purchase, the starting point is the following thermoeconomic correlation:

$$Z = c \cdot A \, [\text{EUR}] \tag{74}$$

where $c \, [\text{EUR}/\text{m}^2]$ represents the cost of the heat-exchange surface unit; $A \, [\text{m}^2]$, its surface.

To highlight the decision-making parameter that imposes the exergetic performance of the device, the heat-exchange surface is explained according to the minimum temperature difference in the device. The minimum temperature difference, ΔT_{min}, determines the amount of exergy destruction in the device. This decision-making parameter balances the operating expense (exergy consumption) and acquisition cost (heat-exchange surface) of the device as follows:

$$A = \frac{\dot{Q}}{U \cdot \Delta T_m} = \frac{\dot{m} \cdot \Delta h}{U \cdot \Delta T_m} \tag{75}$$

where $U \, [\text{W}/(\text{m}^2 \cdot \text{K})]$ represents the global heat-exchange coefficient of the device, and ΔT_m is the average logarithmic temperature difference in the device.

d.1 The refrigeration unit evaporator

The temperature variation is plotted as a function of the heat-transfer surface in Figure 2c as follows:

$$\Delta T_{min}^V = t_8 - t_7 \tag{76}$$

$$\Delta T_m^V = \frac{t_\gamma - t_\delta}{\ln \frac{t_\gamma - t_\nu}{t_\delta - t_\nu}} = \frac{t_\gamma - t_\delta}{\ln \frac{t_\gamma - (t_\delta - \Delta T_{min}^V)}{\Delta T_{min}^V}} = \frac{t_\gamma - t_\delta}{\ln \frac{t_\gamma - t_\delta + \Delta T_{min}^V}{\Delta T_{min}^V}} \tag{77}$$

Considering the temperature drop on the cooled fluid side as imposed design data,

$$t_\gamma - t_\delta = a \tag{78}$$

relationship (77) can be written as follows:

$$\Delta T_m^V = \frac{a}{\ln \frac{a + \Delta T_{min}^V}{\Delta T_{min}^V}} \tag{79}$$

Taking (79) into account, Equation (74) is as follows:

$$Z_V = c_V \frac{\dot{Q}_V \ln \frac{a + \Delta T_{min}^V}{\Delta T_{min}^V}}{a \cdot U_V} \tag{80}$$

Considering $c_V = 87 \, [\text{EUR}/\text{m}^2]$ and $U_V = 500 \, [\text{W}/(\text{m}^2 \cdot \text{K})]$ and choosing $a = 16$ K, Equation (80) can be written as follows:

$$Z_V[\text{EUR}] = 10.87 \cdot \dot{Q}_V[\text{kW}] \ln \left(16 \cdot \left(\Delta T_{min}^V \right)^{-1} + 1 \right) \tag{81}$$

d.2 Condenser

By applying to the condenser the same scheme used in the case of the VCRC evaporator, the following exergoeconomic correlation is obtained:

$$Z_{Cd} = c_{cd} \frac{\dot{Q}_{cd} \ln \frac{b + \Delta T_{min}^{Cd}}{\Delta T_{min}^{Cd}}}{b \cdot U_{Cd}} \tag{82}$$

where from Figure 2b

$$b = t_\tau - t_\xi \tag{83}$$

is imposed by the design and

$$\Delta T_{min}^{Cd} = t_c - t_\tau \tag{84}$$

Considering that $c_{cd} = 254 \left[EUR/m^2 \right]$, $U_{cd} = 1750 \left[W/(m^2 \cdot K) \right]$, and b = 7 K, Equation (82) can be written as follows:

$$Z_{Cd}[EUR] = 20.73 \cdot \dot{Q}_{Cd}[kW] \ln \left(7 \cdot \left(\Delta T_{min}^{Cd} \right)^{-1} + 1 \right) \tag{85}$$

d.3 ORC boiler

In the case of the ORC boiler, the exergoeconomic correlation for calculating the purchase cost is of the following form:

$$Z_B[EUR] = c_B \left[\frac{EUR}{m^2} \right] \cdot A_B \left[m^2 \right] \tag{86}$$

where the heat-transfer surface is the sum of the surfaces corresponding to the processes of heating the subcooled liquid (A_h), boiling (A_{vb}), and overheating (A_{oh}) as follows:

$$A_B = A_h + A_{vb} + A_{oh} \tag{87}$$

The heat-exchange surfaces of each functional area of the boiler are calculated as follows:

- Liquid heating area

$$A_h = \frac{\dot{Q}_h}{U_h \cdot \Delta T_m^h} \tag{88}$$

where from Figure 2a

$$\dot{Q}_h = \dot{m}_{ORC}(h_{20} - h_2) \tag{89}$$

$U_h = 70 \left[W/(m^2 \cdot K) \right]$ is the global heat-transfer coefficient for the liquid heating process as follows:

$$\Delta T_h = \frac{(t_\beta - t_2) - \left(t_{h,g} - t_{20} \right)}{\ln \frac{(t_\beta - t_2)}{\left(t_{h,g} - t_{20} \right)}} \tag{90}$$

The temperature, $t_{h,g}$, on the heating-gas side results from the energy balance of the liquid-heating zone (Figure 2a) as follows:

$$t_{h,g} = t_\beta + \frac{\dot{m}_{ORC}(h_{20} - h_2)}{\dot{m}_g \cdot c_g} \tag{91}$$

where $t_\beta = 140\ °C$.

- Boiling area

$$A_{vb} = \frac{\dot{Q}_{vb}}{U_{vb} \cdot \Delta T_m^{vb}} = \frac{\dot{m}_{ORC}(h_{30} - h_{20})}{U_{vb} \cdot \Delta T_m^{vb}} \tag{92}$$

where $U_{vb} = 90 \left[W/\left(m^2 \cdot K\right)\right]$ is the global heat-transfer coefficient for the boiling process as follows:

$$\Delta T_{vb} = \frac{\left(t_{vb,g} - t_{h,g}\right)}{\ln \frac{\left(t_{vb,g} - t_V\right)}{\left(t_{h,g} - t_V\right)}} \tag{93}$$

The temperature, $t_{vb,g}$, on the heating-gas side results from the energy balance of the boiling zone as follows:

$$t_{vb,g} = t_{h,g} + \frac{\dot{Q}_{vb}}{\dot{m}_g \cdot c_g} \tag{94}$$

- Overheating area

$$A_h = \frac{\dot{Q}_{oh}}{U_{oh} \cdot \Delta T_m^{oh}} = \frac{\dot{m}_{ORC}(t_3 - t_V)}{U_{oh} \cdot \Delta T_m^{oh}} \tag{95}$$

where $U_{oh} = 50 \left[W/\left(m^2 \cdot K\right)\right]$ is the global heat-transfer coefficient for the overheating process as follows:

$$\Delta T_{oh} = \frac{(t_\alpha - t_3) - \left(t_{vb,g} - t_{30}\right)}{\ln \frac{(t_\alpha - t_3)}{\left(t_{vb,g} - t_{30}\right)}} \tag{96}$$

where $t_\alpha = 420\,°C$.

3.4. Exergoeconomic Performance Indicators

3.4.1. Monetary Cost of Zonal Exergy Destruction

To calculate the monetary cost of the exergy destruction in each functional area, the value of the zonal exergy destruction is multiplied by the unitary monetary cost of the local zonal fuel (Table 5).

Table 5. Unit cost of zonal fuel and monetary cost rate of zonal exergy destruction.

Equipment	Zonal Fuel Unit Cost $\left[\frac{EUR}{kJex}\right]$	Monetary Cost Rate of Zonal Exergy Destruction, $\dot{C}_I \left[\frac{EUR}{s}\right]$
Boiler	c_{gMAI}	$c_{gMAI}\left(\dot{Ex}_\alpha - \dot{Ex}_3 + \dot{Ex}_2 - \dot{Ex}_\beta\right)$
Expander	c_3	$c_3\left(\dot{Ex}_3 - \dot{Ex}_4 - \dot{W}_D\right)$
Compressor	c_{10}	$c_{10}\left(\left|\dot{W}_{Cp}\right| - \dot{Ex}_8 - \dot{Ex}_7\right)$
Condenser	c_5	$c_5\left(\dot{Ex}_5 - \dot{Ex}_1 - \dot{Ex}_\tau + \dot{Ex}_\xi\right)$
Throttling Valve	c_1	$c_1\left(\dot{Ex}_A - \dot{Ex}_6\right)$
Evaporator	c_6	$c_6\left(\dot{Ex}_6 - \dot{Ex}_7 - \dot{Ex}_\delta + \dot{Ex}_\gamma\right)$
ORC Pump	c_{11}	$c_{11}\left(\left|\dot{W}_P\right| + \dot{Ex}_{1p} - \dot{Ex}_2\right)$

3.4.2. Exergoeconomic Factor

To appreciate how the investment expense saves the monetary cost of the zonal exergy destruction, the exergoeconomic factor [31] is calculated for each piece of equipment, where the index, k, represents the piece of equipment as follows:

$$f_k = \frac{\dot{Z}_k}{\dot{Z}_k + \dot{C}_{I,k}} < 1 \tag{97}$$

If f_k approaches 1, the operational cost (of exergy destruction, $\dot{C}_{I,k}$) is too low and the investment cost, $\dot{Z}_k$, is too high. In this case, the exergetic performance of the device must be relaxed, and a higher exergy destruction should be accepted; the total cost, consisting of the operating and investment costs, will decrease.

If f_k is too low, the monetary value of the exergy destruction is too high, and the purchase of better equipment, which is obviously more expensive, is recommended.

3.4.3. Zonal Coefficients of Performance

(a) Exergetic Efficiency

For each functional area, the product and fuel are specified (Table 5). Their ratio gives the zonal exergetic efficiency [31] as follows:

$$\eta_{ex,k} = \left(\frac{P}{F}\right)_k 100 \tag{98}$$

For the global ORC-CVRC system, the product is the heat exergy extracted from the chilled water in the VCRC evaporator (evaporator product (Table 4)), and the fuel is the exergy of the combustion gases, $\dot{Ex}_\alpha$; therefore, Equation (98) can be written as follows:

$$\eta_{ex} = \frac{P_{Ev}}{\dot{Ex}_\alpha} 100 \tag{99}$$

(b) The relative weight of the exergy destruction in the fuel consumption of the global ORC-VCRC system

The weight percentage of the zonal exergy destruction in the fuel consumption of the global system is calculated using the following relationship:

$$\psi_k = \frac{\dot{I}_k}{\dot{Ex}_\alpha} 100 \tag{100}$$

The search procedure for the optimal functional and constructive ORC-VCRC system will be guided by the values of the exergoeconomic performance indicators.

3.5. Results of Exergoeconomic Optimization

The following tables show the simulation results for the working fluid, R1224yd(Z), depending on the variation in several decision-making parameters, including the minimum temperature difference in the condenser (ΔT_{minCd}), the minimum temperature difference in the evaporator (ΔT_{minV}), the minimum temperature difference in the boiler (ΔT_{minB}), as well as the variation in the efficiencies of the compressor (η_{sCp}) or the expander (η_{sE}). The unit cost of diesel was considered, $c_{Diesel} = 1.7\ [EUR/l]$.

For $\Delta T_{minCd} = 8$ K, $\Delta T_{minV} = 8$ K, $\Delta T_{minB} = 30$ K, $\eta_{sCp} = 0.8$, $\eta_{sP} = 0.8$ and $\eta_{sE} = 0.8$, the results of the simulation are given in Table 6.

Table 6. The results obtained for $\Delta T_{minCd} = 8$ K; $\Delta T_{minV} = 8$ K; $\Delta T_{minB} = 30$ K; $\eta_{sCp} = 0.8$; $\eta_{sP} = 0.8$; $\eta_{sE} = 0.8$.

Zone	η_{ex} (%)	$\dot{I}$ (kW)	Ψ (%)	$\dot{Z}$ (EUR/h)	CI (EUR/kJ)	$\dot{CI} = CI \cdot \dot{I}$ (EUR/h)	$\dot{CI} + \dot{Z}$ (EUR/h)	f
Boiler	31.43	4.211	37.17	0.01067	$4.07 \cdot 10^{-5}$	0.617	0.6277	0.01701
Expander	82.25	0.4939	4.359	0.01204	$2.021 \cdot 10^{-4}$	0.3593	0.3713	0.03242
Compressor	81.65	0.4035	3.561	0.005774	$2.471 \cdot 10^{-4}$	0.359	0.3648	0.01583
Condenser	-	1.2	13.63	0.00695	$2.535 \cdot 10^{-4}$	1.095	1.102	0.006306
VCRC Evaporator	41.37	0.5094	4.496	0.002576	$5.432 \cdot 10^{-4}$	0.9962	0.9988	0.002579
ORC Pump	81.03	0.01719	0.1517	0.002539	$2.471 \cdot 10^{-4}$	0.0153	0.01783	0.1423

For the above-mentioned decisional variables, the cost of one unit of exergy of the product, c_R [EUR/kJ]] (refrigerating power of the VCRC evaporator) is as follows:

$$c_R = 13.15 \cdot 10^{-4} [\text{EUR/kJ}]$$

Examining Table 6, one observes that the lowest exergetic efficiency, η_{ex}, and the highest relative destruction, ψ, are in the boiler. The low exergoeconomic factor, f, requires the choice of a larger heat-exchange surface (a smaller minimum temperature difference) to reduce the amount of exergy destruction.

Dropping the minimum temperature difference in the boiler to $\Delta T_{minB} = 20$ K, the unitary cost of the refrigerating power becomes, $c_R = 12.35 \cdot 10^{-4}$ [EUR/kJ], and further, for $\Delta T_{minB} = 10$ K, this cost decreases to $c_R = 11.73 \cdot 10^{-4}$ [EUR/kJ].

The same Table 6, shows that the VCRC evaporator has the lowest exergoeconomic factor, f = 0.002579, which indicates a high cost of exergy destruction compared to the cost for amortizing the invested capital. The suggestion is to reduce the temperature difference and, thus, increase the heat-exchange surface area of the evaporator. The minimum temperature difference in the evaporator was reduced to $\Delta T_{minV} = 3$ K.

The results of the simulation for $\Delta T_{minV} = 3$ K, are shown in Table 7.

Table 7. The results obtained for $\Delta T_{minCd} = 8$ K; $\Delta T_{minB} = 10$ K; $\eta_{sCp} = 0.8$; $\eta_{sP} = 0.8$; $\eta_{sE} = 0.8$ when $\Delta T_{minV} = 3$ K is imposed.

Zone	ηex (%)	I (kW)	Ψ (%)	$\dot{Z}$ (EUR/h)	CI (EUR/kJ)	$\dot{CI} = CI \cdot \dot{I}$ (EUR/h)	$\dot{CI} + \dot{Z}$ (EUR/h)	f
Boiler	34.59	3.852	34	0.01137	$4.07 \cdot 10^{-5}$	0.5645	0.5758	0.01975
Expander	82.43	0.5559	4.906	0.01372	$1.723 \cdot 10^{-4}$	0.3448	0.3585	0.03826
Compressor	81.61	0.4547	4.013	0.005171	$2.105 \cdot 10^{-4}$	0.3446	0.3498	0.01479
Condenser	-	1.376	15.64	0.007987	$2.228 \cdot 10^{-4}$	1.103	1.111	0.00718
VCRC Evaporator	56.41	0.3789	3.344	0.005873	$5.443 \cdot 10^{-4}$	0.7425	0.7484	0.00784
ORC Pump	81.05	0.02545	0.2246	0.004271	$2.105 \cdot 10^{-4}$	0.01928	0.02356	0.1813

After reducing the minimum temperature difference in the evaporator to $\Delta T_{minV} = 3$ K, the cost of the product exergy unit became $c_R = 9.681 \cdot 10^{-4}$ [EUR/kJ].

Continuing the optimum search, one observes from Table 7, that now, the condenser has the lowest exergoeconomic factor, f = 0.00718. The conclusion is that the amount of exergy destruction, in the condenser, must be reduced.

Although the condenser is an eminently dissipative area with the role of transferring to the environment the energy generated and transferred in the system and, therefore, does not have an exergetic product, by specifying the condensation temperature, the condenser has an important role in establishing the operation efficiency.

In the aim of decreasing the exergy destruction in the condenser, the minimum temperature difference in this heat exchanger was reduced to $\Delta T_{minCd} = 3$ K. The results of the simulation, in this last case, are presented in Table 8.

Table 8. The results obtained for $\Delta T_{minV} = 3$ K; $\Delta T_{minB} = 10$ K; $\eta_{sCp} = 0.8$; $\eta_{sP} = 0.8$; $\eta_{sE} = 0.8$ when $\Delta T_{minCd} = 3$ K is imposed.

Zone	ηex (%)	I (kW)	Ψ (%)	$\dot{Z}$ (EUR/h)	CI (EUR/kJ)	$\dot{CI} = CI \cdot \dot{I}$ (EUR/h)	$\dot{CI} + \dot{Z}$ (EUR/h)	f
Boiler	33.91	3.93	34.68	0.01117	$4.07 \cdot 10^{-5}$	0.5758	0.5869	0.01904
Expander	82.26	0.5872	5.182	0.01432	$1.615 \cdot 10^{-4}$	0.3415	0.3558	0.04024
Compressor	81.33	0.4837	4.269	0.004574	$1.979 \cdot 10^{-4}$	0.3445	0.3491	0.0131
Condenser	-	0.9541	12.4	0.01742	$2.147 \cdot 10^{-4}$	0.7374	0.7548	0.02307
VCRC Evaporator	56.38	0.4878	4.305	0.007549	$4.258 \cdot 10^{-4}$	0.7476	0.7552	0.009996
ORC Pump	80.73	0.02535	0.2237	0.0053	$1.979 \cdot 10^{-4}$	0.01805	0.02335	0.2269

The cost of one product exergy unit becomes $c_R = 7.585 \cdot 10^{-4}$ [EUR/kJ].

The cost of the exergy destruction in the pump, as well as the cost for amortizing the purchase of the pump, are the lowest in the system (Table 8).

The highest value of the exergoeconomic factor of the pump area indicates the possibility for reducing the quality of the pump by reducing its isentropic efficiency, η_{sP}; but the effect of this reduction on the thermodynamic parameters after pumping does not decrease the unit cost of the global-system product represented by the cold obtained in the evaporator.

Table 8 shows that the compressor has a low acquisition cost, $\dot{Z}$, but induces a high cost of exergy destruction, $\dot{CI}$, which suggests the choice of a more expensive compressor with a higher isentropic compression efficiency, η_{sCp}. The results of the simulation when the isentropic efficiency of the compressor is increased to $\eta_{sCp} = 0.85$, are shown in Table 9.

Table 9. The results obtained for ΔT_{minCd} = 3 K; ΔT_{minV} = 3 K; ΔT_{minB} = 10 K; η_{sP} = 0.8; η_{sE} = 0.8 when η_{sCp} = 0.85 is imposed.

Zone	ηex (%)	I (kW)	Ψ (%)	$\dot{Z}$ (EUR/h)	CI (EUR/kJ)	$\dot{CI} = CI \cdot \dot{I}$ (EUR/h)	$\dot{CI} + \dot{Z}$ (EUR/h)	f
Boiler	33.91	3.93	34.68	0.01117	$4.07 \cdot 10^{-5}$	0.5758	0.5869	0.01904
Expander	82.26	0.5872	8.182	0.01432	$1.583 \cdot 10^{-4}$	0.3347	0.349	0.04102
Compressor	85.97	0.3636	3.209	0.009721	$1.939 \cdot 10^{-4}$	0.2538	0.2635	0.03688
Condenser	-	0.972	12.7	0.01801	$2.04 \cdot 10^{-4}$	0.7139	0.7319	0.02461
VCRC Evaporator	56.38	0.5183	4.574	0.008021	$4.021 \cdot 10^{-4}$	0.7501	0.7582	0.01058
ORC Pump	80.73	0.02535	0.2237	0.0053	$1.939 \cdot 10^{-4}$	0.0177	0.023	0.2305

The cost of the product exergy unit is $c_R = 7.165 \cdot 10^{-4}$ [EUR/kJ]; and if one considers η_{sCp} = 0.89, the following results are obtained (Table 10).

Table 10. The results obtained for ΔT_{minCd} = 3 K; ΔT_{minV} = 3 K; ΔT_{minB} = 10 K; η_{sP} = 0.8; η_{sE} = 0.8 when η_{sCp} = 0.89 is imposed.

Zone	ηex (%)	I (kW)	Ψ (%)	$\dot{Z}$ (EUR/h)	CI (EUR/kJ)	$\dot{CI} = CI \cdot \dot{I}$ (EUR/h)	$\dot{CI} + \dot{Z}$ (EUR/h)	f
Boiler	33.91	3.93	34.68	0.01117	$4.07 \cdot 10^{-5}$	0.5758	0.5869	0.01904
Expander	82.26	0.5872	5.182	0.01432	$1.571 \cdot 10^{-4}$	0.3322	0.3465	0.04132
Compressor	89.69	0.2761	2.357	0.05089	$1.925 \cdot 10^{-4}$	0.185	0.2359	0.2157
Condenser	-	0.9868	12.94	0.01849	$2.005 \cdot 10^{-4}$	0.7123	0.7308	0.0253
VCRC Evaporator	56.38	0.5426	4.789	0.008398	$3.933 \cdot 10^{-4}$	0.7683	0.7767	0.01081
ORC Pump	80.73	0.02535	0.2237	0.0053	$1.925 \cdot 10^{-4}$	0.01756	0.02286	0.2318

For the decisional variables of Table 10, the cost of one unit of product exergy is $c_R = 7.009 \cdot 10^{-4}$ [EUR/kJ].

In the area of the expander, the cost of the exergy destruction is high, which indicates the possibility for reducing the exergy-destruction expense by choosing a more efficient expander. The results of the simulation when the isentropic efficiency is increased to η_{sE} = 0.85, are given in Table 11.

In this case, one unit of product exergy costs $c_R = 6.632 \cdot 10^{-4}$ [EUR/kJ].

With a higher-performance expander (η_{sE} = 0.9), one obtains the results listed in Table 12.

Under these conditions exergy cost of the product unit becomes $c_R = 6.478 \cdot 10^{-4}$ [EUR/kJ]. The further increase in the isentropic efficiency of the expander is tempered by the expander's cost which became higher than, for example the cost of the compressor (Table 12).

Table 11. The results obtained for $\Delta T_{minCd} = 3$ K; $\Delta T_{minV} = 3$ K; $\Delta T_{minB} = 10$ K; $\eta_{sCp} = 0.89$; $\eta_{sP} = 0.8$ when $\eta_{sE} = 0.85$ is imposed.

Zone	ηex (%)	I (kW)	Ψ (%)	$\dot{Z}$ (EUR/h)	CI (EUR/kJ)	$\dot{CI} = CI \cdot \dot{I}$ (EUR/h)	$\dot{CI} + \dot{Z}$ (EUR/h)	f
Boiler	33.91	3.93	34.68	0.01117	$4.07 \cdot 10^{-5}$	0.5758	0.5869	0.01904
Expander	86.74	0.4422	3.903	0.02454	$1.54 \cdot 10^{-4}$	0.2451	0.2697	0.09101
Compressor	89.69	0.2846	2.512	0.05423	$1.799 \cdot 10^{-4}$	0.1843	0.2385	0.2274
Condenser		1.009	13.3	0.01919	$1.908 \cdot 10^{-4}$	0.6934	0.7125	0.02693
VCRC Evaporator	56.38	0.5783	5.104	0.008949	$3.72 \cdot 10^{-4}$	0.7745	0.7835	0.01142
ORC Pump	80.73	0.02535	0.2237	0.0053	$1.799 \cdot 10^{-4}$	0.01641	0.02171	0.2441

Table 12. The results obtained for $\Delta T_{minCd} = 3$ K; $\Delta T_{minV} = 3$ K; $\Delta T_{minB} = 10$ K; $\eta_{sCp} = 0.89$; $\eta_{sP} = 0.8$ when $\eta_{sE} = 0.9$ is imposed.

Zone	ηex (%)	I (kW)	Ψ (%)	$\dot{Z}$ (EUR/h)	CI (EUR/kJ)	$\dot{CI} = CI \cdot \dot{I}$ (EUR/h)	$\dot{CI} + \dot{Z}$ (EUR/h)	f
Boiler	33.91	3.93	34.68	0.01117	$4.07 \cdot 10^{-5}$	0.5758	0.5869	0.01904
Expander	91.19	0.2961	2.613	0.0859	$1.527 \cdot 10^{-4}$	0.1627	0.2486	0.3455
Compressor	89.69	0.3021	2.667	0.05757	$1.753 \cdot 10^{-4}$	0.1906	0.2482	0.232
Condenser	-	1.033	13.66	0.01989	$1.874 \cdot 10^{-4}$	0.6967	0.7166	0.02775
VCRC Evaporator	56.38	0.3109	5.418	0.009501	$3.634 \cdot 10^{-4}$	0.8031	0.8126	0.01169
ORC Pump	80.73	0.02535	0.2237	0.0053	$1.753 \cdot 10^{-4}$	0.01599	0.02129	0.2489

Following the exergoeconomic procedure (Tables 6–12), the optimal functional and constructive solution was obtained. The minimum cost of the cold unit, becomes $c_R = 6.478 \cdot 10^{-4} [\text{EUR/kJ}]$, is reached when the compound system ORC-VCRC operates with the decisional parameters considered in Table 12.

The cost of the cold unit decreased by half compared to the initial situation (Table 6) when $c_R = 13.15 \cdot 10^{-4} [\text{EUR/kJ}]$.

The impact of the fuel cost of the global ORC-VCRC system (diesel fuel) on the cost of the cold unit is presented in Figure 10.

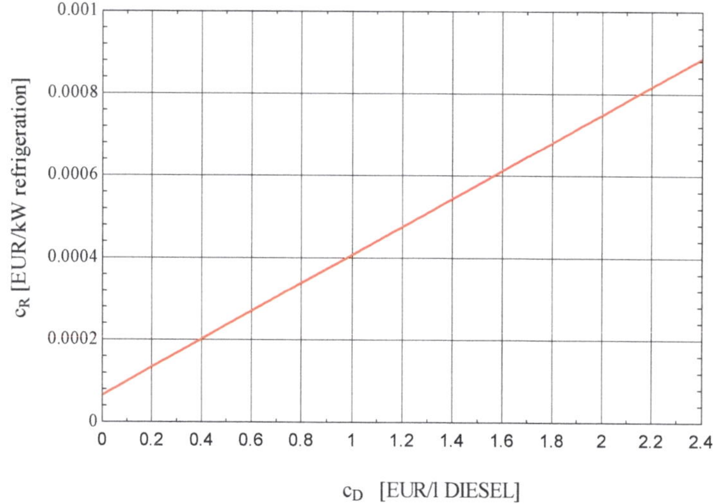

Figure 10. Influence of the diesel fuel unitary cost on the unitary cost of the refrigeration $\Delta T_{minCd} = 3$ K; $\Delta T_{minV} = 3$ K; $\Delta T_{minB} = 10$ K; $\eta_{sCp} = 0.89$; $\eta_{sE} = 0.9$; $\eta_{sP} = 0.8$.

4. Discussion

The presented study refers to the heat recovery of the combustion gases of a stationary internal combustion engine through the cogeneration of mechanical power and cold.

The heat recovery was considered as a theme of the project, for which the combustion gases were considered as an unused part of the fuel used in the engine to produce mechanical power and represent the fuel for the refrigeration plant driven by the ORC system.

If the project did not require the recovery of the energy of the combustion gases, this would be considered as an accepted loss, a situation in which the combustion gases are assigned a zero cost ($c_{gMAI} = 0$).

It is interesting to note that in this case, contrary to the impression that if the cost of the fuel does not matter, the optimal economic solution is obtained for the cheapest and, therefore, non-performing equipment; however, the actual situation is not like that because the lower performance of the equipment affects the product of the system, i.e., the cold product. This is due to the conditions of this case, namely, that the potential of the recoverable energy from the combustion gases, although free, is limited. The solution to the optimization problem is to obtain the maximum amount of product (minimum unit cost of the product) from a limited resource at the minimum monetary cost.

To find the optimal operating and design conditions, the exergoeconomic analysis and optimization method were used, which is the only method that looks for the optimal solution, offering users clear rules for improving the studied system, for which the effects can be followed step by step.

Unlike exergoeconomic analysis, any other optimization procedure based on statistical, evolutionary, or mathematical algorithms represents black boxes for the user, and the results must be accepted on faith. In addition, any mathematical optimization method performs the search within the limits of the specified scheme without providing any hints on its structural improvement.

As proof of the power of exergy analysis to suggest structural changes in the system and reduce the amount of internal consumption of usable energy (reduction in the amount of exergy destruction), an exergy analysis is presented to highlight the high amount of exergy destruction in the ORC boiler and suggests that to reduce the amount exergy destruction, an internal heat exchanger should be used.

But the value of the exergy destruction quantified only from a thermodynamic point of view does not define the conditions where in addition to the interaction with the physical environment, the interaction of the system with its economic environment must also be considered. For this purpose, to assign an economic cost to each exergy destruction, a strategy was followed to find the unit cost of each substance and energy stream as a part of the cost of the resources purchased from outside the system and the cost for amortizing the equipment components.

The system was divided into functional areas for which the exergetic resource (generically, the fuel), product, and destruction were highlighted.

For each piece of equipment, an exergoeconomic correlation was built to provide a connection between the purchase cost, size of the exergetic product, and exergetic performance parameter.

Unlike thermoeconomic correlations, which calculate costs based on material consumption, exergoeconomic correlations provide an image of the sensitivity of the monetary cost for equipment depending on its exergetic performance (that is, the exergy destruction induced by the magnitude of the irreversibility of internal processes).

The modification of the exergetic and exergoeconomic performance coefficients to the changes made in the system guided the optimization procedure.

Owing to the high cost of the exergy destruction associated with the processes in the system, the exergoeconomic optimization procedure seeks to reduce them by increasing the cost of the investment in more efficiently performing equipment.

The cost of the fuel required for driving the thermal engine (the fuel of the global system) has a substantial influence on the cost of the cold unit.

5. Conclusions

The exergoeconomic optimization procedure takes into account the interaction of the system with its physical and economic environments, looking for the functional and constructive conditions for which the unit cost of the product is minimal. The total cost of the system over a period of time is taken into account and calculated as the sum of the operating cost and the amortization cost of the invested capital.

The analysis proposes that the optimal solution is to invest in larger or higher-performance equipment that will reduce the amount of exergy consumption in the system.

Despite increasing the investment cost with larger or higher-performance pieces of equipment, the higher rate of decrease in the operational cost leads to a reduction in the monetary cost of the final product toward the optimal constructive and functional solution.

When reducing the temperature differences in the system heat exchangers (ORC boiler, condenser, and VCRC evaporator), the unitary cost of the refrigeration drops by 44%. The increase in the isentropic efficiency of the ORC expander or VCRC compressor further reduces the unitary cost of refrigeration by another 15%.

As expected from the initial exergy analysis, the ORC boiler had increased influence for decreasing the amount of exergy destruction by increasing the heat-transfer surface area, which reduced the unitary cost of the final product by 26%, followed by the evaporator at 21% and the condenser at 6%. This makes sense because the lower the temperature level at which exergy is destroyed (consumed) the higher is its cost. The increase in the isentropic efficiency and cost of the expander and compressor is accompanied by a reduction in the unitary cost of the refrigeration (the final product of the combined system) by 8% for the expander and 2% for the compressor. This reduced contribution of the compressor and expander is due to their rapid purchase cost increase with the demand for higher isentropic efficiency. Although for an increase of 10% in the isentropic efficiency of the compressor, its purchase cost almost doubles, for a decrease of 66% in the temperature difference in the boiler, its purchase cost increases by only 7%.

Following the optimization procedure, the cost of the cooling unit drops by half. The cost of diesel fuel has a major influence on the unit cost of cooling. A doubling of the cost of diesel fuel leads to an 80% increase in the cost of the cold unit.

The exergoeconomic analysis is the only one that offers research engineers a methodology to search for the optimal conditions step by step and shows the immediate effects of the functional and performance changes in the equipment on the final product. In this way, the level of understanding of the processes that take place in the system, the connections between them, and the design of the equipment increases.

The originality of the proposed exergoeconomic optimization, compared to other mathematical approaches, consists of conducting the optimal constructive and parametric search in open view, providing permanent insights into the changes made to the system.

Author Contributions: Conceptualization, D.T., C.D. and A.D.; methodology, L.G., M.C.B. and V.A.; software, D.T., H.P. and A.D.; validation, V.A., C.D. and H.P.; formal analysis, M.C.B. and V.A.; investigation, D.T. and C.D.; resources, H.P.; data curation, M.C.B.; writing—original draft preparation, D.T., C.D. and A.D.; writing—review and editing, L.G. and C.D.; visualization, D.T. and H.P.; supervision, L.G. and A.D.; project administration, V.A.; funding acquisition, C.D. All authors have read and agreed to the published version of the manuscript.

Funding: This research was funded by the Romanian Ministry of Education and the National University of Science and Technology Politehnica Bucharest through the PubArt Programme.

Institutional Review Board Statement: Not applicable.

Data Availability Statement: The study did not report any data.

Conflicts of Interest: The authors declare no conflict of interest.

Nomenclature

B	ORC boiler
$\dot{C}$	current of monetary cost, EUR/h
c	unitary monetary cost, EUR/unit
Cd	condenser
CI	exergy destruction unit cost, EUR/kJ
$\dot{CI}$	current of exergy destruction cost, EUR/h
Cp	compressor
E	expander
$\dot{Ex}$	current of exergy, kW
Ev	VCRC evaporator
h	mass enthalpy, kJ/kg
$\dot{I}$	current of exergy destruction due to internal irreversibility, kW
ICE	internal combustion engine
$\dot{L}$	current of exergy loss, kW
$\dot{m}$	mass flow rate, kg/s
$\dot{Q}$	current of heat, kW
P	pump exergetic product, kW
s	mass entropy, kJ/(kg K)
T	temperature, K
TV	throttling valve
$\dot{Z}$	rate of amortization of the investment cost, EUR/h
$\dot{W}$	mechanical power, kW
Subscripts	
0	environment, in equilibrium with the environment
B	ORC boiler
Cd	condenser
Cp	compressor
c	condensation
D	diesel fuel
E	expander
Ev	VCRC evaporator
h	heating zone of the boiler
i	inlet
L	exergy loss
m	mean temperature difference
min	minimum temperature difference
o	outlet
oh	overheating
ORC	ORC fluid
P	pump
Q	heat
R	refrigeration
TV	throttling valve
v	evaporation in the VCRC evaporator, evaporator
vb	evaporation in the ORC boiler
VCRC	vapor compression refrigeration cycle
w	water
Superscript	
T	thermodynamic temperature, K
Greek Symbols	
Ψ	share of an exergetic loss or destruction in the fuel consumption

References

1. Radulovic, J. Organic Rankine Cycle: Effective Applications and Technological Advances. *Energies* **2023**, *16*, 2329. [CrossRef]
2. Daniarta, S.; Kolasiński, P.; Rogosz, B. Waste Heat Recovery in Automotive Paint Shop via Organic Rankine Cycle and Thermal Energy Storage System—Selected Thermodynamic Issues. *Energies* **2022**, *15*, 2239. [CrossRef]
3. Aphornratana, S.; Sriveerakul, T. Analysis of a combined Rankine-vapor compression refrigeration cycle. *Energy Convers. Manag.* **2010**, *51*, 2557–2564. [CrossRef]
4. Wang, H.; Peterson, R.; Harada, K.; Miller, E.; Ingram-Goble, R.; Fisher, L.; Yih, J.; Ward, C. Performance of a combined organic Rankine cycle and vapour, compression cycle for heat activated cooling. *Energy* **2011**, *36*, 447–458. [CrossRef]
5. Hu, K.; Zhang, Y.; Yang, W.; Liu, Z.; Sun, H.; Sun, Z. Energy, Exergy, and Economic (3E) Analysis of Transcritical Carbon Dioxide Refrigeration System Based on ORC System. *Energies* **2023**, *16*, 1675. [CrossRef]
6. Arora, C.P. *Refrigeration and Air Conditioning*; Tata McGraw-Hill Education: New York, NY, USA, 2013.
7. Li, T.; Wang, J.; Zhang, Y.; Gao, R.; Gao, X. Thermodynamic Performance Comparison of CCHP System Based on Organic Rankine Cycle and Two-Stage Vapor Compression Cycle. *Energies* **2023**, *16*, 1558. [CrossRef]
8. Yue, C.; You, F.; Huang, Y. Thermal and economic analysis of an energy system of an ORC coupled with vehicle air conditioning. *Int. J. Refrig.* **2016**, *64*, 152–167. [CrossRef]
9. Bounefour, O.; Ouadha, A. Performance improvement of combined organic Rankine-vapor compression cycle using serial cascade. *Energy Procedia* **2017**, *139*, 248–253. [CrossRef]
10. Ochoa, G.V.; Santiago, Y.C.; Forero, J.D.; Restrepo, J.B.; Arrieta, A.R.A. A Comprehensive Comparative Analysis of Energetic and Exergetic Performance of Different Solar-Based Organic Rankine Cycles. *Energies* **2023**, *16*, 2724. [CrossRef]
11. Bett, A.K.; Jalilinasrabady, S. Optimization of ORC Power Plants for Geothermal Application in Kenya by Combining Exergy and Pinch Point Analysis. *Energies* **2021**, *14*, 6579. [CrossRef]
12. Camero, A.B.; Vanegas, J.C.; Forero, J.D.; Ochoa, G.V.; Herazo, R.D. Advanced Exergo-Environmental Assessments of an Organic Rankine Cycle as Waste Heat Recovery System from a Natural Gas Engine. *Energies* **2023**, *16*, 2975. [CrossRef]
13. Kim, K.H.; Perez-Blanco, H. Performance analysis of a combined organic Rankine cycle and vapor compression cycle for power and refrigeration cogeneration. *Appl. Therm. Eng.* **2015**, *91*, 964–974. [CrossRef]
14. Saleh, B. Energy and exergy analysis of an integrated organic Rankine cycle–vapor compression refrigeration system. *Appl. Therm. Eng.* **2018**, *141*, 697–710. [CrossRef]
15. Nazari, N.; Mousavi, S.; Mirjalili, S. Exergo-economic analysis and multi-objective multi-verse optimization of a solar/biomass-based trigeneration system using externally-fired gas turbine, organic Rankine cycle and absorption refrigeration cycle. *Appl. Therm. Eng.* **2021**, *191*, 116889. [CrossRef]
16. Wang, Q.; Wang, J.; Li, T.; Meng, N. Techno-economic performance of two-stage series evaporation organic Rankine cycle with dual-level heat sources. *Appl. Therm. Eng.* **2020**, *171*, 115078. [CrossRef]
17. Elmorsy, L.; Morosuk, T.; Tsatsaronis, G. Comparative exergoeconomic evaluation of integrated solar combined-cycle (ISCC) configurations. *Renew. Energy* **2022**, *185*, 680–691. [CrossRef]
18. Ibrahim, A.G.M.; Rashad, A.M.; Dincer, I. Exergoeconomic analysis for cost optimization of a solar distillation system. *Solar Energy* **2017**, *151*, 22–32. [CrossRef]
19. Rangel-Hernández, V.H.; Belman-Flores, J.M.; Rodríguez-Valderrama, D.A.; Pardo-Cely, D.; Rodríguez-Muñoz, A.P.; Ramírez-Minguela, J.J. Exergoeconomic performance comparison of R1234yf as a drop-in replacement for R134a in a domestic refrigerator. *Int. J. Refrig.* **2019**, *100*, 113–123. [CrossRef]
20. Fergani, Z.; Morosuk, T.; Touil, D. Exergy-Based Multi-Objective Optimization of an Organic Rankine Cycle with a Zeotropic Mixture. *Entropy* **2021**, *23*, 954. [CrossRef]
21. Luo, J.; Morosuk, T.; Tsatsaronis, G.; Tashtoush, B. Exergetic and Economic Evaluation of a Transcritical Heat-Driven Compression Refrigeration System with CO2 as the Working Fluid under Hot Climatic Conditions. *Entropy* **2019**, *21*, 1164. [CrossRef]
22. Tashtoush, B.; Morosuk, T.; Chudasama, J. Exergy and Exergoeconomic Analysis of a Cogeneration Hybrid Solar Organic Rankine Cycle with Ejector. *Entropy* **2020**, *22*, 702. [CrossRef] [PubMed]
23. Salim, M.S.; Kim, M. Multi-objective thermo-economic optimization of a combined organic Rankine cycle and vapor compression refrigeration cycle. *Energy Convers. Manag.* **2019**, *199*, 112054. [CrossRef]
24. Tiwari, D.; Sherwani, A.F.; Muqeem, M.; Goyal, A. Parametric optimization of organic Rankine cycle using TOPSIS integrated with entropy weight method. *Energy Sources Part A Recover. Util. Environ.* **2019**, *44*, 2430–2447. [CrossRef]
25. Bademlioglu, A.H.; Canbolat, A.S.; Kaynakli, O. Multi-objective optimization of parameters affecting Organic Rankine Cycle performance characteristics with Taguchi-Grey Relational Analysis. *Renew. Sustain. Energy Rev.* **2020**, *117*, 109483. [CrossRef]
26. Lozano, M.A.; Valero, A. Theory of the exergetic cost. *Energy* **1993**, *18*, 939–960. [CrossRef]
27. Serra, L. Optimización Exergoeconómica de Sistemas Térmicos. Ph.D. Thesis, University of Zaragoza, Zaragoza, Spain, 1994.
28. Gutierrez, J.C.; Ochoa, G.V.; Duarte-Forero, J. A comparative study of the energy, exergetic and thermo-economic performance of a novelty combined Brayton S-CO$_2$-ORC configurations as bottoming cycles. *Heliyon* **2020**, *6*, e04459. [CrossRef]
29. Dobrovicescu, A.; Grosu, L. Exergoeconomic Optimization of a Gas Turbine. *Oil Gas Sci. Technol.—Rev. IFP Energ. Nouv.* **2012**, *67*, 661–670. [CrossRef]
30. Grosu, L. Contribution à l'Optimisation Thermodynamique et Economique des Machines a Cycle Inverse a Deux et Trois Reservoirs de Chaleur. Ph.D. Thesis, Institut National Polytechnique de Lorraine, Lorraine, France, 2000.

31. Bejan, A.; Tsatsaronis, G.; Moran, M.J. *Optimization and Thermal Design*; John Wiley & Sons, Inc.: Hoboken, NJ, USA, 1996; pp. 432–448.
32. Darvish, K.; Ehyaei, M.A.; Atabi, F.; Rosen, M.A. Selection of optimum working fluid for organic Rankine cycles by exergy and exergy-economic analyses. *Sustainability* **2015**, *7*, 15362–15383. [CrossRef]

Article

Advanced Exergy-Based Optimization of a Polygeneration System with CO_2 as Working Fluid

Jing Luo, Qianxin Zhu and Tatiana Morosuk *

Institute for Energy Engineering, Technische Universität Berlin, Marchstr. 18, 10587 Berlin, Germany
* Correspondence: tetyana.morozyuk@tu-berlin.de

Abstract: Using polygeneration systems is one of the most cost-effective ways for energy efficiency improvement, which secures sustainable energy development and reduces environmental impacts. This paper investigates a polygeneration system powered by low- to medium-grade waste heat and using CO_2 as a working fluid to simultaneously produce electric power, refrigeration, and heating capacities. The system is simulated in Aspen HYSYS® and evaluated by applying advanced exergy-based methods. With the split of exergy destruction and investment cost into avoidable and unavoidable parts, the avoidable part reveals the real improvement potential and priority of each component. Subsequently, an exergoeconomic graphical optimization is implemented at the component level to improve the system performance further. Optimization results and an engineering solution considering technical limitations are proposed. Compared to the base case, the system exergetic efficiency was improved by 15.4% and the average product cost was reduced by 7.1%; while the engineering solution shows an increase of 11.3% in system exergetic efficiency and a decrease of 8.5% in the average product cost.

Keywords: polygeneration; carbon dioxide; supercritical cycle; advanced exergy-based analysis; optimization

Citation: Luo, J.; Zhu, Q.; Morosuk, T. Advanced Exergy-Based Optimization of a Polygeneration System with CO_2 as Working Fluid. *Entropy* **2024**, *26*, 886. https://doi.org/10.3390/e26100886

Academic Editors: Daniel Flórez-Orrego, Meire Ellen Ribeiro Domingos and Rafael Nogueira Nakashima

Received: 30 August 2024
Revised: 10 October 2024
Accepted: 18 October 2024
Published: 21 October 2024

1. Introduction

According to the *World Energy Outlook* published by the International Energy Agency [1], energy efficiency improvement is one of the most important elements for achieving sustainable development. Polygeneration systems, which simultaneously generate three or more energy products in a single integrated process, can effectively increase the system efficiency and have a large potential for decreasing the cost of the products. Therefore, increased attention is being paid to the design and optimization of polygeneration systems [2,3].

In this paper, a polygeneration system using CO_2 as the working fluid is optimized. The system can simultaneously produce electricity, heating, and refrigeration capacities, and it is designed to be powered by low- to medium-grade waste heat. Because of the unique thermophysical properties of CO_2 near its critical point [4], the system is expected to be compact and thermodynamically very efficient while having a low product cost.

Figure 1 shows a flow diagram of the evaluated polygeneration system and the corresponding thermodynamic cycle in a pressure–enthalpy diagram. The initial idea of this system design was inspired by a heat-driven compression refrigeration machine, which couples a closed-cycle gas-turbine cycle (closed direct/power cycle) with a transcritical vapor-compression refrigeration cycle (inverse cycle) via a gas cooler (GC), a mixer (MIX), and a splitter (SPLIT):

- The power cycle consists of a compressor for the power cycle (CM–P), a heat exchanger (HE), and an expander (EX). The "driving energy" is a medium-temperature heat source.
- The refrigeration cycle consists of a throttling valve (TV), an evaporator (EVAP), and a compressor for the refrigeration cycle (CM–R). The refrigeration capacity is generated within the EVAP.

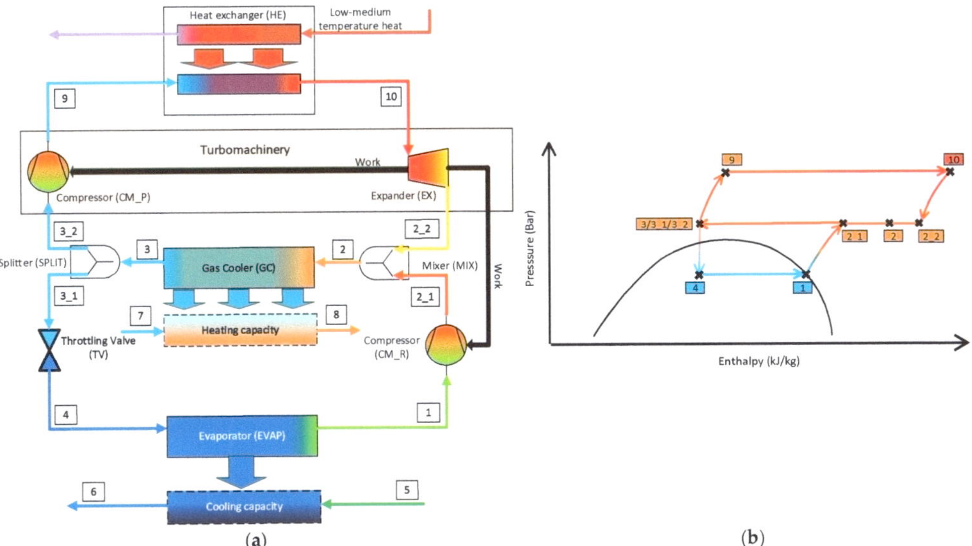

Figure 1. Process flow diagram (**a**) and pressure–enthalpy diagram (**b**) of the proposed polygeneration system.

The power produced by the power cycle is used to drive the refrigeration cycle.

The polygeneration system (Figure 1) has been evaluated using different methods based on the exergy-based analysis [5] and using single-objective and multi-objective optimizations [6]. In this paper, a graphical method of advanced exergy-based optimization is applied. The novelty of this research is the application of this method, for the first time, to the system consisting of the direct and reverse thermodynamic cycles.

2. Methodology

In this paper, the advanced exergy-based methods (reported in detail and applied for power [7] and refrigeration [8] cycles) are applied and modified for evaluating and optimizing the polygeneration system. Compared to a conventional exergetic analysis, an advanced exergy-based analysis reveals the real improvement potential of each component (the kth component) as well as the interdependencies among components [7].

2.1. Advanced Exergy-Based Analysis

Equations (1) and (2) present the basic idea of an advanced exergy-based analysis by dividing both the exergy destruction rate $\dot{E}_{D,k}$ and the associated investment cost $\dot{Z}_{D,k}$ of the kth component into unavoidable (superscript *UN*) and avoidable (superscript *AV*) parts to identify the thermodynamic and economic potential for improvement:

$$\dot{E}_{D,k} = \dot{E}_{D,k}^{UN} + \dot{E}_{D,k}^{AV} \tag{1}$$

$$\dot{Z}_{D,k} = \dot{Z}_{k}^{UN} + \dot{Z}_{k}^{AV} \tag{2}$$

The unavoidable part of the exergy destruction ($\dot{E}_{D,k}^{UN}$ and the corresponding cost of the unavoidable part of the exergy destruction $\dot{C}_{D,k}^{UN}$) cannot be reduced because of the availability and cost of materials, manufacturing methods, and other technological limitations.

The unavoidable investment cost $(\dot{Z}_{D,k}^{UN})$ for each system component can be calculated by assessing the minimum values of $\left(\frac{\dot{Z}_k}{\dot{E}_{P,k}}\right)^{UN}$ [7].

In Table 1, the definitions of fuel and product for each component and for the overall system are given. For the compressor (CM_R) and the throttling valve (TV), the separate consideration of the thermal (superscript T) and mechanical (superscript M) parts of the physical exergy is required [9]. The advanced exergy-based analysis is initially conducted for a workable design called "base case". Then, a "best case" as well as a "worst case" are assumed for the kth component to compute its unavoidable exergy destruction $\dot{E}_{D,k}^{UN}$ and the unavoidable investment cost $\dot{Z}_k^{UN}$, respectively. The parameters selected for these three cases are listed in Table 2. Moreover, the "overall-system approach" [8] is applied for calculating the unavoidable parts of each component by simulating the entire system with all selected parameter values for that corresponding case only once, which has the advantage of less computation time compared to the "component approach" [7] that needs to simulate each component separately for the "best" and "worst" cases.

Table 1. Definition of the exergetic fuel and product for each component and the overall system.

Component	$\dot{E}_F$	$\dot{E}_P$
HE	$\dot{E}_{11} - \dot{E}_{12}$	$\dot{E}_{10} - \dot{E}_9$
EX	$\dot{E}_{10} - \dot{E}_{2\_2}$	$\dot{W}_{EX}$
GC	$\dot{E}_2 - \dot{E}_3$	$\dot{E}_8 - \dot{E}_7$
CM_P	$\dot{W}_{CM\_P}$	$\dot{E}_9 - \dot{E}_{3\_2}$
CM_R	$\dot{W}_{CM\_R} + \dot{E}_1^T$	$\dot{E}_{2\_1}^M - \dot{E}_1^M + \dot{E}_{2\_1}^T$
EVAP	$\dot{E}_4 - \dot{E}_1$	$\dot{E}_6 - \dot{E}_5$
TV	$\dot{E}_{3\_1}^M - \dot{E}_4^M + \dot{E}_{3\_1}^T$	$\dot{E}_4^T$
MIX	Dissipative Component: $\dot{E}_D = \dot{E}_{2\_1} + \dot{E}_{2\_2} - \dot{E}_2$	
SPLIT	-	-
Overall System	$\dot{E}_{11} - \dot{E}_{12}$	$\left(\dot{E}_6 - \dot{E}_5\right) + \dot{W}_{net} + \left(\dot{E}_8 - \dot{E}_7\right)$

Table 2. Values of parameters assumed for the splitting of exergy destructions and investment costs into avoidable/unavoidable parts.

Component	Parameter [Unit]	"Best" Case	Base Case	"Worst" Case
HE	$\Delta T_{HE}[K]$	5	20	40
EX	$\eta_{EX}[\%]$	98	90	70
GC	$\Delta T_{GC}[K]$	1	5	10
CM_P	$\eta_{CM_P}[\%]$	95	85	70
EVAP	$\Delta T_{EVAP}[K]$	1	5	10
CM_R	$\eta_{CM\_R}[-]$	95	85	70

Detailed economic and conventional exergoeconomic analyses for the poligeneration system are reported by the authors in [4,5].

In addition, a modified exergetic efficiency ε_k^{AV} and a modified exergoeconomic factor f_k^{AV} are computed by Equations (3) and (4), respectively. These indicators provide design engineers with more information for the further evaluation and improvement of the system at the component level [7].

$$\varepsilon_k^{AV} = \frac{\dot{E}_{P,k}}{\dot{E}_{F,k} - \dot{E}_{D,k}^{UN}} = 1 - \frac{\dot{E}_{D,k}^{AV}}{\dot{E}_{F,k} - \dot{E}_{D,k}^{UN}} \tag{3}$$

$$f_k^{AV} = \frac{\dot{Z}_k^{AV}}{\dot{Z}_k^{AV} + \dot{C}_{D,k}^{AV}} = \frac{\dot{Z}_k^{AV}}{\dot{Z}_k^{AV} + c_{F,k}\dot{E}_{D,k}^{AV}} \tag{4}$$

2.2. Advanced Exergy-Based Graphical Optimization

As discussed in [7], the relation of $\frac{\dot{Z}_k^{CI}}{\dot{E}_{P,k}}$ and $\frac{\dot{C}_{D,k}}{\dot{E}_{P,k}}$ of the kth component could be presented by a curve having a horizontal and a vertical asymptote (Figure 2). The horizontal asymptote indicates the unavoidable investment cost rate per unit of product exergy $\left(\frac{\dot{Z}_k^{CI}}{\dot{E}_{P,k}}\right)^{UN}$ calculated with the parameters of the component corresponding to the "worst case", while the vertical asymptote represents the cost rate associated with the unavoidable exergy destruction within the component per unit of product exergy $\left(\frac{\dot{C}_{D,k}}{\dot{E}_{P,k}}\right)^{UN}$ calculated with the parameters given for the "best case". The optimal design point of the component can be found at the point where the derivative of the curve y = f(x) shown in Figure 2 equals to -1, $\frac{dy}{dx} = -1$.

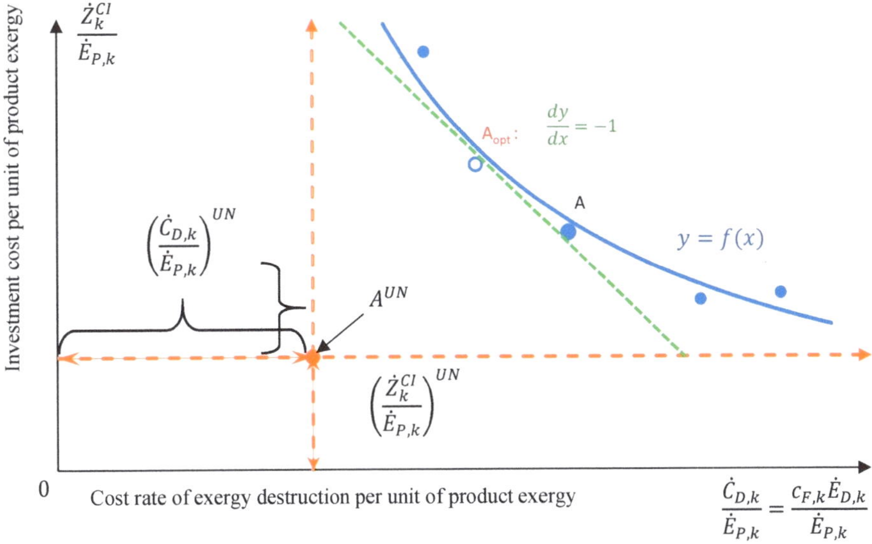

Figure 2. Advanced exergy-based graphical optimization by minimizing the sum of the associated investment cost rate and the exergy destruction cost rate for the kth component, adapted from [7].

In this work, the aforementioned optimization is slightly modified by calculating the intersection point A^{UN} of two asymptotes; then, this is set as the new zero point of a modified x-y diagram of $\left(\frac{\dot{C}_{D,k}}{\dot{E}_{P,k}}\right)^{AV}$ to $\left(\frac{\dot{Z}_k^{CI}}{\dot{E}_{P,k}}\right)^{AV}$, as shown in Figure 3a. In this newly modified diagram, the fitted function can be expressed as $y = ax^b$ with b < 0. In Figure 3b, the fitted function is linearized and simplified by taking the logarithms of both sides. The easier the function is, the simpler the process of curve fitting is. Now, with the linear function $lny = blnx + ln\,a$, the problem can be defined as a linear regression problem, and the goodness-of-fit can be shown by the coefficient of determination R^2 of the regression line [10]. An R^2 of 1 indicates that the regression predictions fit the data perfectly. After the values of a and b are obtained from the linear curve fitting process, the optimal point $\left(A_{opt}\right)^{AV}$ of the new curve with the consideration of only avoidable parts, similarly, can be found by the point with $\frac{dy}{dx} = -1$, and the unavoidable part needs to be added to the avoidable optimal results to compute the final A_{opt}.

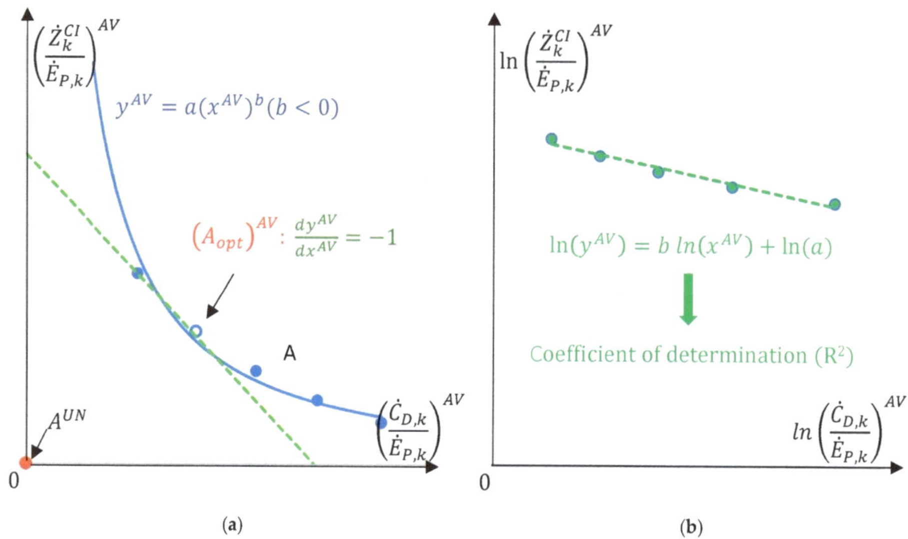

Figure 3. Modified graphical optimization: (**a**) minimizing the cost objective with the newly set zero point; (**b**) linearization of the fitted curve along with the least squares regression line.

3. Results and Discussion

In this section, the advanced exergy-based results of the proposed polygeneration system are given and discussed in detail, which include the results of advanced exergetic and exergoeconomic analyses for evaluating the component potential improvement and the optimization results based on the advanced exergy-based graphical optimization.

3.1. Results of Advanced Exergy-Based Analyses

In Table 3, the results obtained from the advanced exergy-based analyses are presented. The absolute value of the avoidable exergy destruction $\dot{E}_{D,k}^{AV}$, which reveals the real potential of improvement within each component, is in descending order of magnitude: TV, GC, HE, EX, CM_R, MIX, CM_P, and EVAP. If we compare the unavoidable to the avoidable parts of the components, the heat exchangers show the tendency of $\dot{E}_{D,k}^{UN} \gg \dot{E}_{D,k}^{AV}$; for the turbomachinery, $\dot{E}_{D,k}^{UN} \ll \dot{E}_{D,k}^{AV}$. However, regarding the ε_k^{AV}, the turbomachinery shows its highest efficiency (94.4% for the EX, 90.5% for the CM_P, and 90.3% for the CM_R), which indicates that the space available for technical modifications of the turbomachinery is rather small. The avoidable exergy destruction within the turbomachines may be caused more by the irreversibility occurring in the other components, an assumption that could be further proven by an advanced exergy-based method for splitting the exergy destruction into endogenous and exogenous parts. The endogenous and exogenous parts will be discussed in a future publication. Moreover, as the TV and MIX cannot be improved by themselves, one can conclude that the GC has the highest potential for improvement with the highest $\dot{E}_{D,k}^{AV}$ value and the HE with the second-highest $\dot{E}_{D,k}^{AV}$ comes next. Structural optimization needs to be carried out to improve the performance of TV and MIX further and the overall system to determine the best topology for the proposed polygeneration system.

Table 3. Results of advanced exergetic and exergoeconomic analyses for the base case.

Component	$\dot{E}_{D,k}^{UN}[kW]$	$\dot{E}_{D,k}^{AV}[kW]$	$\varepsilon_k^{AV}[\%]$	$\left(\dot{Z}_k^{CI}\right)^{AV}[\$/h]$	$f_k^{AV}[\%]$
HE	62.21	5.27	95.5	0.044	62.3
EX	1.02	4.49	94.4	6.967	95.0
GC	20.60	6.31	77.9	0.394	38.3
CM_P	1.12	2.56	90.5	2.333	87.7
EVAP	2.78	1.06	86.8	0.075	39.5
CM_R	1.93	4.45	90.3	0.851	60.1
TV	13.93	8.73	61.5	0.002	0.5
MIX	9.58	2.87	–	0.000	0.0

3.2. Results of Advanced Exergy-Based Graphical Optimization

The curve fitting of the avoidable parts, $\left(\dfrac{\dot{C}_{D,k}}{\dot{E}_{P,k}}\right)^{AV}$ to $\left(\dfrac{\dot{Z}_k^{CI}}{\dot{E}_{P,k}}\right)^{AV}$, for each component with its fitted function ($y = ax^b$ with $b < 0$) is illustrated in Figure 4. In addition, the coefficient of determination R^2 is also calculated for each curve to show how well the curve fits the original simulation data. The curves for the components HE, GC, and CM_R all fitted well with their R^2 values being above 0.9, while the curves for the CM_P and EVAP have relatively poor values of R^2, which may, to some extent, affect the identification of their optimal points.

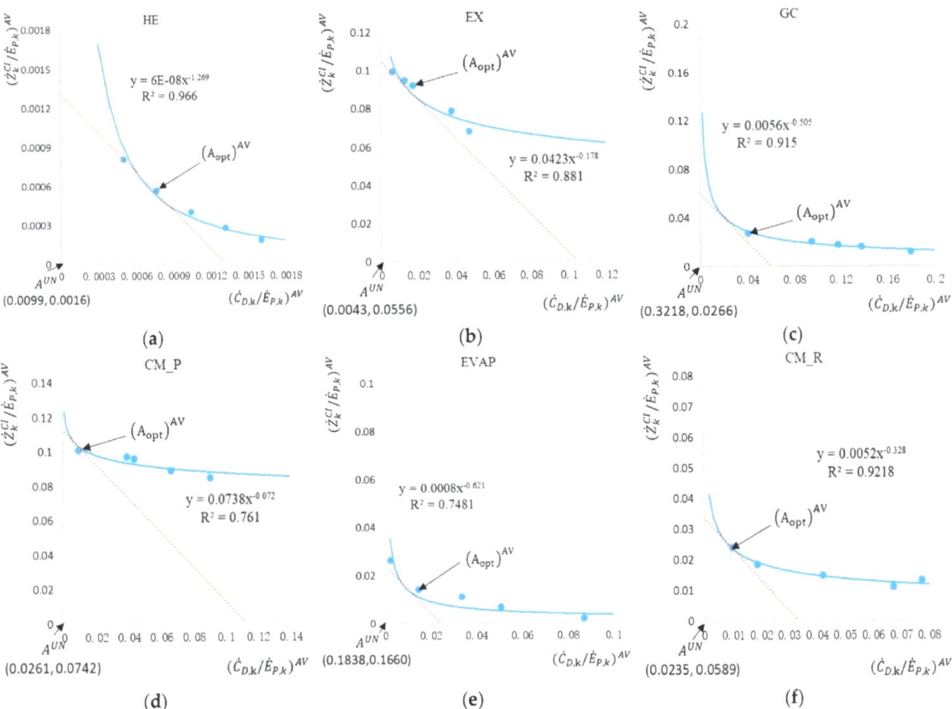

Figure 4. Exergoeconomic graphical component optimization based on the avoidable parts for the components of the polygeneration system: (**a**) heater (HE); (**b**) expander (EX); (**c**) gas cooler (GC); (**d**) compressor for the power cycle (CM_P); (**e**) evaporator (EVAP); and (**f**) compressor for the refrigeration cycle (CM_R).

Table 4 summarizes the optimal results obtained for each component based on the graphical optimization. The GC requires an improvement of its pinch point temperature difference from 5 K to 2 K, and the pinch point temperature difference for HE, similarly, needs to be reduced to 15 K from its initial setting of 20 K in the base case. For the EX, the optimal value of η_{EX} remains the same, which can be confirmed by the limited capabilities for improving this component with its high value for the exergetic efficiency based on avoidable values: $\varepsilon_{EX}^{AV} = 94.4\%$. The components CM_P and CM_R require an improvement in their isentropic efficiencies: from 85% in the base case to $\eta_{CM\_P} = 90\%$ and $\eta_{CM\_R} = 92\%$, respectively.

Table 4. Results for the overall system in the base case, optimal case, and engineering solution case.

	Base Case	Optimization Results	Engineering Solution	Improvement Potential
Operating parameters for each component				
HE	$\Delta T = 20$ K	$\Delta T = 15$ K	$\Delta T = 15$ K	high
EX	$\eta = 0.9$	$\eta = 0.9$	$\eta = 0.9$	relatively low
GC	$\Delta T = 5$ K	$\Delta T = 2$ K	$\Delta T = 2$ K	highest (possible but difficult)
CM_P	$\eta = 0.85$	$\eta = 0.90$	$\eta = 0.85$	relatively low
EVAP	$\Delta T = 5$ K	$\Delta T = 4$ K	$\Delta T = 4$ K	lowest
CM_R	$\eta = 0.85$	$\eta = 0.92$	$\eta = 0.85$	low
Exergetic and exergoeconomic analysis				
$\varepsilon_{Overall}(\%)$	16.5	19.1	18.4	
$\dot{W}_{Electricity}[kW]$	0.00	0.00	0.00	
$\dot{Q}_{Cooling}[kW]$	6.95	7.14	7.14	
$\dot{Q}_{Heating}[kW]$	22.21	17.21	18.21	
$c_{P,Electricity}[\$/kWh]$	0.47	0.31	0.31	
$c_{P,Cooling}[\$/kWh]$	0.82	0.57	0.58	
$c_{P,Heating}[\$/GJ]$	0.84	0.86	0.84	
$c_{P,Overall}[\$/GJ]$	0.83	0.78	0.76	
Relative change to the base case				
Overall exergetic efficiency		15.4%	11.3%	
Overall average product cost		−7.1%	−8.5%	

However, all optimal parametric values calculated for each component could only be considered by design engineers as theoretical optimization results. Engineers should also consider the current technical development and the additional costs of implementing the combination of these optimal values in the real design. Thus, an engineering solution for this system is also presented in Table 4 based on the current economic and technical background. In the engineering solution scenario, no modifications are required for the two compressors based on their current high values of ε^{AV} over 90%; otherwise, a further improvement will result in a high penalty associated with the purchased equipment cost. On the contrary, modifications of the minimum temperature differences in heat exchangers are relatively less costly and easier to achieve. However, we should mention that it may be difficult to operate the GC with a $\Delta T_{GC} = 2$ K, which requires special materials and techniques.

Compared to the base case, the exergetic efficiency for the overall system ($\varepsilon_{Overall}$) increases by 15.4% and 11.3% for the cases with optimization results and with the engineering solution, respectively. Simultaneously, the overall average product cost decreases by 7.1% for the optimization results case and by 8.5% for the engineering solution case. Thus, a "cost optimum" is obtained by the so-called engineering solution. From these results, we conclude that optimizing single components in isolation does not, in general, lead to the system optimal design and that the design engineers must critically review the results of any theoretical optimization before implementation.

Figure 5 shows the distribution of the total cost $\left(\dfrac{\dot{Z}_k^{ci}+C_{D,k}}{\dot{E}_{P,k}}\right)$ associated with each system component for the base case and the optimizations. The most significant difference can be observed for the GC.

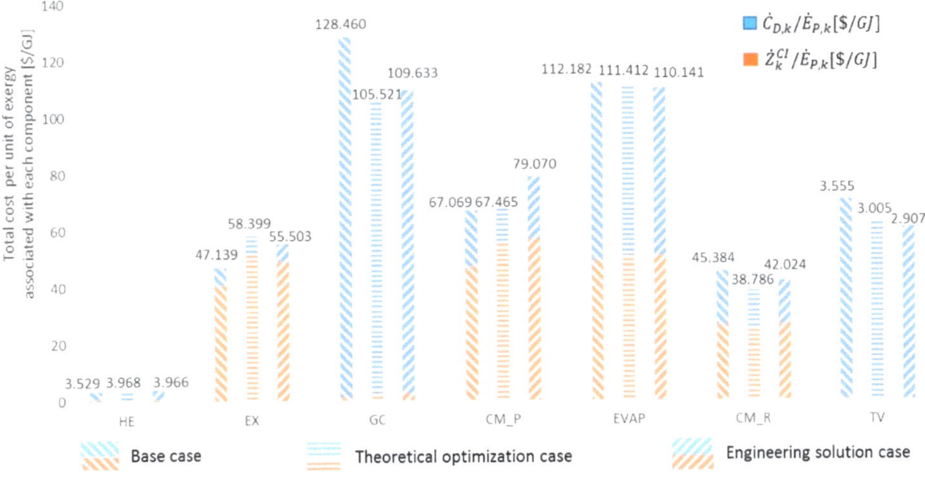

Figure 5. Total cost $\left(\dfrac{\dot{Z}_k^{ci}+C_{D,k}}{\dot{E}_{P,k}}\right)$ associated with each system component (USD/GJ).

4. Conclusions

In this work, advanced exergy-based analyses and optimizations were conducted for the polygeneration system using CO_2 as the working fluid at the assumption of a refrigeration capacity of 100 kW with heat recovery of the generation of hot water at the temperature of 65 °C. By applying advanced exergy-based analyses, the avoidable inefficiencies within the system components were identified. This information assists designers in further improving the system performance from the thermodynamic and cost viewpoints. Conventional exergetic analysis [5,6] showed that the improvement priorities for the components in the overall system should be in the order of the heater (HE), the gas cooler (GC), the throttling valve (TV), the compressor for the refrigeration cycle (CM_R), the expander (EX), the evaporator (EVAP) and the compressor for the power cycle (CM_P). However, the advanced exergetic analysis suggested that the priority of technical modification for the components should be given to the gas cooler (GC), followed by the heater (HE) as the throttling process of the throttling valve (TV) with the highest avoidable exergy destruction value could not be improved by itself. A total amount of 45.2 kW, 30.8% of the overall exergy destruction rate, could be lowered with the consideration of only the overall system avoidable part (calculated by setting all parametric variables in the "best" condition with maximum efficiency for the system).

An exergoeconomic graphical optimization focusing only on the avoidable parts of components was carried out. The optimization results revealed an improvement in terms of system exergetic efficiency by more than 15%, with a reduction of more than 7% in the average product cost. However, no interactions among components were included in the advanced exergoeconomic analysis. Thus, these "optimization" results can be further improved. This fact is demonstrated by the results of the engineering solution presented here. It should be noted, however, that single-component optimization is an easily implemented and practical approach for improving the system performance with less computation time; it is especially user-friendly for non-programmers [10].

However, there is still potential for improvement. As indicated through the exergetic analysis, the optimization results, which might require further modification of the structure, especially the throttling valve (TV), need to be further investigated. Moreover, for the turbomachine (EX and CM_P), which showed an increase in the overage total cost in the optimization results case, sensitivity analyses regarding the turbine inlet temperature and turbine inlet pressure might also be necessary for further research.

Author Contributions: Conceptualization, J.L.; methodology, T.M.; software, Q.Z.; validation, J.L. and Q.Z.; formal analysis, J.L. and Q.Z.; investigation, J.L.; resources, J.L.; data curation, J.L.; writing— original draft preparation, J.L. and Q.Z.; writing—review and editing, T.M.; visualization, Q.Z.; supervision, T.M.; project administration, T.M.; funding acquisition, T.M. All authors have read and agreed to the published version of the manuscript.

Funding: This research received no external funding.

Institutional Review Board Statement: Not applicable.

Data Availability Statement: Data is contained within the article.

Conflicts of Interest: The authors declare no conflicts of interest.

Nomenclature

$c_{F,k}$	fuel cost per unit of exergy of kth component, USD/kWh
$c_{P,k}$	product cost per unit of exergy of kth component, USD/kWh
$\dot{C}_{D,k}$	destruction cost rate associated with kth component, USD/h
$\dot{E}_i^M$	mechanical exergy rate of ith stream, kW
$\dot{E}_i^T$	thermal exergy rate of ith stream, kW
$\dot{E}_{F,k}$	fuel exergy rate of kth component, kW
$\dot{E}_{P,k}$	product exergy rate of kth component, kW,
$\dot{E}_{D,k}$	exergy destruction rate of kth component, kW
$\dot{E}_{D,k}^{AV}$	avoidable exergy destruction rate of kth component, kW
$\dot{E}_{D,k}^{UN}$	unavoidable exergy destruction rate of kth component, kW
f_m	material factor for the calculation of purchased equipment cost, -
f_k	exergoeconomic factor of kth component, %
f_k^{AV}	modified exergoeconomic factor of kth component, %
T_{HS}	thermodynamic average temperature of the stream of matter providing the low to medium-grade heat, K
$\dot{Z}_k^{CI}$	capital investment cost rate of kth component, USD/h
$\dot{Z}_k^{AV}$	avoidable capital investment cost rate of kth component, USD/h
$\dot{Z}_k^{UN}$	unavoidable capital investment cost rate of kth component, USD/h
ΔT	pinch point temperature difference, K
ε_k	exergetic efficiency of kth component, %
ε_k^{AV}	modified exergetic efficiency of kth component, %
η	isentropic efficiency, %

Abbreviations

CM–P	compressor in power sub-cycle
CM–R	compressor in refrigeration sub-cycle
EVAP	evaporator
EX	expander
GC	gas cooler
HE	heat exchanger
MIX	mixer
SPLIT	splitter
TV	throttling valve

References

1. IEA. World Energy Outlook 2023. Available online: https://www.iea.org/reports/world-energy-outlook-2023 (accessed on 5 August 2024).
2. Liu, P.; Gerogiorgis, D.I.; Pistikopoulos, E.N. Modeling and optimization of polygeneration energy systems. *Catal. Today* **2007**, *127*, 347–359. [CrossRef]
3. Chicco, G.; Mancarella, P. Trigeneration primary energy saving evaluation for energy planning and policy development. *Energy Policy* **2007**, *35*, 6132–6144. [CrossRef]
4. Lorentzen, G. Revival of carbon dioxide as a refrigerant. *Int. J. Refrig.* **1994**, *17*, 292–300. [CrossRef]
5. Luo, J.; Morosuk, T.; Tsatsaronis, G. Exergoeconomic investigation of a multi-generation system with CO_2 as the working fluid using waste heat. *Energy Convers. Manag.* **2019**, *197*, 111882. [CrossRef]
6. Tashtoush, B.; Luo, J.; Morosuk, T. Exergy-Based Optimization of a CO_2 Polygeneration System: A Multi-Case Study. *Energies* **2024**, *17*, 291. [CrossRef]
7. Morosuk, T.; Tsatsaronis, G. Advanced exergy-based methods used to understand and improve energy-conversion systems. *Energy* **2019**, *169*, 238–246. [CrossRef]
8. Tsatsaronis, G.; Morosuk, T.V. Advanced Exergoeconomic Evaluation and its Application to Compression Refrigeration Machines. In Proceedings of the ASME International Mechanical Engineering Congress and Exposition, IMECE 2007, Seattle, WA, USA, 11–15 November 2007. IMECE2007-412202.
9. Morosuk, T.; Tsatsaronis, G. Splitting physical exergy: Theory and application. *Energy* **2019**, *167*, 698–707. [CrossRef]
10. Cain, J.W. Mathematics of fitting scientific data. In *Molecular Life Sciences: An Encyclopedic Reference*; Bell, E., Ed.; Springer: New York, NY, USA, 2014; pp. 1–7.

Article

Comparative Exergy and Environmental Assessment of the Residual Biomass Gasification Routes for Hydrogen and Ammonia Production

Gabriel Gomes Vargas [1,*], Daniel Alexander Flórez-Orrego [2,3,*] and Silvio de Oliveira Junior [1]

[1] Polytechnic School, University of São Paulo, Av. Luciano Gualberto 380, São Paulo 05508-010, Brazil; soj@usp.br

[2] Industrial Process and Energy Systems Engineering, École Polytechnique Fédérale de Lausanne EPFL, Sion, 1950 Valais, Switzerland

[3] Faculty of Mines, National University of Colombia, Av. 80 #65-223, Medellin 050034, Colombia

* Correspondence: gabrielvargas@usp.br (G.G.V.); daniel.florezorrego@epfl.ch (D.A.F.-O.)

Abstract: The need to reduce the dependency of chemicals on fossil fuels has recently motivated the adoption of renewable energies in those sectors. In addition, due to a growing population, the treatment and disposition of residual biomass from agricultural processes, such as sugar cane and orange bagasse, or even from human waste, such as sewage sludge, will be a challenge for the next generation. These residual biomasses can be an attractive alternative for the production of environmentally friendly fuels and make the economy more circular and efficient. However, these raw materials have been hitherto widely used as fuel for boilers or disposed of in sanitary landfills, losing their capacity to generate other by-products in addition to contributing to the emissions of gases that promote global warming. For this reason, this work analyzes and optimizes the biomass-based routes of biochemical production (namely, hydrogen and ammonia) using the gasification of residual biomasses. Moreover, the capture of biogenic CO_2 aims to reduce the environmental burden, leading to negative emissions in the overall energy system. In this context, the chemical plants were designed, modeled, and simulated using Aspen plus™ software. The energy integration and optimization were performed using the OSMOSE Lua Platform. The exergy destruction, exergy efficiency, and general balance of the CO_2 emissions were evaluated. As a result, the irreversibility generated by the gasification unit has a relevant influence on the exergy efficiency of the entire plant. On the other hand, an overall negative emission balance of -5.95 kg$_{CO2}$/kg$_{H2}$ in the hydrogen production route and -1.615 kg$_{CO2}$/kg$_{NH3}$ in the ammonia production route can be achieved, thus removing from the atmosphere 0.901 t$_{CO2}$/t$_{biomass}$ and 1.096 t$_{CO2}$/t$_{biomass}$, respectively.

Keywords: biomass gasification; decarbonization; bioproducts; exergy analysis; energy integration

check for **updates**

Citation: Vargas, G.G.; Flórez-Orrego, D.A.; de Oliveira Junior, S. Comparative Exergy and Environmental Assessment of the Residual Biomass Gasification Routes for Hydrogen and Ammonia Production. *Entropy* **2023**, *25*, 1098. https://doi.org/10.3390/e25071098

Academic Editor: Jean-Noël Jaubert

Received: 23 May 2023
Revised: 14 July 2023
Accepted: 18 July 2023
Published: 22 July 2023

1. Introduction

Biomass is an important source of renewable energy that may help reduce fossil fuel dependence and CO_2 emissions in the chemical sector. This is especially applicable in the case of Brazil, considering its substantial biomass potential. In recent years, biofuels have accounted for almost 70% of global renewable energy production [1], and biomass was responsible for 25.5% of Brazilian domestic energy supply [2]. This contribution could be boosted further if biomass wastes were converted into valuable energy products such as hydrogen and ammonia. In this way, fossil energy consumption and greenhouse gas emissions could be reduced, while the costs and environmental impact of waste disposal could be relieved. In Brazil, for example, sugarcane and orange bagasses are the primary residues of the sugarcane and juice industries, respectively, which are, in turn, the primary suppliers of bioenergy and juice in the country [2,3]. Typically, bagasse provides combined heat and power production for sugarcane mills. Even though they are well-established procedures in the industry, they are still reasonably inefficient and could be replaced by improved energy conversion processes [4,5]. On the other hand, sewage sludge is a by-product of wastewater treatment that contains various organic and inorganic materials and needs to be disposed of appropriately. In recent years, conventional sewage sludge disposal methods, including landfills and anaerobic digestion, have been adopted. However, these methods take a long time to digest and require large amounts of land. Moreover, they tend to cause environmental issues such as undesirable emissions (e.g., odor and leachate) and the accumulation of heavy metals in soils [6,7].

Biomass gasification has a large potential to simultaneously deal with the treatment of cumbersome wastes while increasing the effectiveness of the utilization of the chemical processes' byproducts. Several studies have explored the potential of different biomass feedstocks for power generation and biofuel production. Promising results were reported for corn cobs, with a gas production yield of about 2 m^3/kg, a heating value of 5.6 to 5.8 MJ/m^3, and a cold gas efficiency between 66% and 68% [8]. Other studies focused on converting olive tree pruning and olive pits into electricity and heat, achieving satisfactory cold gas efficiency (70.7–75.5%) and a favorable calorific value (4.8 to 5.4 MJ kg^{-1}) [9]. Vine pruning showed promising results in a 350 kW downdraft gasifier, with a syngas heating value of 5.7 MJ/m^3, cold gas efficiency of 65%, and power efficiency of 21% [10]. Biohydrogen production from the gasification of agricultural waste through dark fermentation is reportedly an environmentally friendly solution [11]. Hydrogen production from coconut coir and palm kernel shell through air gasification showed substantial hydrogen gas production potential [12]. In this regard, the gasification of agricultural residual biomasses is recognized as a promising method for achieving a sustainable bioeconomy and reducing dependence on fossil fuels, averaging 67% efficiency in energy conversion [13].

Over the last five decades, extensive research has been carried out on biomass gasification, mainly focusing on syngas production [14]. Comprehensive research is underway for the development of cost-effective and energy-efficient gasifiers. Gasifiers can be broadly categorized based on [15]:

I.   Fluid dynamics (updraft, downdraft),

II.   Modes of heat transfer to the gasification process (auto thermal or directly heated gasifiers and allothermal or indirectly heated gasifiers),

III. Gasification agents (air, oxygen, or steam blown), and

IV.   Pressure (atmospheric or pressurized).

In the above categorization of gasifiers, the classification based on fluid dynamics and modes of heat transfer is of prime importance. Fluid dynamics primarily determines the characteristics of the gases/solids in contact during the gasification process and plays a vital role in influencing the performance of a gasifier. In addition to fluid dynamics, the modes of heat transfer in a gasifier are also important aspects of the study. In the case of a directly heated gasifier, the entire gasification process occurs in a single reactor, and heat evolved from the exothermic reactions is used to carry out endothermic gasification reactions. These gasifiers exhibit several configurations, e.g., fixed bed, fluidized bed, or circulating fluidized bed (operated at temperatures below 900 °C) and entrained flow gasifiers (operated at a higher temperature range of 1200–1500 °C) [16]. The heating values of the product gas from these gasifiers using air and oxygen as gasification agents are in the range of 4–7 MJ/Nm^3 and 10–12 MJ/Nm^3, respectively [17].

In contrast, an indirectly heated gasifier consists of two reactors. The heat required for the gasification process is produced by a separate combustion reactor and transported to the reduction reactor using heat-carrying material, such as sand. Syngas obtained from this type of gasifier is rich in CO and H_2, as the flue gas that is released from the combustion reactor flows separately from the product gas, thus preventing its dilution. This fact results in a higher heating value for the gas (12–20 MJ/Nm^3) compared to an indirectly heated gasifier. Also, since no oxygen separation unit is necessary and a smaller amount of gas cleaning equipment is installed, a lower capital investment is required [18]. In addition, as the two reactors operate separately, it is easy to control and scale up [19]. The dual fluidized bed (DFB) reactor is a type of indirectly heated gasifier. DFB gasification systems have been studied at laboratory and pilot scales over the last two decades. A DFB gasification plant has been running successfully in Güssing, Austria (8 MWth) since 2001, along with industrial-scale operations in Oberwart, Austria (8.5 MWth) and Ulm, Germany (15 MWth) [20]. Apart from this, the Gothenburg Biomass Gasification (GoBiGas) project in Göteborg, Sweden, has been recently commissioned to produce substitute natural gas (SNG) using wood pellets as feedstock. Finally, the comparative analysis conducted by Florez-Orrego et al. [21] shed light on the emissions profile of fossil fuel and biomass pathways for chemical production. As a result, a negative CO_2 emissions balance is achieved, indicating a favorable global impact on mitigating atmospheric CO_2 levels. Notably, the study revealed that for each ton of ammonia produced, approximately 1.7 to 2.3 tons of CO_2 are effectively sequestered from the environment. Furthermore, the research emphasized the advantageous aspects of utilizing bagasse, despite its indirect emissions. These emissions are offset not only by the captured biogenic emissions but also by the utilization of "greener" electricity imports.

In that regard, gasification is a prominent research topic among the available techno-logical routes in a residual biomass conversion context [22]. The technology could lead to higher energy conversion and production yields [23,24] and reduced sizes for treatment plants [25] and costs [26]. Previous studies have already investigated the use of biomass for synthetic natural gas [27], hydrogen [4], ammonia [21], nitrogen fertilizers [28], and electricity production [29,30]. Some conversion routes are shown in the literature for residual biomass [5,31,32]. However, while different options have already been proposed for each biomass waste, there is a lack of studies dedicated to analyzing the performance of waste upgrade systems and comparing the utilization of all those resources using a common basis defined by thermodynamic and environmental indicators. Thus, this work proposes alternative routes for the conversion of biomass wastes into hydrogen and ammonia, in addition to the optimization and hierarchization of these energy conversion routes. For this purpose, residual biomass with low or no added value will be used, such as sugarcane bagasse, sewage sludge, and orange bagasse. This fact reduces the risk of the perception of biomass utilization as a competitor for food and land resources.

2. Process Description

The considered approach integrates a biomass gasification system for agricultural or human wastes, a synthesis gas purification unit, and a conditioning system to produce hydrogen and ammonia. Data was collected through a bibliographic review, in addition to data provided by the Basic Sanitation Company of the State of São Paulo [33]. For the sake of comparison, it was assumed a biomass mass flow rate of 26,400 kg/h. The analysis is focused on determining the minimum energy requirements of those facilities; thus, the composite curves of the chemical systems will be further discussed. Depending on the waste heat available, all heat and electricity requirements should be imported, or part of the energy produced in the form of fuel can be internally consumed. Also, if the amount of waste heat from the exothermic reactions exceeds the domestic demands, power could be internally generated using a waste heat recovery steam network.

2.1. Biomass Drying and Chipping Process

In the gasification section, moisture is first removed in a rotary dryer with a specific power consumption of 15 kWh per wet ton of biomass [34]. In this process, the water content of the biomass is reduced to 7% [35]. Furthermore, electricity is used in the chipping process for grinding bagasse to obtain 0.5 mm diameter particles. The power consumption is estimated at 3% of the lower heating value of the biomass input [36]. In order to conduct mass and energy balances for the biomass pre-treatment processes, a FORTRAN subroutine was developed within the Aspen® Plus software [37]. The subroutine was utilized to calculate the quantity of moisture removed in the rotary dryer, denoted as $m_{H_2O,removed}$ (kg/h) in Equation (1). This calculation is based on the initial moisture content of the biomass, represented as $\psi_{H_2O,moist-bio}$ (%), the desired moisture content of the biomass at the gasifier inlet, denoted as $\psi_{H_2O,dry-bio}$ (%), and the mass flow rate of the wet biomass feed, indicated as $m_{H_2O,moist-bio}$ (kg/h).

$$m_{H_2O,removed} = \left(\psi_{H_2O,moist-bio} - \frac{1 - \psi_{H_2O,moist-bio}}{1 - \psi_{H_2O,dry-bio}} \times \psi_{H_2O,dry-bio} \right) \times m_{H_2O,moist-bio} \quad (1)$$

2.2. Gasification Process

After the chipping and drying processes, biomass is fed to the gasification unit. A Battelle Columbus Laboratory (BCL) indirect-heated gasifier is adopted (Figure 1) [4,12,18,38]. The gasifier separates the solids from the syngas (i.e., sand and char) and transfers them to a combustion chamber. In this latter case, air is blown to burn the char, which provides heat to the reduction zone. To this end, the hot particles (sand) are separated from the flue gas through a cyclone and recycled back to the reduction bed, ensuring the provision of heat for the endothermic reactions (drying, pyrolysis, and reduction). This approach separates the combustion reactions from the reduction reactions, preventing the dilution of the produced syngas with nitrogen [39]. The temperature in the combustion column reaches approximately 950 °C, while the gasification column operates at a temperature of around 850 °C [4,12,18]. The gasifier operates at atmospheric pressure.

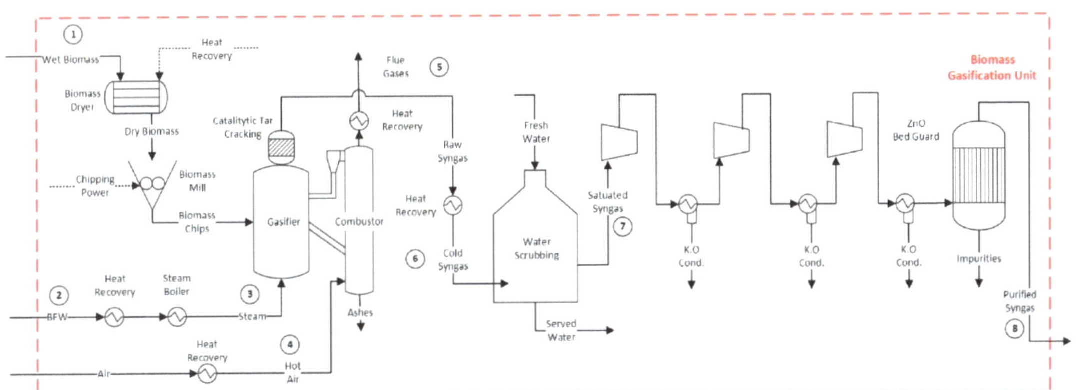

Figure 1. Flowsheet of the biomass pre-treatment and gasification unit. See Supplementary Material for numbered stream properties. Flow properties (1–8) can be found in the Supplementary Material, Tables S1–S3.

The ultimate and proximate analyses of the dry biomass residues are shown in Table 1. For the calculation of the enthalpy and density of solids, the HCOALGEN and DCOALGEN models are chosen [40]. The gasification process consumes saturated steam as the gasification agent, with a steam-to-biomass ratio of 0.50 [41]. Additionally, combustion air is preheated up to 400 °C [42] to facilitate the combustion of a portion of the char generated. Essentially, in the gasification process, it is crucial to maintain a balance between the heat supplied by the combustion zone and the heat required for the drying, pyrolysis, and reduction stages.

Following the decomposition in the pyrolysis process, the reduction reactions occur in the presence of steam and can be summarized as shown in (R. 1–R. 9) in Table 2.

Table 1. Proximate and ultimate analyses of different waste biomass used in the gasification process (%). db stand for dry basis.

Parameter	Sugar Cane Bagasse [43]	Sewage Sludge [44]	Orange Bagasse [45]
Proximate analysis			
Fixed Carbon	50.00	18.40	9.23
Volatile Material	14.32	7.60	73.20
Moisture	83.54	64.90	20.60
Ash	2.14	27.50	6.20
Ultimate Analysis (%) [db]			
Carbon	46.70	33.90	46.40
Hydrogen	6.02	6.30	5.54
Oxygen	44.95	25.50	40.15
Nitrogen	0.17	5.88	1.70
Sulphur	0.02	0.67	0.00
Chlorine	0.00	0.21	0.00

Table 2. General reactions of the gasification process.

Reaction	ΔH^0_{298K} (kJ/mol)	Name	No.
$C + O_2 \rightarrow CO_2$	−394	Complete combustion	(R. 1)
$C + CO_2 \rightarrow 2CO$	+173	Boudouard reaction	(R. 2)
$C + H_2O \rightarrow CO + H_2$	+131	Char steam gasification	(R. 3)
$C + 2\,H_2 \rightarrow CH_4$	−75	Char gasification	(R. 4)
$CO + \frac{1}{2}\,O_2 \rightarrow CO_2$	−283	Carbon oxidation	(R. 5)
$H_2 + \frac{1}{2}\,O_2 \rightarrow H_2O$	−242	Hydrogen oxidation	(R. 6)
$CH_4 + 2\,O_2 \rightarrow CO_2 + 2H_2O$	−283	Methane oxidation	(R. 7)
$CO + H_2O \rightarrow CO_2 + H_2$	−41	Water-gas shift reaction	(R. 8)
$6CO + 9H_2 \rightarrow 6H_2O + C_6H_6$	−1583	Tar formation	(R. 9)

The syngas produced exits the gasifier and goes through a thermal catalytic cracking process, which converts the produced tar into more desirable compounds [43]. Subsequently, the synthesis gas is cooled down to a temperature of 400 °C. To eliminate impurities that could potentially impact downstream equipment, the gas is subjected to a scrubbing process using water. Following this, the syngas is compressed to 35 bar. To ensure the removal of sulphur compounds, a zinc oxide guard bed is utilized. More information regarding the properties of the mass flows identified in tags 1 to 8 of Figure 1 can be found in the Supplementary Material, Tables S1–S3.

2.3. Syngas Conditioning Process

Upon exiting the gasifier, the syngas undergoes the necessary treatment and adjustment to its composition. This critical process is performed in the syngas treatment unit. In the hydrogen production route, syngas can be directly sent to the water gas shift reactors (Figure 2). However, for ammonia production, it is crucial to achieve an H_2:N_2 molar ratio of 3:1. To this end, an autothermal reformer (ATR) is employed, followed by water gas shift reactors, as shown in Figure 3. In the ATR reactor, the partial combustion of the syngas with air enables the introduction of the necessary nitrogen, which provides the energy for the reforming reactions [44]. The reforming reactions consume saturated steam and occur in the presence of a high-temperature-resistant nickel catalyst [45].

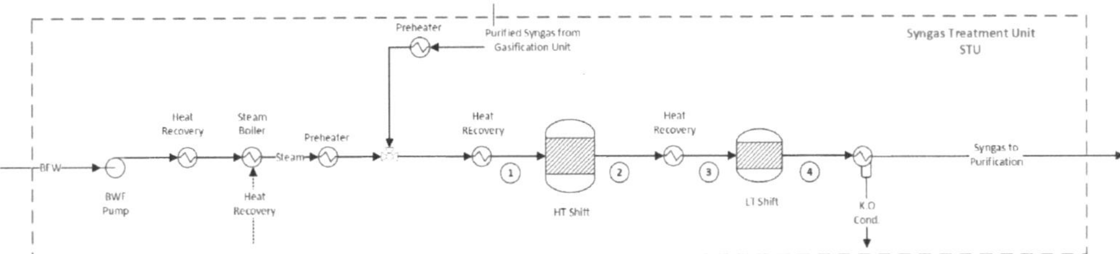

Figure 2. Flowsheet of the syngas conditioning unit for hydrogen production. Flow properties (1–4) can be found in the Supplementary Material, Tables S4–S6.

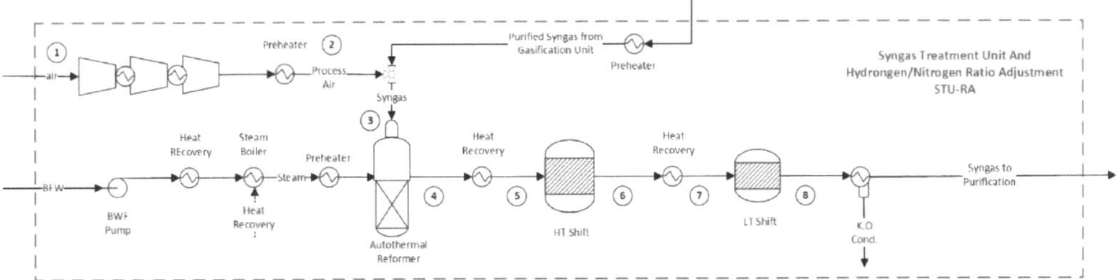

Figure 3. Flowsheet of the syngas treatment unit for ammonia production. Flow properties (1–8) can be found in the Supplementary Material, Tables S7–S9.

The reactions occurring in the ATR involve reactions (R. 2) in Table 2 and (R. 10–R. 11) in Table 3, as well as the combustion of methane, hydrogen, and carbon monoxide. As a result, the molar fraction of the methane slip in the syngas at the ATR outlet is around 0.45% mol [45]. Next, the synthesis gas is cooled to reach an appropriate feed temperature for the downstream high- and low-temperature shift reactors (HT/LT Shift). The recovery of the residual heat is typically achieved by generating high-pressure saturated steam [45].

Table 3. Reforming and water gas shift reactions in the ATR.

Reaction	ΔH^0_{298K} (kJ/mol)	Name	
$CH_4 + H_2O \rightarrow CO + 3H_2$	+206	Steam reform	(R. 10)
$CO + H_2O \rightarrow CO_2 + H_2$	−41	Water gas shift reaction	(R. 11)

In the HT shift reactor, an iron-chrome catalyst is used to increase the production of hydrogen by reacting the remaining CO and water in the syngas (R. 11) [46]. The exothermic WGS reaction is limited by equilibrium, which results in a residual CO concentration of approximately 3% mol [45]. To further enhance the conversion of CO, a second WGS reactor is used at a lower temperature (LT Shift) in the presence of a copper–zinc catalyst. The residual CO content at the outlet is typically around 1% [45]. More information about the properties of the mass flows in the syngas conditioning unit can be found in the Supplementary Material, Tables S4–S6 for hydrogen identified in tags 1 to 4 of Figure 2 and Tables S7–S9 for ammonia production routes identified in tags 1 to 8 of Figure 3, respectively. Finally, the cooled syngas (35 °C) continues to the syngas purification unit, described in the next section.

2.4. Carbon Dioxide Capture and Methanation Processes

A syngas purification unit is needed to remove the carbon compounds produced in the previous sections; otherwise, they could poison the ammonia catalyst. This unit encompasses a CO_2 capture unit, a methanator, and a dryer, as shown in Figure 4. However, for the hydrogen production route, the methanation unit is spared since no catalyst protection is required, as shown in Figure 5.

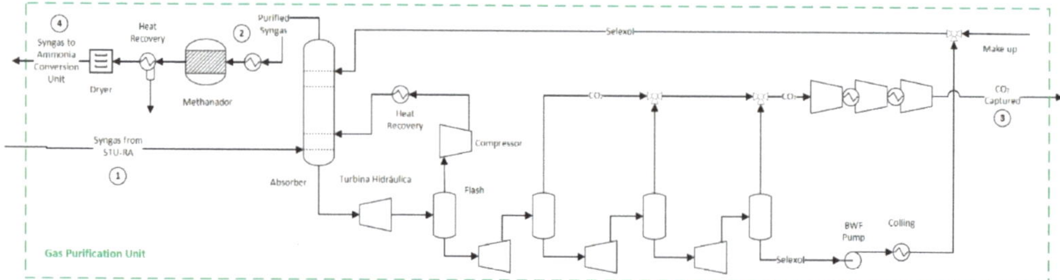

Figure 4. Flowsheet of the syngas purification unit for the ammonia production route. Flow properties (1–4) can be found in the Supplementary Material, Tables S10–S12.

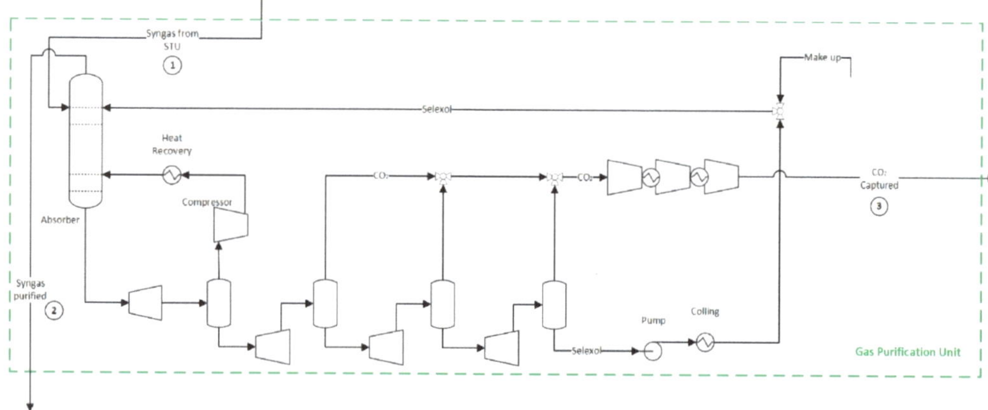

Figure 5. Flowsheet of the syngas purification unit for hydrogen production. Flow properties (1–3) can be found in the Supplementary Material, Tables S13–S15.

In the CO_2 capture unit of both ammonia and hydrogen production routes, the syngas enters the CO_2 absorber (35 bar) and is brought into contact with a physical solvent (dimethyl ethers of polyethylene glycols) to form a CO_2-rich bottom solution. The purified syngas, primarily composed of H_2 and CO, exits from the top of the absorber column and is sent to the downstream processes. Meanwhile, the pure CO_2 is gradually released by pressure let-downs through a series of flash drums and expanders, which recover the expansion energy. CO_2 is sent for transport and disposal at high purity. The lean solvent is recycled back to the absorber [47]. After the CO_2 removal step, there may still be residual amounts of CO and CO_2 in the syngas, which need to be eliminated to meet the purity requirements for the ammonia synthesis loop. For this reason, a methanation unit converts those compounds into inert methane by consuming a fraction of the hydrogen over a nickel catalyst. More information about those streams and their properties can be found in the Supplementary Material, Tables S10–S12 for the ammonia production route identified in tags 1 to 4 of Figure 3 and Tables S13–S15 for the hydrogen production route identified in tags 1 to 3 of Figure 5.

2.5. Pressure Swing Adsorption and Hydrogen Compression in Hydrogen Production Route

The final hydrogen purification is commonly accomplished using pressure swing adsorption (PSA) technology. PSA operates by modulating the internal pressure of adsorption columns, which allows for selective retention of gases (Figure 6). By removing CO_2 prior to hydrogen purification, the size of the PSA system is reduced, resulting in cost and space savings. The PSA unit operates at 30 bar and 35 °C [4]. The hydrogen recovery efficiency is 95% mol [48]. Pure hydrogen is obtained as the product of the PSA system, while purge gas containing impurities is typically burned to recover the energy. The purified hydrogen is compressed to 200 bar [49]. Supplementary Material provides detailed information on the streams involved in this unit; see Tables S16–S18 identified in tags 1 to 5 of Figure 6.

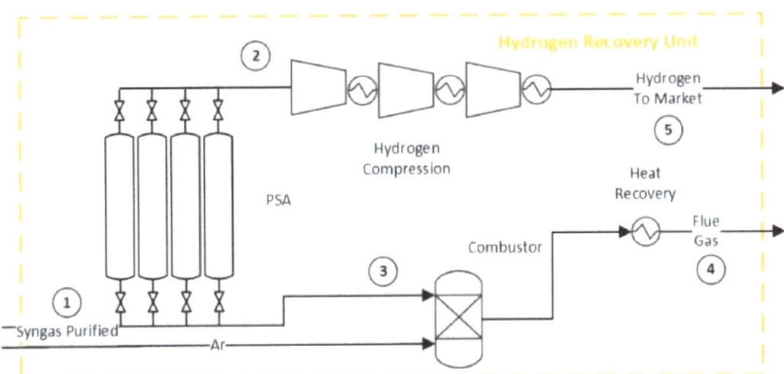

Figure 6. Flowsheet of the pressure swing adsorption for hydrogen recovery and purge gas combustion. Flow properties (1–5) can be found in the Supplementary Material, Tables S16–S18.

2.6. Ammonia Synthesis Loop

In the integrated route for ammonia production, the final purification stage (Figure 3) results in a syngas with the desired N_2/H_2 ratio and a small amount of inert gases. This purified syngas is then compressed at 200 bar and fed to the ammonia synthesis loop (Figure 7). Since reactants are not completely converted in one pass, the unreacted mixture is recompressed and recycled to the ammonia converter. A mixture of fresh and recycled syngas at 200 bar and 35 °C is preheated and introduced into the converter. In the converter, the ammonia synthesis reaction (R. 12) takes place in the presence of an iron-based catalyst, with a fractional conversion that typically ranges between 10% and 30%. The process design and operational parameters are based on refs. [21,45].

$$N_2 + 3H_2 \rightarrow 2NH_3 \qquad (\Delta h^0_{298k} = 92 \text{ kJ/kmol}) \tag{R. 12}$$

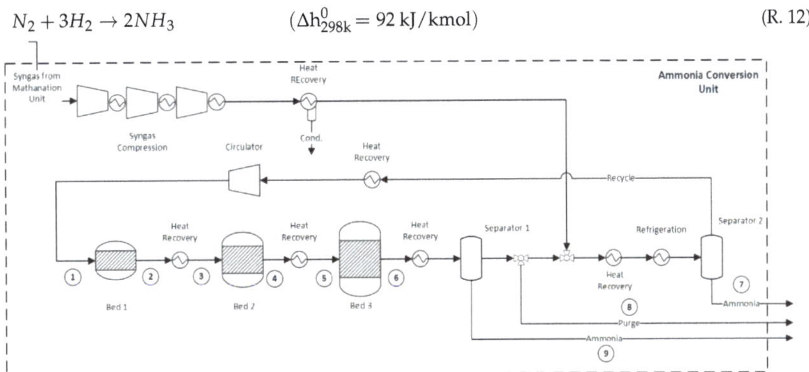

Figure 7. Flowsheet of the ammonia synthesis loop. Flow properties (1–9) can be found in the Supplementary Material, Tables S19–S21.

The ammonia synthesis is highly exothermic. As a result, moderate temperatures ranging from 350 °C to 550 °C are commonly employed [45] to achieve appropriate equilibrium conversion and an acceptable reaction rate. To effectively manage the temperature and optimize the performance of the ammonia synthesis process, three or more sequential catalytic beds with an intercooling system are adopted. By dividing the reaction into multiple catalytic beds and incorporating intercooling, the temperature can be better managed, reducing the risk of catalyst deactivation and improving overall process performance. Also, this setup enables higher per-pass conversions. After the ammonia synthesis, a significant portion of the produced ammonia is initially condensed using a water-cooling system. However, relying solely on water cooling does not provide satisfactory ammonia condensation. Thus, the unreacted mixture is further cooled to approximately −20 °C to increase the ammonia condensation and the overall efficiency of the ammonia loop. Finally, as an excessive build-up of methane (inert) has negative effects on the reaction conversion and circulation rate, a portion of the hydrogen-rich recycled gas is purged from the system. In this way, the overall inert concentration, including methane, is kept below a suitable threshold, typically 8% mol [45]. The characteristics of the streams associated with this unit are shown in detail in the Supplementary Material identified in tags 1 to 9 of Figure 7, Tables S19–S21.

2.7. Integrated Flowsheets of the Ammonia and Hydrogen Production Routes Using Residual Biomass

Figures 8 and 9 summarize the flowcharts of the hydrogen and ammonia production routes using residual biomass. The distinct features of the two production routes will be responsible for different energy demands, CO_2 emissions, and chemical production per unit of biomass consumed.

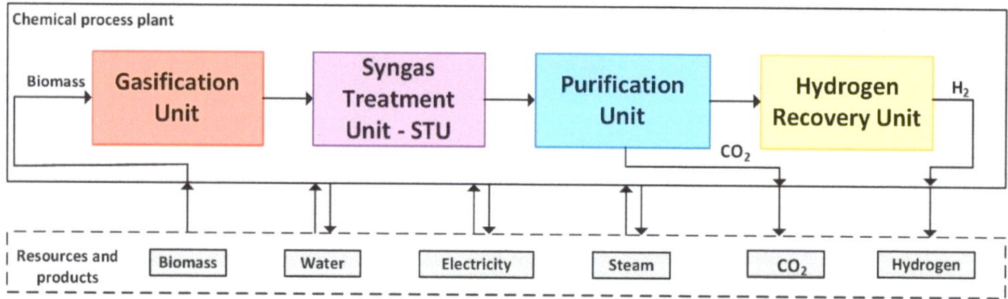

Figure 8. Flowcharts of the hydrogen production route from residual biomass.

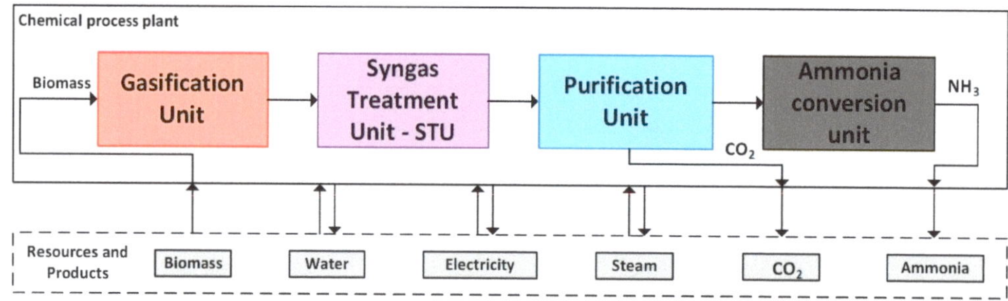

Figure 9. Flowcharts of the ammonia production route from residual biomass.

3. Materials and Methods

The mass, energy, and exergy balances for each unit operation of the chemical plants are carried out in this work. Indicators based on the exergy concept, namely, the plantwide and extended exergy efficiencies, as well as the CO_2 balance, are used to assess the hydrogen and ammonia production performance.

3.1. Process Modeling

The evaluation of the thermodynamic properties of each process flow, as well as the mass, energy, and exergy balances of each operation unit, is performed using Aspen Plus® V8.8 software [37]. The thermodynamic model used in gasification, treatment, hydrogen production, and ammonia synthesis is Peng-Robinson EOS with Boston–Mathias modifications [50]. On the other hand, for the simulation of the CO_2 capture unit using DPEG, the thermodynamic model of the theory of the statistical association of the chain (PC-SAFT) is adopted [21,47,51]. The gasification model is composed of sequential pre-treatment (dryer and chipping), pyrolysis, reduction, and combustion processes. Moisture removal is simulated by using a FORTRAN subroutine [21]. The calculation of the mass fractions of volatiles (x_j), condensables, and solids in the pyrolysis reaction step, as well as the gas volume fractions (v_i) of hydrogen, carbon monoxide, carbon dioxide, and methane produced, is carried out using a set of empirical correlations [52]. These correlations are functions of the reaction temperature (T) and are represented by Equations (2)–(8) [52]. A subroutine in MS Excel® integrated into the Aspen simulator performs the atomic balance of species (C, H, O, N, and S, Ash) present in the volatiles, condensables, char, and ash generated during the pyrolysis section.

$$x_{Gas} = 311.10 - 351.45\left(\frac{T}{500}\right) + 121.43\left(\frac{T}{500}\right)^2 \qquad \textit{Gases (\% mass of dry biomass)} \quad (2)$$

$$x_{Char} = -15.03 + 50.58\left(\frac{T}{500}\right) - 18.09\left(\frac{T}{500}\right)^2 \qquad \textit{Char (\% mass of dry biomass)} \quad (3)$$

$$x_{Tar} = -196.07 + 300.86\left(\frac{T}{500}\right) - 103.34\left(\frac{T}{500}\right)^2 \ \text{Tar(\% mass of dry biomass)} \quad (4)$$

$$y_{CO} = 240.53 - 225.12\left(\frac{T}{500}\right) + 67.50\left(\frac{T}{500}\right)^2 \ \text{CO (\% mole of gas)} \quad (5)$$

$$y_{CO_2} = -206.86 + 267.66\left(\frac{T}{500}\right) - 77.50\left(\frac{T}{500}\right)^2 \ CO_2(\% \text{ mole of gas}) \quad (6)$$

$$y_{CH_4} = -168.64 + 214.47\left(\frac{T}{500}\right) - 62.51\left(\frac{T}{500}\right)^2 \ CH_4(\% \text{ mole of gas}) \quad (7)$$

$$y_{H_2} = 234.97 - 257.01\left(\frac{T}{500}\right) + 72.50\left(\frac{T}{500}\right)^2 \ H_2(\% \text{ mole of gas}) \quad (8)$$

Compressors and pumps are modeled with 60% and 80% isentropic efficiencies, respectively. The PSA has a hydrogen recovery efficiency of 95% mol [48]. The determination of the chemical exergy adopts the standard environment model proposed by Szargut et al. [53] with reference conditions at T_0 = 298.15 K and P_0 = 101.3 kPa. The ratio of specific chemical exergy to lower heating value is calculated using the correlation proposed by ref. [53] for solid fuels with specified mass ratios, Equation (1).

$$\beta = \frac{b^{ch}}{LHV} = \frac{1.0438 + 0.1882\frac{H}{C} - 0.2509\left(1 + 0.7256\frac{H}{C}\right)}{1 - 0.3035\frac{O}{C}} \quad (9)$$

whereas the biomass lower heating value (LHV, MJ/kg) is estimated according to Equation (10) [54]:

$$LHV = 349.1C + 1178.3H + 100.5S - 103.4O - 15.1N - 21.5ASH - 0.0894h_{lv}H \tag{10}$$

The mass fractions of carbon (C), hydrogen (H), sulphur (S), oxygen (O), nitrogen (N), and ashes (A) in the dry biomass are reported in Table 1. In addition, h_{lv} is the enthalpy of evaporation of water at standard conditions (2442.3 kJ/kg). The calculated LHV and chemical exergy of biomasses are summarized in Table 4.

Table 4. Calculated lower heating value (LHV) and specific chemical exergy (b^{CH}) for the selected waste streams used in the gasification process.

Biomass	LHV (MJ/kg)	b^{CH} (MJ/kg)
Sugar cane bagasse	17.39	19.50
Sewage sludge	19.25	16.13
Orange bagasse	25.24	20.26

The chemical exergy of a mixture can be calculated using Equation (11) [55]:

$$\bar{b}_{ch,\,mist} = \sum_i x_i\bar{b}_{ch,\,i} + RT_0\sum_i x_i\ln\gamma_i x_i \tag{11}$$

where $\bar{b}_{ch,\,i}$ represents the standard chemical exergy of the substance i at P_0 and T_0; x_i is the molar fraction of the component i; R is the universal gas constant; and γ_i is the activity coefficient.

3.2. CO_2 Emissions

The general balance of CO_2 emissions (GBE) is performed according to Equation (3)

$$GBE = CO_{2\,Direct}^{\,Biogenic} + CO_{2\,indirect}^{\,Fossil} - CO_{2\,Avoided}^{\,Biogenic} \tag{12}$$

where direct biogenic CO_2 corresponds to direct emissions derived from the biomass conversion, such as the reactions in the gasifier. Since biomass-derived emissions could be considered circular emissions, the captured biogenic CO_2 emissions may improve the overall emissions balance by reducing the amount of CO_2 in the atmosphere (i.e., negative emissions). The indirect fossil CO_2 emissions consider those emissions that arise from the upstream supply chains of the electricity (62.09 gCO_2/kWh) [56], the sugarcane bagasse and the orange bagasse (0.0043 gCO_2/kJ$_{biomass}$) [57], as well as the sludge (0.0106 gCO_2/kJ$_{sludge}$) [58].

3.3. Exergy Efficiency

The overall exergy efficiency of the chemical production routes is evaluated using two performance indicators [4,56], namely the rational and the relative exergy efficiency. The rational efficiency, Equation (13), considers that all the outlets (incl. CO_2 and purge gas) of the chemical plant are products, while the relative exergy efficiency, Equation (14), is a measure of the deviation from the theoretical exergy consumption when only bio-products are produced in the plant. Thus, the second definition is more conservative, adopting lower values for those processes that produce less useful products.

$$\eta_{rational} = \frac{B_{useful,output}}{B_{input}} = 1 - \frac{B_{Dest}}{B_{input}} = 1 - \frac{B_{Dest}}{B_{biomass} + W_{imported}} \tag{13}$$

$$\eta_{relative} = \frac{B_{consumed,ideal}}{B_{consumed,actual}} = \frac{B_{bioproduct}}{B_{biomass} + W_{imported}} \tag{14}$$

where B is the exergy flow rate (kW) and B_{Dest} represents the exergy destruction rate. W is the electrical power imported from the grid. The bio-product refers either to hydrogen or ammonia.

3.4. Definition of the Optimization Problem

The minimum energy requirement (MER) is calculated using the OSMOSE Lua platform developed at the IPESE group of the Federal Polytechnique School of Lausanne—EPFL, in Switzerland [59]. To calculate the MER, each hot and cold stream contribution to the overall heat balance is considered and incorporated into the respective hot and cold composite curves. The minimum temperature difference (ΔT_{min}) concept is employed to ensure reasonable heat transfer rates, and its value varies depending on the characteristics of the heat flow. For gaseous, liquid, and two-phase flows, a respective temperature difference contribution of 8 °C, 5 °C, and 2 °C is adopted [60]. The objective function and the associated constraints of the MER optimization problem are shown in Equations (15)–(17):

$$Min_{R_r} R_{N_r+1} \tag{15}$$

Subject to heat balance of each interval of temperature r

$$\sum_{i=1}^{N} Q_{i,r} + R_{i,r} - R_r = 0 \forall_r = 1 \ldots N \tag{16}$$

$$\text{Feasibility of the solution } R_r \geq 0 \tag{17}$$

where N is the number of temperature intervals defined by considering the supply and the target temperatures of the entire set of streams, and Q is the heat exchanged between the process streams ($Q_{i,r} > 0$ hot streams, <0 cold streams). Finally, R is the heat cascaded from higher ($r + 1$) to lower (r) temperature intervals (kW).

4. Results and Discussion

Due to its impact on global process energy efficiency and chemical yield, the gasification system is considered the most important unit. Thus, the results obtained from the simulation of the gasification system were validated using the study conducted by Marcantonio et al. [41] using walnut husk (M_{db}: 12%, Ash_{db}: 1.2%, VM_{db}: 80.6%, FC_{db}: 18.2%, C: 47.9%, H: 6.3%, N: 0.32%, O: 44.27%, S: 0.015%). The comparative results shown in Figure 10 show good agreement with the reported study. The most significant deviation was found for CO_2 (5%), whereas, for the other substances, the error of the simulation was less than 3%. It can be attributed to the inherent complexities of the gasification reactions and the uncertainties associated with biomass composition.

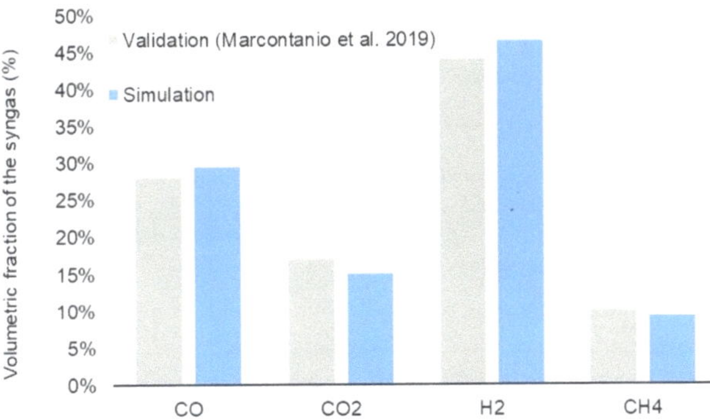

Figure 10. Comparison of the biomass gasification modeling results between the simulation in this work and the literature data reported by Marcantonio et al. [41] for walnut husk.

According to Figure 11, among the investigated biomass residues, orange peel gasification exhibits the highest cold gas efficiency (80.66%), which implies that a substantial portion of the energy content in the waste material is effectively converted into syngas. Conversely, sugarcane bagasse gasification shows the highest carbon conversion efficiency (92.88%), indicating that a major proportion of the carbon in the residual biomass is successfully converted into syngas components. However, it is also important to mention that sugarcane bagasse conversion exhibits the lowest cold gas efficiency (77.62%) among the studied configurations.

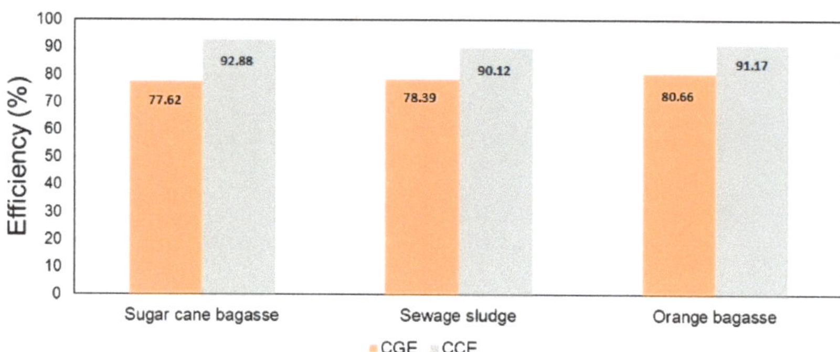

Figure 11. Cold gas efficiency (CGE) and carbon conversion efficiency (CCE) in the gasification process of biomass residues.

Tables 5 and 6 present a breakdown of the exergy destruction among the main equipment and processes of the different biomass conversion routes. As expected, the gasifier contributes the largest share of exergy destruction in the plants. The biomass grinding and drying and syngas scrubbing and compression are also accounted for as part of the gasification unit, which is in agreement with other studies [21,38].

Table 5. Breakdown of the exergy destruction in the biomass to ammonia conversion routes.

	Sugar Cane Bagasse	Sewage Sludge	Orange Bagasse
Gasification (%)	68.1	69.8	73.6
Chipping (%)	2.0	2.1	2.2
Dryer (%)	3.5	2.6	1.6
Scrubber (%)	3.6	3.9	2.7
ATR (%)	2.3	2.5	2.1
Shift reactors (%)	0.9	0.9	0.8
Physical absorption (%)	3.4	2.9	3.3
Methanator (%)	0.2	0.2	0.3
Compression (%)	4.3	3.3	2.4
Ammonia reactors (%)	4.0	4.0	3.1
Others (%)	7.7	7.8	7.9

Table 6. Breakdown of the exergy destruction in the biomass to hydrogen conversion routes.

	Sugar Cane Bagasse	Sewage Sludge	Orange Bagasse
Gasification (%)	56.7	57.8	63
Chipping (%)	4.6	4.6	4.3
Dryer (%)	2.6	1.9	1.2

Table 6. *Cont.*

	Sugar Cane Bagasse	Sewage Sludge	Orange Bagasse
Scrubber (%)	2.7	2.9	2.1
Compression (%)	2.4	2.7	1.9
Shift reactors (%)	0.3	0.3	0.2
Physical absorption (%)	7.0	5.9	6.0
PSA combustor (%)	19.9	19.8	17.6
Others (%)	3.8	4.1	3.7

Considering the different residual biomass conversion routes, the process that presents the highest fraction of exergy destruction in relation to the total exergy destruction in the plant was the orange bagasse gasifier for ammonia conversion (73.6%), as shown in Table 5. On the other hand, gasification via sugarcane bagasse has the lowest exergy destruction share (68.1%). According to Table 5, the ATR only contributes 2.0% to the total exergy destruction in the plant, despite the partial combustion of the produced syngas. Compression systems have relatively high participation in the irreversibility of the whole energy conversion system (5%), especially in the case of ammonia production via sugarcane bagasse. This circumstance is due to the fact that a large amount of syngas compression entails the loss of valuable energy in the form of waste heat. A way to help reduce the amount of exergy destroyed in Thomass-based production plants is to employ better technologies to remove bagasse moisture as well as implement hot catalytic cleaning of the syngas, thus avoiding the waste heat in the water scrubbing section. An increase in the gasifier pressures would also help avoid excessive compression power consumption [61].

The difference in the exergy destruction in the hydrogen (Table 5) and ammonia (Table 6) production routes can be partly explained by the irreversible combustion process of the purge gas in the former route, along with higher power consumption by the hydrogen compression and export system. The gasifier's relative contribution to the overall exergy destruction is thus smaller in the context of the hydrogen conversion routes. It should also be kept in mind that the amount of exergy recovered per ton of ammonia produced is higher than in the case of hydrogen production routes, even though the latter route has a smaller number of unit operations.

The plantwide exergy efficiency (i.e., without considering the supply chain efficiency), shown in Figure 12, also exhibits this trend. The performance of the hydrogen production route is lower than that of the ammonia production route due to a more stringent purification system and higher compression levels. A large production of offgas and its flaring impairs further its exergy efficiency [29]. Compression and intercooling also require a significant amount of energy input per unit of hydrogen produced. In contrast, when liquid ammonia is expanded, energy can be harnessed through expansion, thus partially recovering the compression power [61,62].

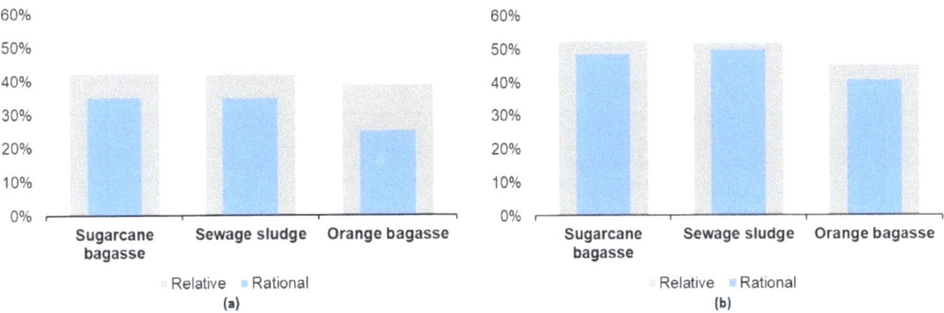

Figure 12. Comparison of the exergy efficiencies of (**a**) hydrogen and (**b**) ammonia production routes using different types of residual biomass.

4.1. Energy Integration Analysis and Power Generation Potential

The energy integration approach relies on pinch analysis methodology to maximize waste heat recovery throughout the plant. It allows for calculating the minimum energy requirements (MER) of the chemical processes. From the analysis of the composite curves presented in Figure 13, enough waste heat is available from the biomass conversion routes of agricultural waste and sewage sludge, avoiding the need for additional fuel imports. It will still need an additional cooling requirement, such as that provided by a cooling water system.

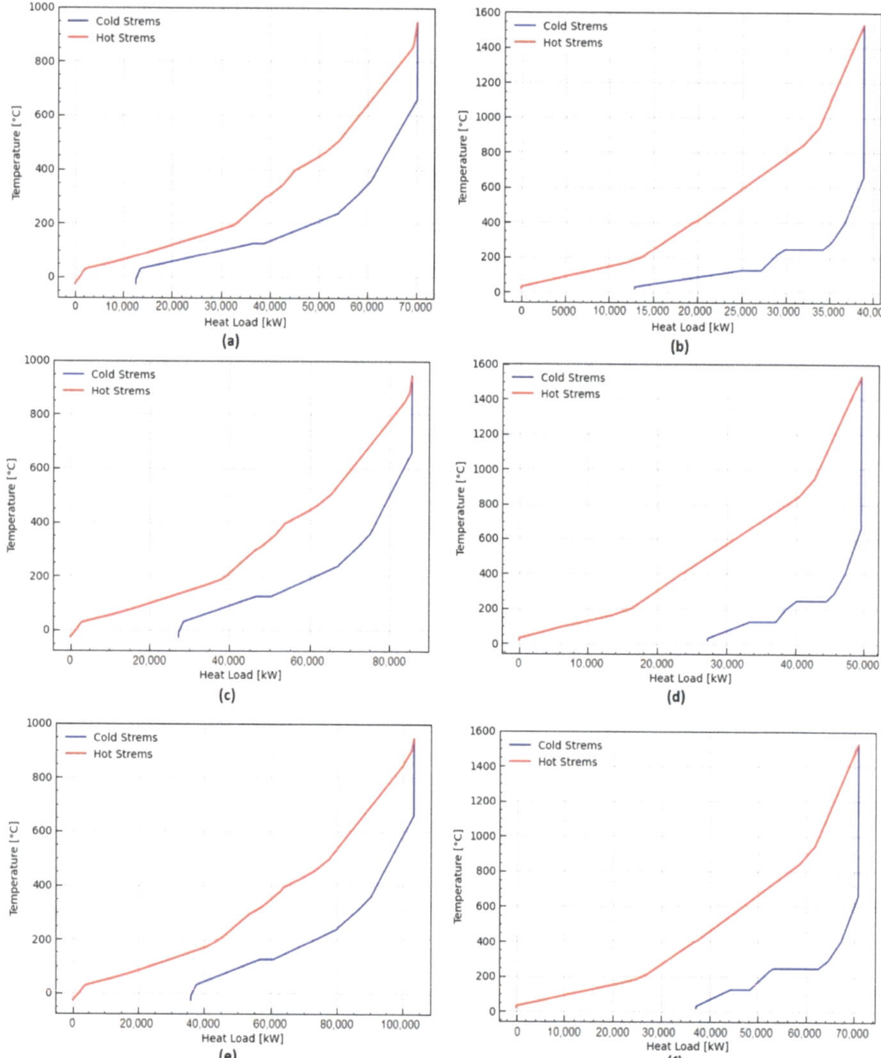

Figure 13. Cold and hot composite curves for different waste biomass conversion processes exhibiting no need for external heating requirements, but showing the need for further cooling requirements: (**a**) sugarcane bagasse to hydrogen; (**b**) sugarcane bagasse to ammonia; (**c**) sewage sludge to hydrogen; (**d**) sewage sludge to ammonia; (**e**) orange bagasse to hydrogen; (**f**) orange bagasse to ammonia.

On the other hand, the electricity requirements, such as compression, refrigeration, pumping, and grinding, could be satisfied either by importing renewable electricity from the electricity mix or by self-generating some power using a Rankine cycle. Waste heat available from the chemical systems also suggests opportunities for providing waste heat to a nearby urban settlement or, depending on the temperature levels, using the waste heat to produce refrigeration using absorption refrigeration systems. The total amount of waste heat cascading available at a high temperature can be better appreciated from Figure 14a–f. In order to quantify the potential power generation using a Rankine cycle, a temperature of waste heat recovery steam generation of 400 °C and a condensation temperature of 25 °C are adopted. Assuming a realistic Carnot efficiency of 50% and considering the waste heat cascade shown in Figure 14a–f, the power generated in a Rankine cycle-based power plant operating at the mentioned temperatures can be calculated and is reported in Table 7. Major differences between the power generation potential of the ammonia and the hydrogen production routes are observed. The potential for power generation in the ammonia production routes is higher than in hydrogen production.

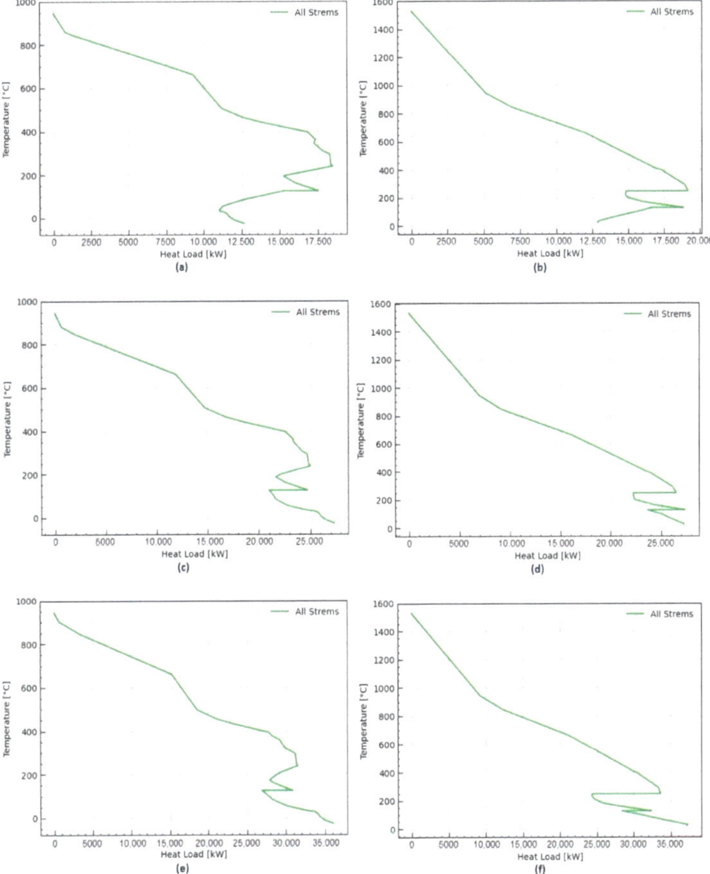

Figure 14. Grand composite curves for different waste biomass conversion processes exhibiting no need for external heating requirements, but showing the need for further cooling requirements: (**a**) sugarcane bagasse to hydrogen; (**b**) sugarcane bagasse to ammonia; (**c**) sewage sludge to hydrogen; (**d**) sewage sludge to ammonia; (**e**) orange bagasse to hydrogen; (**f**) orange bagasse to ammonia.

Table 7. Power generation potential using a Rankine cycle-based power plant with a waste heat recovery steam generator at 400 °C and a condenser at 25 °C that recovers heat throughout the chemical production plants.

Chemical Plant	Power Generated
Sugarcane bagasse to hydrogen	6208 kW
Sugarcane bagasse to ammonia	7259 kW
Sewage sludge to hydrogen	11,835 kW
Sewage sludge to ammonia	13,147 kW
Orange bagasse to hydrogen	13,735 kW
Orange bagasse to ammonia	15,171 kW

The exothermic reactions involved in the ammonia synthesis contribute to the higher power output observed in the ammonia production route. Among the biomass residues studied, the orange peel biomass exhibits the highest potential for power output (15,171 kW), while the sugarcane biomass conversion route shows the lowest potential for power output (7259 kW), which can be attributed to the properties of the biomass used. The hydrogen production routes show lower power generation potential. The orange peel conversion has the highest potential for power generation (13,735 kW) when used to produce hydrogen, while the sugarcane conversion route has a small power generation potential (6208 kW) in a Rankine cycle-based power plant when used to produce hydrogen. These results highlight the influence of biomass composition and its conversion process on the r generation potential.

4.2. General CO_2 Emissions Balance

Figure 15a,b, and Tables 8 and 9 summarize the results of the balance of CO_2 emissions for each biomass-based chemical production route. As it can be seen, the indirect fossil contributions to the emissions balance are not negligible, which reveals environmental burdens that might otherwise remain hidden if imported electricity or biomass were considered emission-free inputs. The indirect emissions from the sewage sludge supply chain (3.89 kg_{CO2}/kg_{H2}) are the largest among all the chemical production routes. Nevertheless, all the hydrogen production routes using any residual biomass present an overall negative balance of emissions, as the avoided emissions offset the effect of the indirect ones. Among the hydrogen production routes, the conversion route of orange peel showed the best performance in terms of negative emissions. The biomass conversion plants captured a significant amount of emissions along the supply chain, thus making a positive contribution to the environmental impact. Considering the CO_2 emissions balances for the chemical processes of ammonia production (Figure 8b), the conversion route using sewage sludge presents the worst performance in terms of emissions balance, although negative emissions can still be obtained (-0.448 kg_{CO2}/kg_{NH3}). The biomass conversion route of orange peel to produce ammonia proved to be the best solution for the improvement of the global emissions balance (-1.615 kg_{CO2}/kg_{NH3}). It is worth mentioning that the conversion routes using sugarcane bagasse have shown excellent performance in terms of CO_2 emissions reduction. The utilization of these biomasses for producing hydrogen and ammonia as value-added products shows negative values for the emission balance for all the conversion routes.

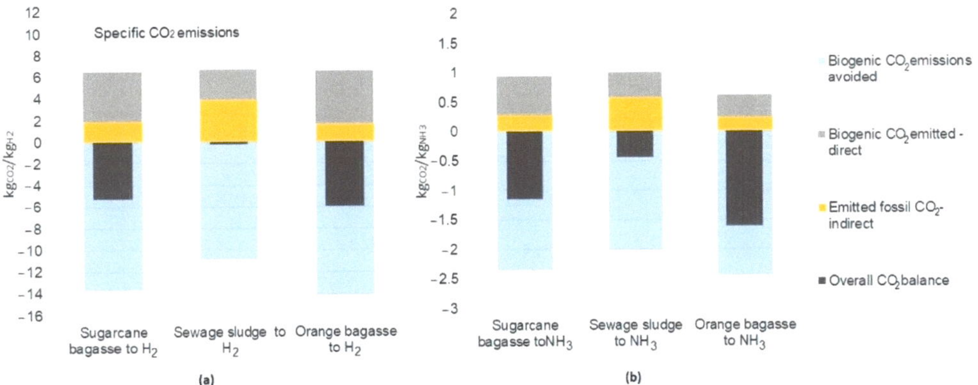

Figure 15. General and detailed emissions (biogenic and fossil, emitted directly, indirectly, and avoided) for the conversion process of (**a**) hydrogen and (**b**) ammonia for different types of selected biomass.

Table 8. CO_2 emissions and other exergy consumption remarks for hydrogen production using different types of waste biomass.

Process Parameter	Sugarcane Bagasse	Sewage Sludge	Orange Bagasse
Biomass Consumption ($t_{biomass}/t_{H2}$)	27.39	20.54	15.86
Syngas produced in the gasifier (MJ/kg_{H2})	188.82	187.96	237.24
Hydrogen Produced (t_{H2}/day)	23.32	31.13	40.32
Heating requirement [1] (GJ/t_{H2})	0.00	0.00	0.00
Cooling requirement [1] (GJ/t_{H2})	47.60	75.66	83.64
Captured CO_2 ($t_{CO2}/t_{biomass}$)	0.503	0.534	0.901
Fossil CO_2 emitted—indirect [2] (kg_{CO2}/kg_{H2})	1.919	3.896	1.629
Indirect emitted CO_2—EE (%)	0.081	0.078	0.072
Indirect emitted CO_2—Biomass (%)	0.919	0.922	0.928
Total fossil CO_2 emitted (kg_{CO2}/kg_{H2})	1.919	3.896	1.629
Biogenic CO_2 emissions avoided [3] ($kg_{CO2}/kg\ H2$)	13.682	10.869	14.166
Biogenic CO_2 emitted—direct ($kg_{CO2}/kg\ H2$)	6.527	6.747	6.584
Total atmospheric emissions ($kg_{CO2}/kg\ H2$)	8.447	10.643	8.213
General balance of CO_2 emissions [4] ($kg_{CO2}/kg\ H2$)	−5.235	−0.226	−5.953

[1]—Chemical process heating requirements (energy basis) determined from the composite curves. [2]—Considers indirect emissions due to sewage sludge (0.0106 gCO_2/kJ_{sludge}) [58], electricity (62.09 gCO_2/kWh), and residual bagasse (0.0043 $gCO_2/kJ_{biomass}$) supply chains [46,56]; [3]—CO_2 emissions captured through the physical absorption system; [4]—considers the total CO_2 emitted (fossil or biogenic) minus the biogenic CO_2 captured.

Table 9. CO_2 emissions and other exergy consumption remarks for ammonia production using different types of waste biomass.

Process Parameter	Sugarcane Bagasse	Sewage Sludge	Orange Bagasse
Biomass Consumption ($t_{biomass}/tN_{H3}$)	3.93	3.05	2.26
Syngas produced in the gasifier (MJ/kg_{NH3})	28.02	27.93	34.99
Ammonia produced (t_{NH3}/day)	157.16	209.47	273.43
Heating requirement [1] (GJ/t_{NH3})	0.00	0.00	0.00
Cooling requirement [1] (GJ/t_{NH3})	6.72	11.28	11.36
Captured CO_2 ($t_{CO2}/t_{biomass}$)	0.603	0.668	1.096
Fossil CO_2 emitted—indirect [2] (kg_{CO2}/kg_{NH3})	0.272	0.572	0.230
Indirect emitted CO_2—EE (%)	0.101	0.093	0.091
Indirect emitted CO_2—Biomass (%)	0.899	0.907	0.909
Total fossil CO_2 emitted (kg_{CO2}/kg_{NH3})	0.272	0.572	0.230
Biogenic CO_2 emissions avoided [3] (kg_{CO2}/kg_{NH3})	2.351	2.022	2.450

Table 9. *Cont.*

Process Parameter	Sugarcane Bagasse	Sewage Sludge	Orange Bagasse
Biogenic CO_2 emitted—direct (kg_{CO2}/kg_{NH3})	0.936	1.003	0.606
Total atmospheric emissions (kg_{CO2}/kg_{NH3})	1.208	1.574	0.835
General balance of CO_2 emissions [4] (kg_{CO2}/kg_{NH3})	−1.142	−0.448	−1.615

[1]—Chemical process heating requirements (energy basis) determined from the composite curves. [2]—Considers indirect emissions due to sewage sludge (0.0106 g_{CO2}/kJ_{sludge}) [58], electricity (62.09 g_{CO2}/kWh), and residual bagasse (0.0043 $g_{CO2}/kJ_{biomass}$) supply chains [46,56]; [3]—CO_2 emissions captured through the physical absorption system; [4]—considers the total CO_2 emitted (fossil or biogenic) minus the biogenic CO_2 captured.

5. Conclusions

In this work, the use of residual biomass gasification in integrated chemical production plants is presented. The energy integration and extended exergy analyses allowed us to point out the opportunities to maximize the recovery of available waste heat exergy throughout the plant. As a result, the implementation of a Rankine cycle allowed the recovery of residual heat from biomass conversion in the ammonia production route, resulting in a potential power output of approximately 15,171 kW. Similarly, in the hydrogen production route, the power generation potential reached 13,735 kW. The sugarcane bagasse-based route shows the highest hydrogen yield rate (40.32 t H_2 per day) and the largest ammonia production rate (237.43 t NH_3 per day). The exergy efficiencies calculated ranged from 39% to 43% for hydrogen production routes and from 46% to 57% for ammonia production routes. The overall emission balances ranged from −0.226 to −5.953 kg_{CO2}/kg_{H2} and −0.448 to −1.615 kg_{CO2}/kg_{NH3}, respectively. Negative values point towards the environmental benefit of producing chemical products through residual biomass by depleting CO_2 from the atmosphere. Many efforts in the research and development of technologies for more efficient conversion of renewable energy sources should aim to boost alternative routes of production of chemicals at larger scales. It should be noted that by defining the extended plant consumption and the extended efficiency concepts, the real effect of the production process, including the upstream supply chain inefficiencies, can be assessed. In this way, the results proved to be strongly dependent on the indirect fossil emissions of those supply chains. In fact, the contribution to atmospheric emissions is not negligible, and it reveals environmental issues that might otherwise remain hidden if imported electricity or biomass were considered emission-free energy inputs.

Supplementary Materials: The following supporting information can be downloaded at: https://www.mdpi.com/article/10.3390/e25071098/s1.

Author Contributions: Conceptualization, G.G.V. and D.A.F.-O.; Methodology, G.G.V. and D.A.F.-O.; Formal analysis, G.G.V., D.A.F.-O. and S.d.O.J.; Writing—original draft, G.G.V.; Writing—review & editing, D.A.F.-O. and S.d.O.J.; Supervision, S.d.O.J. All authors have read and agreed to the published version of the manuscript.

Funding: This study was financed in part by the Coordenação de Aperfeiçoamento de Pessoal de Nível Superior—Brasil (CAPES)—Finance Code 001. The first author acknowledges CAPES for his Ph.D. grant. The second author acknowledges the Colombian Ministry of Science, Technology, and Innovation—MINCIENCIAS (grant no. 1.128.416.066-646/2014). The third author acknowledges CNPq (the Brazilian National Council for Scientific and Technological Development) for the grant 306484/2020-0.

Institutional Review Board Statement: Not applicable.

Data Availability Statement: Not applicable.

Conflicts of Interest: The authors declare no conflict of interest.

Nomenclature

Latin symbols

M	moisture content (%)
VM	volatile matter content (%)
FC	fixed carbon content (%)
C	carbon (%)
H	hydrogen (%)
N	nitrogen (%)
S	sulphur (%)
Cl	chlorine (%)
O	oxygen (%)
b	specific chemical exergy (kJ/kg)
h	enthalpy (kJ/kg)
B	chemical exergy flow rate (kW)
W	electrical power (kW)
R	cascaded heat transfer rate (kW)
Q	heat exchanged (kJ)
y	molar fraction (-)
T	Temperature ($^\circ$C, K)
N	number of intervals (-)

Superscript

CH	chemical exergy

Subscripts

db	dry basis
r	interval of temperature

Greek symbols

β	ratio of specific chemical exergy
η	exergy efficiency
ψ	Moisture (%)

Abbreviations

DEPG	dimethyl ethers of polyethylene glycols
PSA	pressure swing adsorption
GBE	general balance of emissions ($t_{CO2}/t_{product}$)
LHV	lower heating value (kJ/kg)
MER	minimum energy requirement (kW)

References

1. Thraen, D.; Shaubach, K.; Global Wood Pellet Industry and Trade Study 2017. IEA Bioenergia 2017. Available online: https://www.ieabioenergy.com/blog/publications/two-page-summary-global-wood-pellet-industry-and-trade-study-2017/ (accessed on 5 September 2022).
2. EPE. Brazilian Energy Balance Year 2016. 2016. Available online: https://www.epe.gov.br/en/publications/publications/brazilian-energy-balance/brazilian-energy-balance-2016 (accessed on 2 March 2023).
3. Marcio, R.; Santos, D.; De Alencar Nääs, I.; Neto, M.M.; Vendrametto, O. An overview on the brazilian orange juice production chain 1 uma visão da produção brasileira de suco de laranja. *Rev. Bras. Frutic. Jaboticabal SP* **2013**, *35*, 218–255.
4. Nakashima, R.; Flórez-Orrego, D.; de Oliveira Junior, S. Integrated anaerobic digestion and gasification processes for upgrade of ethanol biorefinery residues. *J. Power Technol.* **2019**, *99*, 104–114.
5. Pellegrini, L.F.; Deoliveirajr, S. Exergy analysis of sugarcane bagasse gasification. *Energy* **2007**, *32*, 314–327. [CrossRef]
6. González, D.; Colón, J.; Gabriel, D.; Sánchez, A. The effect of the composting time on the gaseous emissions and the compost stability in a full-scale sewage sludge composting plant. *Sci. Total Environ.* **2019**, *654*, 311–323. [CrossRef]
7. Swati, A.; Hait, S. Fate and bioavailability of heavy metals during vermicomposting of various organic wastes—A review. *Process Saf. Environ. Prot.* **2017**, *109*, 30–45. [CrossRef]
8. Biagini, E.; Barontini, F.; Tognotti, L. Gasification of agricultural residues in a demonstrative plant: Corn cobs. *Bioresour. Technol.* **2014**, *173*, 110–116. [CrossRef] [PubMed]
9. Vera, D.; Jurado, F.; Margaritis, N.K.; Grammelis, P. Experimental and economic study of a gasification plant fuelled with olive industry wastes. *Energy Sustain. Dev.* **2014**, *23*, 247–257. [CrossRef]

10. Biagini, E.; Barontini, F.; Tognotti, L. Gasification of agricultural residues in a demonstrative plant: Vine pruning and rice husks. *Bioresour. Technol.* **2015**, *194*, 36–42. [CrossRef]

11. Guo, X.M.; Trably, E.; Latrille, E.; Carrère, H.; Steyer, J.-P. Hydrogen production from agricultural waste by dark fermentation: A review. *Int. J. Hydrog. Energy* **2010**, *35*, 10660–10673. [CrossRef]

12. Ghani, W.A.W.A.K.; Moghadam, R.A.; Salleh, M.A.M.; Alias, A.B. Air Gasification of Agricultural Waste in a Fluidized Bed Gasifier: Hydrogen Production Performance. *Energies* **2009**, *2*, 258–268. [CrossRef]

13. Durán-Sarmiento, M.A.; Del Portillo-Valdés, L.A.; Rueda-Ordonez, Y.J.; Florez-Rivera, J.S.; Rincón-Quintero, A.D. Study on the gasification process of the biomass obtained from agricultural waste with the purpose of estimating the energy potential in the Santander region and its surroundings. *IOP Conf. Ser. Mater. Sci. Eng.* **2022**, *1253*, 012001. [CrossRef]

14. Widjaya, E.R.; Chen, G.; Bowtell, L.; Hills, C. Gasification of non-woody biomass: A literature review. *Renew. Sustain. Energy Rev.* **2018**, *89*, 184–193. [CrossRef]

15. Göransson, K.; Söderlind, U.; He, J.; Zhang, W. Review of syngas production via biomass DFBGs. *Renew. Sustain. Energy Rev.* **2011**, *15*, 482–492. [CrossRef]

16. Perez, D.L.; Brown, A.; Mudhoo, A.; Timko, M.; Rostagno, M.; Forster-Carneiro, T. Applications of subcritical and supercritical water conditions for extraction, hydrolysis, gasification, and carbonization of biomass: A critical review. *Biofuel Res. J.* **2017**, *4*, 611–626. [CrossRef]

17. Koppatz, S.; Pfeifer, C.; Rauch, R.; Hofbauer, H.; Marquard-Moellenstedt, T.; Specht, M. H_2 rich product gas by steam gasification of biomass with in situ CO_2 absorption in a dual fluidized bed system of 8 MW fuel input. *Fuel Process Technol.* **2009**, *90*, 914–921. [CrossRef]

18. Leckner, B. Developments in fluidized bed conversion of solid fuels. *Therm. Sci.* **2016**, *20*, 135. [CrossRef]

19. Kraussler, M.; Binder, M.; Hofbauer, H. Behavior of GCMS tar components in a water gas shift unit operated with tar-rich product gas from an industrial scale dual fluidized bed biomass steam gasification plant. *Biomass Convers. Biorefinery* **2016**, *7*, 69–79. [CrossRef]

20. Thunman, H.; Seemann, M.; Vilches, T.B.; Maric, J.; Pallares, D.; Ström, H.; Berndes, G.; Knutsson, P.; Larsson, A.; Breitholtz, C.; et al. Advanced biofuel production via gasification—Lessons learned from 200 man-years of research activity with Chalmers' research gasifier and the GoBiGas demonstration plant. *Energy Sci. Eng.* **2018**, *6*, 6–34. [CrossRef]

21. Flórez-Orrego, D.; Maréchal, F.; Junior, S.D.O. Comparative exergy and economic assessment of fossil and biomass-based routes for ammonia production. *Energy Convers. Manag.* **2019**, *194*, 22–36. [CrossRef]

22. Budzianowski, W.M. Low-carbon power generation cycles: The feasibility of CO_2 capture and opportunities for integration. *J. Power Technol.* **2011**, *91*, 6–13.

23. Palacios-Bereche, R.; Mosqueira-Salazar, K.J.; Modesto, M.; Ensinas, A.V.; Nebra, S.A.; Serra, L.M.; Lozano, M.-A. Exergetic analysis of the integrated first- and second-generation ethanol production from sugarcane. *Energy* **2013**, *62*, 46–61. [CrossRef]

24. Ribeiro Domingos, M.; Florez-Orrego, D.; Maréchal, F. Comparison of regression techniques for generating surrogate models to predict the thermodynamic behavior of biomass gasification systems. In Proceedings of the 33rd European Symposium on Computer Aided Process Engineering, Athens, Greece, 18–21 June 2023.

25. Allesina, G.; Pedrazzi, S.; Guidetti, L.; Tartarini, P. Modeling of coupling gasification and anaerobic digestion processes for maize bioenergy conversion. *Biomass Bioenergy* **2015**, *81*, 444–451. [CrossRef]

26. Ribeiro Domingos, M.E.G.; Flórez-Orrego, D.; Teles Dos Santos, M.; Oliveira, S., Jr.; Marechal, F. Incremental financial analysis of black liquor upgraded gasification in integrated kraft pulp and ammonia production plants under uncertainty of feedstock costs and carbon taxes. In Proceedings of the 32nd European Symposium on Computer Aided Process Engineering, Toulouse, France, 12–15 June 2022.

27. Gassner, M.; Maréchal, F. Increasing Efficiency of Fuel Ethanol Production from Lignocellulosic Biomass by Process Integration. *Energy Fuels* **2013**, *27*, 2107–2115. [CrossRef]

28. Telini, R.; Florez-Orrego, D.; Oliveira, S., Jr. Techno-economic and environmental assessment of ammonia production from residual bagasse gasification: A decarbonization pathway for nitrogen fertilizers. *Front. Energy Res.* **2022**, *10*, 881263. [CrossRef]

29. Tock, L.; Maréchal, F. Co-production of hydrogen and electricity from lignocellulosic biomass: Process design and thermo-economic optimization. *Energy* **2012**, *45*, 339–349. [CrossRef]

30. Vargas, G.G.; Oliveira, S. Exergy Assessment of Electricity Generation via Biomass Gasification by Neural Network Algorithm. In Proceedings of the 36th International Conference on Efficiency, Cost, Optimization, Simulation and Environmental Impact of Energy Systems, Las Palmas, Spain, 25–30 June 2023.

31. Ribeiro Domingos, M.E.G.; Flórez-Orrego, D.; Teles Dos Santos, M.; Oliveira Junior, S.; Maréchal, F. Multi-time integration approach for combined pulp and ammonia production and seasonal CO_2 management. *Comput. Chem. Eng.* **2023**, *176*, 108305. [CrossRef]

32. Vargas, G.G.; Florez-Orrego, D.; Oliveira, S. Comparative exergy assessment of residual biomass gasification routes for hydrogen and ammonia production. In Proceedings of the 35th International Conference on Efficiency, Cost, Optimization, Simulation and Environmental Impact of Energy Systems, Copenhagen, Denmark, 3–7 July 2022.

33. SABESP. State Basic Sanitation Company from Sao Paulo. Available online: www.sabesp.gov.br (accessed on 2 March 2023). (In Portugese)

34. Sues Caula, A. *Are European Bioenergy Targets Achievable?: An Evaluation Based on Thermoeconomic and Environmental Indicators*; Technische Universiteit Eindhoven: Eindhoven, The Netherlands, 2011. [CrossRef]

35. Basu, P. *Biomass Gasification and Pyrolysis: Pratical Design and Theory*; Academic Press: Cambridge, MA, USA, 2010.

36. Bergman, P.C.A.; Boersman, A.R.; Kiel, J.H.A.; Prins, M.J.; Ptasinski, K.J.; Janssen, F.J.J.G. Torrefaction for Entrained-Flow Gasification of Biomass Revisions A B Made by. Available online: www.ecn.nl/biomass (accessed on 11 August 2021).

37. ASPENTECH. *Aspen Plus V8.8*; Aspen Technology Inc.: Bedford, MA, USA, 2015.

38. Ribeiro Domingos, M.E.G.; Orrego, D.F.; dos Santos, M.T.; Velásquez, H.I.; Junior, S.D.O. Exergy and environmental analysis of black liquor upgrading gasification in an integrated kraft pulp and ammonia production plant. *Int. J. Exergy* **2021**, *35*, 35. [CrossRef]

39. Kinchin, C.M.; Bain, R.L. *Hydrogen Production from Biomass via Indirect Gasification: The Impact of NREL Process Development Unit Gasifier Correlations*; National Renewable Energy Laboratory: Golden, CO, USA, 2009. [CrossRef]

40. Puig-Gamero, M.; Argudo-Santamaria, J.; Valverde, J.; Sánchez, P.; Sanchez-Silva, L. Three integrated process simulation using aspen plus®: Pine gasification, syngas cleaning and methanol synthesis. *Energy Convers. Manag.* **2018**, *177*, 416–427. [CrossRef]

41. Marcantonio, V.; De Falco, M.; Capocelli, M.; Bocci, E.; Colantoni, A.; Villarini, M. Process analysis of hydrogen production from biomass gasification in fluidized bed reactor with different separation systems. *Int. J. Hydrog. Energy* **2019**, *44*, 10350–10360. [CrossRef]

42. Spath, P.; Aden, A.; Eggeman, T.; Ringer, M.; Wallace, B.; Jechura, J. *Biomass to Hydrogen Production Detailed Design and Economics Utilizing the Battelle Columbus Laboratory Indirectly-Heated Gasifier*; National Renewable Energy Laboratory: New York, NY, USA, 2005. [CrossRef]

43. Spath, P.L.; Dayton, D.C. Preliminary Screening—Technical and Economic Assessment of Synthesis Gas to Fuels and Chemicals with Emphasis on the Potential for Biomass-Derived Syngas. 2003. Available online: http://www.osti.gov/bridge (accessed on 3 May 2022).

44. Gail, E.; Gos, S.; Kulzer, R.; Lorösch, J.; Rubo, A. *Ullmann's Encyclopedia of Industrial Chemistry*, 5th ed.; Wiley: Hoboken, NJ, USA, 2012; pp. 673–710.

45. Flórez-Orrego, D.; Oliveira Junior, S. On the allocation of the exergy costs and CO_2 emission cost for an integrated syngas and ammonia production plant. In Proceedings of the ECOS 2015—28th International Conference on Efficiency Cost, Optimization Simulation and Environmental Impact Energy Systtems, Pau, France, 29 June–3 July 2015.

46. Flórez-Orrego, D.; Oliveira Junior, S. On the efficiency, exergy costs and CO_2 emission cost allocation for an integrated syngas and ammonia production plant. *Energy* **2016**, *117*, 341–360. [CrossRef]

47. Adams, T.A.; Salkuyeh, Y.K.; Nease, J. Chapter 6—Processes and simulations for solvent-based CO_2 capture and syngas cleanup. In *Reactor and Process Design in Sustainable Energy Technology*; Shi, F., Ed.; Elsevier: Amsterdam, The Netherland, 2014; pp. 163–231.

48. Kuo, P.-C.; Illathukandy, B.; Wu, W.; Chang, J.-S. Energy, exergy, and environmental analyses of renewable hydrogen production through plasma gasification of microalgal biomass. *Energy* **2021**, *223*, 120025. [CrossRef]

49. Pires, A.P.B.; Filho, V.F.D.S.; Alves, J.L.F.; Marangoni, C.; Bolzan, A.; Machado, R.A.F. Application of a new pilot-scale distillation system for monoethylene glycol recovery using an energy saving falling film distillation column. *Chem. Eng. Res. Des.* **2019**, *153*, 263–275. [CrossRef]

50. ASPENTECH. *Aspen Physical Property System—Physical Property Methods*; Aspen Technology, Inc.: Cambridge, MA, USA, 2011.

51. Field, R.P.; Brasington, R. Baseline Flowsheet Model for IGCC with Carbon Capture. *Ind. Eng. Chem. Res.* **2011**, *50*, 11306–11312. [CrossRef]

52. Puig-Arnavat, M.; Bruno, J.C.; Coronas, A. Modified Thermodynamic Equilibrium Model for Biomass Gasification: A Study of the Influence of Operating Conditions. *Energy Fuels* **2012**, *26*, 1385–1394. [CrossRef]

53. Szargut, J.; Morris, D.R.; Steward, F.R. *Exergy Analysis of Thermal, Chemical, and Metallurgical Processes*; Hemisphere: New York, NY, USA, 1987.

54. Channiwala, S.; Parikh, P. A unified correlation for estimating HHV of solid, liquid and gaseous fuels. *Fuel* **2002**, *81*, 1051–1063. [CrossRef]

55. Larson, D.C.; Weinberger, C.B.; Lawley, A.; Thomas, D.H.; Moore, T.W. Fundamentals of Engineering Energy. *J. Eng. Educ.* **1994**, *83*, 325–330. [CrossRef]

56. Flórez-Orrego, D.; Silva, J.A.; de Oliveira, S., Jr. Renewable and non-renewable exergy cost and specific CO_2 emission of electricity generation: The Brazilian case. *Energy Convers. Manag.* **2014**, *85*, 619–629. [CrossRef]

57. Flórez-Orrego, D.; da Silva, J.A.M.; Velásquez, H.; de Oliveira, S. Renewable and non-renewable exergy costs and CO_2 emissions in the production of fuels for Brazilian transportation sector. *Energy* **2015**, *88*, 18–36. [CrossRef]

58. Chen, Y.C.; Kuo, J. Potential of greenhouse gas emissions from sewage sludge management: A case study of Taiwan. *J. Clean. Prod.* **2016**, *129*, 196–201. [CrossRef]

59. Yoo, M.-J.; Lessard, L.; Kermani, M.; Maréchal, F. OsmoseLua—An Integrated Approach to Energy Systems Integration with LCIA and GIS. In *Computer Aided Chemical Engineering*; Elsevier: Amsterdam, The Netherlands, 2015. [CrossRef]

60. Linnhoff, B.; Hindmarsh, E. The pinch design method for heat exchanger networks. *Chem. Eng. Sci.* **1983**, *38*, 745–763. [CrossRef]

61. de Souza, A.C.C.; Luz-Silveira, J.; Sosa, M.I. Physical-Chemical and Thermodynamic Analyses of Ethanol Steam Reforming for Hydrogen Production. *J. Fuel Cell Sci. Technol.* **2006**, *3*, 346–350. [CrossRef]
62. da Silva, A.L.; de Fraga Malfatti, C.; Müller, I.L. Thermodynamic analysis of ethanol steam reforming using Gibbs energy minimization method: A detailed study of the conditions of carbon deposition. *Int. J. Hydrog. Energy* **2009**, *34*, 4321–4330. [CrossRef]

Article

Modeling and Optimization of Hydraulic and Thermal Performance of a Tesla Valve Using a Numerical Method and Artificial Neural Network

Kourosh Vaferi [1], Mohammad Vajdi [1,*], Amir Shadian [2], Hamed Ahadnejad [1], Farhad Sadegh Moghanlou [1,*], Hossein Nami [3,*] and Haleh Jafarzadeh [4]

1   Department of Mechanical Engineering, University of Mohaghegh Ardabili, Ardabil 5619913131, Iran; k.vaferi@uma.ac.ir (K.V.); hamed.ahadnejad@outlook.com (H.A.)
2   Department of Mechanical Engineering, University of Tabriz, Tabriz 5166616471, Iran; amir.shadian99@ms.tabrizu.ac.ir
3   SDU Life Cycle Engineering, Department of Green Technology, University of Southern Denmark, Campusvej 55, 5230 Odense M, Denmark
4   Department of Civil Engineering, School of Science and Engineering, Khazar University, Baku 1096, Azerbaijan; haleh_jafarzadeh@yahoo.com
*   Correspondence: vajdi@uma.ac.ir (M.V.); f_moghanlou@uma.ac.ir (F.S.M.); hon@igt.sdu.dk (H.N.)

Citation: Vaferi, K.; Vajdi, M.; Shadian, A.; Ahadnejad, H.; Moghanlou, F.S.; Nami, H.; Jafarzadeh, H. Modeling and Optimization of Hydraulic and Thermal Performance of a Tesla Valve Using a Numerical Method and Artificial Neural Network. *Entropy* **2023**, *25*, 967. https://doi.org/10.3390/e25070967

Academic Editors: Daniel Flórez-Orrego, Meire Ellen Ribeiro Domingos and Rafael Nogueira Nakashima

Received: 10 May 2023
Revised: 17 June 2023
Accepted: 19 June 2023
Published: 22 June 2023

Abstract: The Tesla valve is a non-moving check valve used in various industries to control fluid flow. It is a passive flow control device that does not require external power to operate. Due to its unique geometry, it causes more pressure drop in the reverse direction than in the forward direction. This device's optimal performance in heat transfer applications has led to the use of Tesla valve designs in heat sinks and heat exchangers. This study investigated a Tesla valve with unconventional geometry through numerical analysis. Two geometrical parameters and inlet velocity were selected as input variables. Also, the pressure drop ratio (PDR) and temperature difference ratio (TDR) parameters were chosen as the investigated responses. By leveraging numerical data, artificial neural networks were trained to construct precise prediction models for responses. The optimal designs of the Tesla valve for different conditions were then reported using the genetic algorithm method and prediction models. The results indicated that the coefficient of determination for both prediction models was above 0.99, demonstrating high accuracy. The most optimal PDR value was 4.581, indicating that the pressure drop in the reverse flow direction is 358.1% higher than in the forward flow direction. The best TDR response value was found to be 1.862.

Keywords: Tesla valve; optimization; diodicity; thermo-hydraulic performance; artificial neural network

1. Introduction

Non-moving-part valves (NMPVs) are efficient equipment used as passive fluid controllers. Compared to conventional check valves with moving parts, NMPVs have advantages in terms of manufacturing and do not require external power to operate [1]. One specific type of NMPV is the Tesla valve. The Tesla valve, first introduced by Nikola Tesla in 1920, is a check valve with no moving parts in its structure and is also known as a fluid diode [2–4]. This valve allows fluid to flow easily in the forward direction but prevents fluid from flowing in the reverse direction. Due to its unique construction, the Tesla valve causes a lower pressure drop in the forward direction than in the reverse direction [5]. In fact, Tesla valves can be considered one-way valves. Due to the demand for passive fluid flow control, especially in mini and micro scales, the utilization of Tesla valves is becoming increasingly attractive to researchers. These valves are widely utilized in industries for controlling flow rate and direction in equipment such as internal combustion engines [6], turbines [7], pumps [8], and compressors [9,10]. Tesla valves can also be used in mini and microfluid applications, such as micromixers [11], and for the decompression process in hydrogen fuel cells [12,13]. Furthermore, their thermal characteristics make them suitable for heat transfer applications, especially in heat sinks for battery cooling [14,15].

De Vries et al. [16] designed a new construction for the Tesla valve to improve fluid flow and reduce thermal resistance in a pulsating heat pipe (PHP). They experimentally investigated its diodicity and operation by steady two-phase flow and laminar single-phase modeling. Laminar single-phase modeling demonstrated that the new construction of the Tesla valve generates more diodicity than other conventional Tesla valves at low Reynolds numbers. In addition, they found a 14% decrease in thermal resistance for the PHP with Tesla valves compared to similar PHP without Tesla valves in their structure. Jin et al. [12] numerically investigated the hydrogen decompression process using a Tesla valve with reverse flow direction for a wide range of inlet velocities. The results of their study highlighted that a large valve angle, small inner curve radius, and small hydraulic diameter could offer a high ΔP. Qian et al. [17] investigated the exergy loss and the possibility of aerodynamic noise occurrence in a Tesla valve with hydrogen fluid flow used for decompression. To perform these analyses, they applied changes in the valve inlet and outlet pressure ratio and the number of valve stages. They reported that increasing the pressure ratio raises the Ma and exergy loss. Also, they found that the Ma increases and exergy loss decreases by increasing the stage number of the valve. Monika et al. [15] presented a novel configuration for a multi-stage Tesla valve. They numerically analyzed it to investigate the temperature gradient created in a cold plate with a Tesla channel for the thermal management of Li-ion batteries. They observed that their new design provides more efficient cooling than conventional channels by improving the heat transfer rate of the cold plate. Liu et al. [2] presented a symmetrical design for the bent channel structure of the Tesla valve and investigated the hydraulic characteristics of the fluid using the finite element method (FEM). The results indicated that by enhancing the symmetry of the structure, the hydraulic diodicity performance of the valve increases. Bao and Wang [10] improved the relative pressure drop ratio (RPDR) and absolute pressure drop ratio (APDR) parameters to compensate diodicity performance of the Tesla valve. They designed a novel Tesla valve with special tapering and widening in its body and compared it with other types of Tesla valves. Their results stated that the novel presented design has a better APDR than conventional Tesla valves, and it was also found that this parameter increases linearly with the increase in the number of valve stages while the RPDR gradually reaches a constant value. In this study, they also investigated the thermal diodicity and observed that this parameter increases with increasing velocity and number of stages, but it is independent of wall and inlet temperatures. Lu et al. [14] presented a cold plate cooling system inspired by a Tesla valve for the enhancement of cooling in batteries. The optimization results showed that under specific geometrical conditions and a velocity of 0.83 m/s, the cold plate with Tesla valve channels and reverse flow establishes a good equilibrium between thermal performance and energy consumption. Yang et al. [18], using computational fluid dynamics (CFD) and conducting experiments, designed a new micromixer with a Tesla valve structure to obtain an effective mixing process in microfluidic equipment for biological applications. They observed that the mixing performance is more suitable for Reynolds numbers ranging from 0.1 to 100. Sun et al. [19] numerically analyzed a microchannel heat sink with Tesla valve-shaped channels. By examining the thermo-hydraulic performance, they realized that using the channels with Tesla valve design instead of the smooth channel increases the Nusselt number by 102.3%, and the friction factor increases by 3.21 times.

Artificial neural network (ANN) is a technology inspired by the brain and biological nervous system that mimics their electrical activity. This method forms the core of deep learning algorithms and is a subset of machine learning. One of the main benefits of ANN over other models is its ability to represent a multivariable problem based on the complex interactions between the variables and extract implicit nonlinear correlations among them [20–22]. This model-optimization method is widely used across various fields due to its impressive performance [23–27].

Polat and Cadirci [28] investigated the heat transfer of a microchannel heat sink with a diamond-shaped pin fin array under laminar, steady-state, and incompressible flow boundary conditions. They utilized a multi-layer ANN model coded in Python and trained with CFD outcomes to investigate Nusselt and Poiseuille numbers representing thermal and hydrodynamic features. The results showed that the pin-fin angle has the greatest impact on Nusselt and Poiseuille numbers. Kanesan et al. [29] developed a response model for thermal energy storage heat sinks, which are commonly used to cool electronics. They examined an aluminum heat sink for thermal energy storage (TES) using paraffin as the phase change material (PCM). By combining the trained ANN model with the genetic algorithm (GA) method, they optimized the variables related to the TES heat sink's geometry and the used PCM volume. This research demonstrates that combining ANN with GA creates a more effective optimization tool. Mahmoudabadbozchelou et al. [30] studied the enhancement of the heat transfer rate of impinging jets by adding nanoparticles to the background fluid. They utilized ANN and GA methods to optimize the uniform cooling of a continuously heated surface. The results indicated that the addition of nanoparticles to water led to an increase in heat transfer due to the increased thermal conductivity of the fluid, and larger particle sizes and concentrations caused a further increase. Kuang et al. [31] investigated the heat transfer in the boiling process of hydrogen flow and used ANN to identify the most influential parameter in this process. The study found that the boiling number is one of the most critical factors in determining the boiling heat transfer coefficient. The researchers also observed that the effect of saturation pressure on the flow boiling heat transfer coefficient is more significant than its effect on the flow rate of liquid hydrogen. Yunn Heng et al. [32] proposed a rapid and reliable transient thermal prediction technique for estimating the exit temperature of a parabolic trough collector tube. ANN was applied to analyze the increase in exit temperature produced by a single heat flux pulse. They observed that the outcomes could be utilized for preliminary system planning, heat balance assessment, and systems engineering. The study reported that this method works well with changing and steady solar radiation, making it useful for designing parabolic trough technologies in any weather condition worldwide. Ermis et al. [33] investigated phase change heat transfer in a finned-tube latent heat thermal storage system using an ANN approach. The trained ANN model predicted the total quantity of stored heat with an average error of 5.58%, resulting in a more accurate heat storage estimation than the numerical model results. In another study, Xie et al. [34] evaluated multi-layer neural network designs based on experimental datasets of Nusselt number and friction factor for three heat exchangers. Their findings demonstrated that the ANN method performs well in predicting heat transfer and fluid flow for laminar or turbulent regimes in such heat exchangers, such that the variance between their study's predicted and experimental results was approximately 4%. Their work suggests that ANNs can be utilized for thermal system performance anticipation, particularly heat exchanger modeling for heat transfer assessment. Beigmoradi et al. [35] conducted a study on the aerodynamic optimization of the rear end of a car using ANN. They selected several geometric parameters as input variables and studied the drag coefficient and maximum acoustic power level as responses. The Taguchi method was used to reduce the number of tests, and the GA method was used to optimize the model. The results indicate that the drag coefficient decreases with the increase in the rear box length parameter, but it leads to an increase in the acoustic power response. Li et al. [36] used three ANNs to predict the properties of China RP-3 kerosene at a faster rate. Their results show that the properties predicted by ANN models have high accuracy and are consistent with the calculations of the extended corresponding state principle method. They also observed that the prediction of properties is 10^4 times faster than the calculations, which is a significant achievement. George et al. [37] optimized the design of a multi-layer porous wave absorber by using the ANN method and a data set consisting of 200 combinations. The trained prediction model in their work has a determination coefficient of 0.97, indicating high accuracy in predicting results. They found that the optimal range of design variables for submergence depth was 0.055–0.067, the distance between plates was 0.064–0.080, and porosity was 0.117–0.173. Zhu et al. [38] investigated the total output power due to changes in the configuration of the array of wave energy converters. Using the ANN method, they trained a prediction model for the desired response. They reported that using energy converters with shorter distances improves energy absorption and was also suitable and beneficial for engineering applications.

The present study investigated a two-stage Tesla valve with an unconventional geometry numerically. The parameters of the divider baffle length (L), step length (S), and inlet velocity of the valve (V) were considered as input variables. Also, the diodicity (PDR) and the ratio of the temperature difference (TDR) in reverse and forward directions were selected as the responses. Of course, due to the consistency of the properties and mass flow rate of the passing fluid in both directions, TDR also indicates the heat transfer ratio. Numerous numerical experiments were conducted under different conditions to obtain the prediction models for each response, and the results were used to train artificial neural networks. These models can predict the values of the responses accurately and quickly without the need for complex calculations or additional experiments. Finally, the optimal conditions and designs of the Tesla valve for various applications were determined using the genetic algorithm method and the obtained prediction models. This approach can significantly reduce the time and cost required for designing and optimizing the Tesla valve for specific applications.

2. Methodology

2.1. Tesla Valve Structure

According to the results of the research conducted on Tesla valves and preliminary analysis, a specific structure for this device was proposed, which was based on the design by Bao and Wang [10]. The physical shape and geometrical characteristics of the Tesla valve structure are shown in Figure 1. A three-dimensional view of the intended structure is depicted in Figure 1a. The constant and variable dimensions of the simulated geometry are shown in Figure 1b and Figure 1c, respectively. The most significant difference between this Tesla valve and other conventional Tesla valves is the use of a unique three-way pattern that consists of a divider baffle. In this work, the divider baffle length (L) and step length (S) were considered as the two geometrical variables to investigate the performance of the Tesla valve. It should also be noted that in addition to the variables mentioned, the effects of the velocity of the input flow to the Tesla valve were investigated.

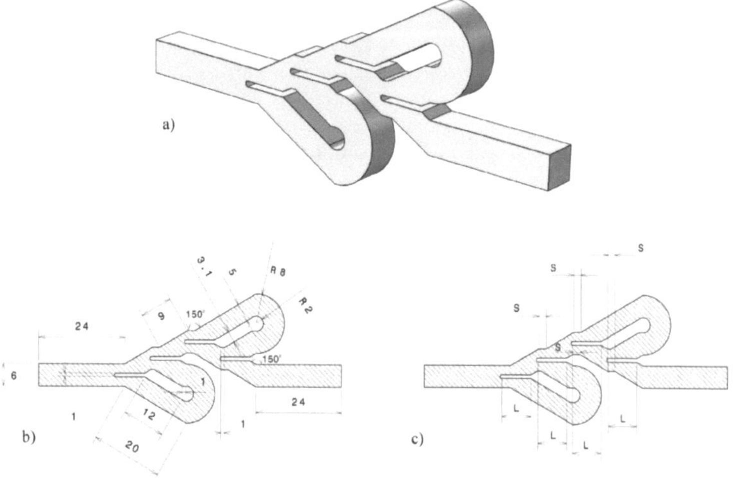

Figure 1. (a) 3D view of the investigated geometry, (b) fixed geometric parameters, (c) variable geometric parameters.

The fluid movement pattern in this unconventional design for the Tesla valve in reverse and forward directions is based on Figure 2a and Figure 2b, respectively. When the fluid enters the valve in the reverse direction, the dividers direct part of the fluid flow into the bent channels. While with the movement of the fluid in the forward direction, the main flow of the fluid can easily pass through the main channel of the valve. Therefore, it is expected that due to more friction and fluid interaction in the reverse direction, a greater ΔP will occur in this direction.

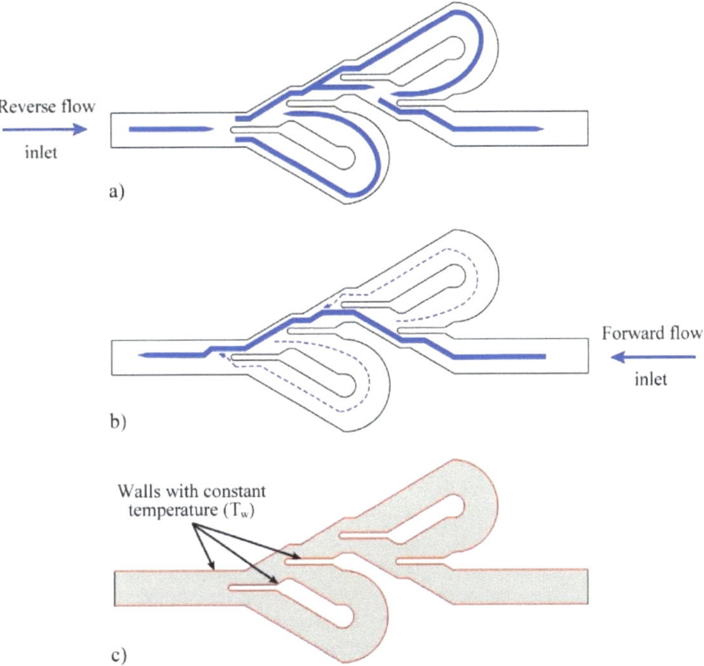

Figure 2. (a) The reverse fluid movement pattern in the Tesla valve, (b) the forward fluid movement pattern in the Tesla valve, (c) the thermal boundary condition in the Tesla valve.

2.2. Assumptions and Boundary Conditions

All numerical analysis and simulation processes were performed using FEM in COMSOL Multiphysics software. Since this study aims to enhance the diodicity and TDR of the Tesla valve, water fluid was simulated under special conditions. Consequently, several boundary conditions and assumptions were applied to achieve an appropriate result. Boundary conditions and assumptions are as follows:

- This study was investigated in a steady state;
- The fluid passing through the Tesla valve was turbulent, single-phase, and incompressible;
- The thermophysical characteristics of the water were considered to be constant, as listed in Table 1;
- No-slip and no-temperature-jump conditions were assumed for the walls in contact with the fluid;
- The wall's temperature around the fluid was adopted constant and equal to 350 K (Figure 2c);
- The water inlet temperature was considered constant and equal to 293.15 K;
- The gauge pressure of the Tesla valve outlet was considered zero.

Table 1. The thermophysical characteristics of the water at 293.15 K [10].

Properties	Value
ρ (kg/m^3)	998.2
μ (Pa·s)	1.003×10^{-3}
C_p (J/kg·K)	4182
λ (W/m·K)	0.6

2.3. Governing Equations

The present study numerically investigated the fluid flow and heat transfer inside the designed Tesla valve using CFD in two dimensions. Governing equations that must be solved for determining thermo-hydraulic parameters of the flow are conservation equations and k-ε turbulence model equations. For an incompressible viscous flow, the first equation is the mass conservation equation, and it can be expressed as follows [10,39]:

$$\rho \nabla \cdot V = 0 \tag{1}$$

The momentum conservation equation is written as follows:

$$\rho V \cdot \nabla V = -\nabla P + \nabla \cdot \left((\mu + \mu_t) \left(\nabla V + (\nabla V)^T \right) \right) \tag{2}$$

Also, the energy conservation equation is given by:

$$\rho C_p V \cdot \nabla T = \nabla \cdot \left(\left(\lambda + \frac{\mu_t}{Pr} \right) \nabla T \right) \tag{3}$$

where V, μ, μ_t, and λ represent the velocity, dynamic viscosity, turbulent viscosity, and thermal conductivity, respectively. In this research, the standard k-ε turbulent model was employed in order to analyze the flow in the valve. This turbulence model is the most prevalent model used in CFD to represent the mean flow characteristics of turbulent flow. It is a two-equation model that uses two transport equations to provide a general description of turbulence. The turbulent kinetic energy and specific dissipation rate for turbulent flow in the standard k-ε model are defined by Equations (4) and (5), respectively [40–42].

$$\rho V \cdot \nabla k = \nabla \cdot \left(\left(\mu + \frac{\mu_t}{\sigma_k} \right) \nabla k \right) + P_k - \rho \varepsilon \tag{4}$$

$$\rho V \cdot \nabla \varepsilon = \nabla \left(\left(\mu + \frac{\mu_t}{\sigma_\varepsilon} \right) \nabla \varepsilon \right) + C_{\varepsilon 1} \frac{\varepsilon}{k} P_k - C_{\varepsilon 2} \rho \frac{\varepsilon^2}{k} \tag{5}$$

$$\mu_t = \rho C_\mu \frac{k^2}{\varepsilon} \tag{6}$$

$$P_k = \mu_t \left(\nabla V : \left(\nabla V + (\nabla V)^T \right) \right) \tag{7}$$

where k and P_k represent the turbulent kinetic energy and the production of this energy due to the mean velocity gradients, respectively, and the parameter ε is the energy dissipation rate obtained in the turbulent flow. Also, the constants related to the turbulence model are presented in Table 2.

Table 2. Constant parameters related to standard k-ε turbulence model.

Constant Parameter	σ_k	σ_ε	$C_{\varepsilon 1}$	$C_{\varepsilon 2}$	C_μ
Value	1	1.30	1.44	1.92	0.09

In the present study, the segregated approach, a pressure-based solver, was used to solve the governing equations. As implemented in COMSOL Multiphysics, this approach solves the velocity and pressure in one step. In contrast, other variables, such as temperature, are solved separately in other steps. This solver configuration allows for the decoupling and independent solution of different variables, which can help improve computational efficiency and convergence [43]. This research used a parallel direct sparse solver (PARDISO) in numerical simulations. The PARDISO solver is a state-of-the-art direct sparse solver in computational science and engineering. This solver is known for its efficiency, scalability, and ability to handle large-scale linear systems from various numerical simulations. It employs advanced algorithms and parallel computing techniques to efficiently handle the matrix factorization and solve the system of equations [44–46].

Diodicity is an essential factor that evaluates the Tesla valve's hydraulic performance. This parameter highlights the effectiveness of the valve based on the ratio of ΔP in reverse flow to forward flow in an identical flow rate [1]. By increasing diodicity, the performance of this device as a check valve will be improved. On the other hand, if this device is used in thermal applications, the thermal parameters of this device should be improved. The hydraulic and thermal diodicity of the Tesla valve were presented as PDR (pressure drop ratio) and TDR (thermal difference ratio), respectively, which were calculated as follows:

$$PDR = \frac{\Delta P_r}{\Delta P_f} = \frac{(P_{in} - P_{out})_r}{(P_{in} - P_{out})_f} \tag{8}$$

$$TDR = \frac{\Delta T_r}{\Delta T_f} = \frac{(T_{out} - T_{in})_r}{(T_{out} - T_{in})_f} \tag{9}$$

2.4. Mesh Independency

COMSOL Multiphysics was used to develop a two-dimensional triangular mesh type, as seen in Figure 3. In this simulation, the average mesh quality is 0.93. According to the statistics in Table 3, the thermal and hydraulic results of the numerical simulation with the number of 50.44 elements per 1 mm^2 are independent of the mesh. As can be seen, by increasing the mesh elements number to 88.58 per 1 mm^2, the time to solve the simulation increases by 56%, while the results related to the temperature difference and pressure drop change by 1.23% and 0.44%, respectively.

a)
b)

Figure 3. (**a**) Overview and (**b**) refined mesh near the divider baffle for the computational domain.

Table 3. Mesh independence validation for tesla valve with six-stage and reverse flow at conditions $V = 0.2$ m/s, $T_{in} = 274.15$ K, and $T_w = 368.15$ K.

Number of Elements per 1 mm^2	Solve Time	ΔT_r		ΔP_r	
		Value (K)	Difference (%)	Value (Pa)	Difference (%)
3.89	00:00:48	24.210	21.45	1905	7.15
22.68	00:03:38	29.320	4.87	1835	0.83
50.44	00:25:13	30.440	1.23	1828	0.44
88.58	00:39:27	30.820	-	1820	-

2.5. Simulation Method Validation

Since this work is a numerical study, it needs to be validated by a reliable experimental reference. Therefore, all processes and numerical simulations in the present study were based on the experimental research conducted by Bao and Wang [10], and the results of this work were compared with their research findings to ensure the accuracy and validity of the analysis done in this work. In the present work, the thermo-hydraulic performance of the equivalent shunts (ES) Tesla valve with six stages was investigated and compared with the reference study [10]. Figure 4 shows that both hydraulic and thermal results obtained from the numerical analysis have a reasonable correlation with the findings provided in the experimental reference study, indicating that the numerical method adopted in this research has reasonable accuracy. It is worth noting that in the subsequent part of this work, a two-stage Tesla valve was designed instead of the six-stage valve to allow for physical changes and further investigations.

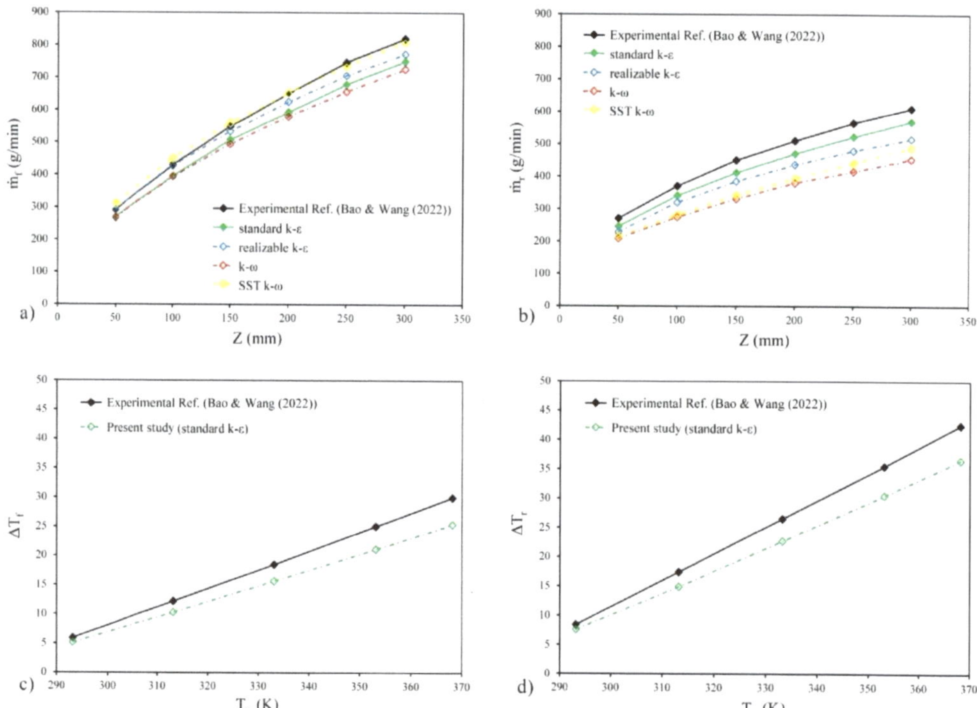

Figure 4. Comparison of results calculated by the numerical method of the present work with the experimental study of Bao and Wang [10]: (**a**) the hydraulic results in the forward flow, (**b**) the hydraulic results in the reverse flow, (**c**) the thermal results in the forward flow, and (**d**) the thermal results in the reverse flow.

2.6. Turbulent Model Validation

In this research, different turbulence models were tested and compared with the experimental results to select the turbulence model used in the numerical analysis, according to Figure 4a,b. Equation (10) was used to calculate the average error of numerical analysis for each turbulence model. According to the results of the numerical analysis using the k-ω turbulence model, the average error of the hydraulic results in the reverse flow was 25.7%, and it was 10.2% in the forward flow. However, when the SST k-ω model was used in the simulations, the average errors for the reverse and forward directions results were reduced to 22.7% and 2.6%, respectively. The realizable *k-ε* model showed an average error of 15% in the reverse flow and 3.3% in the forward flow for the results obtained in the analysis of the Tesla valve. On the other hand, the standard k-ε model gave an average error of 8.3% and 8.4% for hydraulic results in the reverse and forward directions, respectively. Based on these results, it can be concluded that the standard *k-ε* model is the most appropriate turbulence model to use in numerical simulations because it gives suitable and close each other average errors in both directions.

The $y+$ is a dimensionless parameter similar to the local Reynolds number used in CFD to characterize the near-wall flow behavior by quantifying the distance from the solid wall to the nearest mesh element's center. This parameter is used as a criterion to evaluate the appropriateness of the grid element size on the walls' borders. On the other hand, it plays a crucial role in selecting an appropriate wall modeling approach, such as wall functions or low Reynolds number (LRN) models, based on the flow regime and the desired level of accuracy for capturing the near-wall physics. This parameter is calculated according to Equations (11) and (12). In these equations, V_τ and τ_w represent the friction velocity and wall shear stress, respectively. Also, y is the distance between the wall and the center of the nearest mesh element to the wall. Higher $y+$ values (often above 30) indicate fully turbulent flows where wall functions are effective. In comparison, lower $y+$ values suggest laminar or transitional flows requiring more refined modeling techniques to resolve the near-wall region accurately [47]. The calculation of $y+$ was performed across different turbulence models, and the results are presented in Table 4. It is evident from the data that the $y+$ values almost in all cases exceed 30, indicating a predominantly turbulent near-wall flow regime.

Table 4. The value of $y+$ in different turbulence models at Z = 300 mm, T_{in} = 274.15 K, and T_w = 368.15 K conditions.

Turbulence Model	Wall Model	Value of $y+$		Suitability Situation
		Reverse Flow	Forward Flow	
Standard *k-ε*	Wall functions	41.910	40.749	suitable
Realizable *k-ε*	Wall functions	41.784	45.745	suitable
SST k-ω	LRN	29.211	36.539	unsuitable
k-ω	LRN	39.675	42.715	unsuitable

According to the $y+$ parameter results, it can be seen that the use of the wall functions model was suitable for investigating the fluid flow behavior near the walls in the present study. These wall functions are derived from empirical correlations based on experimental data, and *k-ε* models widely use these wall functions to capture the turbulence characteristics near walls. Also, because the values of $y+$ were above 30, using the LRN wall model, which is used in the *k-ω* and *SST k-ω* turbulence models, was unsuitable for use in the present work [48,49]. Further, considering that the numerical results of the standard *k-ε* turbulence model were in better agreement with the experimental results compared to the realizable *k-ε* model, the standard *k-ε* model was used for further investigations in this work.

$$\overline{Error} = 100 \times \frac{1}{m} \sum_{i=1}^{m} \frac{|r_{num} - r_{exp}|}{r_{exp}} \tag{10}$$

$$y^+ = \frac{\rho V_\tau y}{\mu} \tag{11}$$

$$V_\tau = \sqrt{\frac{\tau_w}{\rho}} \tag{12}$$

2.7. Artificial Neural Network Approach

One way to reduce computational time and save financial resources is by using black-box methods, which deal with input and output data without considering the possible physical processes. A practical approach in this field is the artificial neural network (ANN), inspired by the human nervous system. Neurons are the fundamental processors in neural networks, and each neuron may receive multiple inputs from other neurons and have one or more outputs based on its activity [33].

In this study, a separate three-layer structure was designed for each response (TDR and PDR), consisting of input, hidden, and output layers, as shown in Figure 5a. The input layer has three neurons that serve as the network's inputs. The hidden layer has four neurons, and the output layer contains one neuron. Based on the investigations conducted, it was observed that prediction models with fewer than four neurons in the hidden layer did not have satisfactory performance. Conversely, an excessive increase in neurons in this layer led to overfitting in the model's results. The activation functions must be derivable to perform the backpropagation function in model training. Therefore, the tangent-sigmoid activation function was used in hidden layer neurons, and the linear activation function was used in the output layer neuron to obtain the response value. To better understand the performance of each neuron in the hidden and output layers, Figure 5b was presented, which shows the performed calculations.

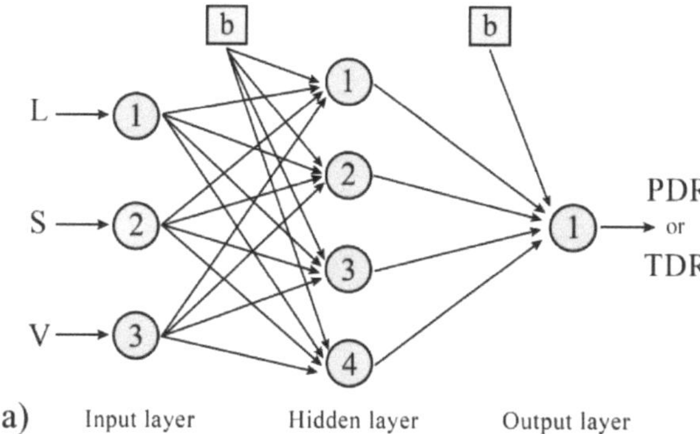

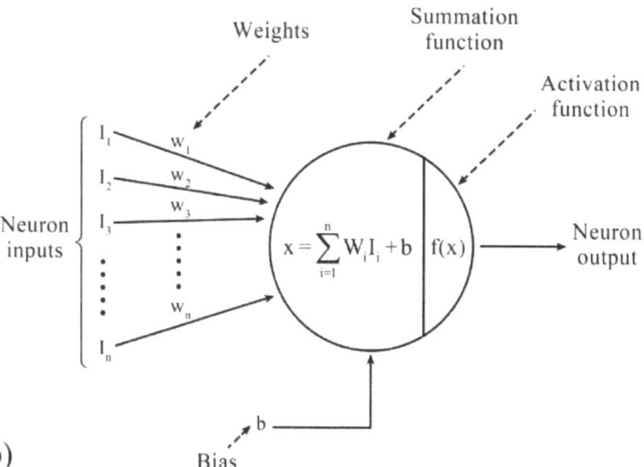

Figure 5. (**a**) The three-layer structure of the ANN network was used in the present work, (**b**) calculations performed in neurons of the hidden and output layers.

MATLAB (2016b) software was applied to optimize models using the ANN approach, and the Levenberg–Marquardt method was employed to train artificial neural networks by utilizing the available data. The variables and their ranges for simulation and optimization were tabulated in Table 5, and these ranges were selected based on the geometric and boundary conditions. Furthermore, Table 6 presents a list of numerical tests selected using the central composite design (CCD) method, which can minimize the number of tests, which is an important aspect, especially in experimental studies [50]. To ensure the development of reliable prediction models, a careful data allocation strategy was employed. As illustrated in Table 6, 70% of the data was allocated for training the models, 15% for validation purposes, and 15% for testing the models. Notably, this classification was performed randomly, ensuring an unbiased data distribution. To assess the validity and accuracy of the predicted results by the ANN method, three error functions were used, including the determination coefficient (R^2), mean absolute error (MAE), and root mean square error (RMSE).

Table 5. Selected variables and levels.

Variable	Variable Levels		
	−1	**0**	**+1**
L (mm)	2	5	8
S (mm)	2	5	8
V (m/s)	0.2	1.1	2

Table 6. Selected numerical experiments to perform numerical simulations.

Numerical Experiment Number	Position Used in Machine Learning	Variables		
		L (mm)	*S* (mm)	*V* (m/s)
1	Train	8	2	0.2
2	Train	5	5	1.1
3	Train	8	8	0.2
4	Validation	5	5	0.2
5	Train	5	5	2
6	Train	5	8	1.1
7	Train	8	8	2
8	Train	5	2	1.1
9	Train	2	2	2
10	Train	2	5	1.1
11	Test	8	2	2
12	Train	2	8	0.2
13	Train	2	8	2
14	Validation	8	5	1.1
15	Test	2	2	0.2

3. Results

3.1. Numerical Results

The Tesla valve functions as a one-way valve and creates a high-pressure drop in one direction compared to the other direction, making the flow of movement easier in one direction. Therefore, the ratio of ΔP in one direction to the other, known as diodicity, is one of the most important parameters in this device. Tesla valves can also be used in thermal applications. In this work, the thermal performance of the device was investigated in addition to its hydraulic performance. Due to the stability of fluid properties and mass flow rate in both flow directions, the ratio of directions temperature difference (TDR) also indicates the heat transfer ratio. The information related to the numerical tests is reported in Table 7, and preliminary analyses can be performed using these data. Based on the results, the maximum values of TDR and PDR are observed in test numbers 8 and 9, respectively.

The highest ΔP in the reverse direction occurred in test number 7, and the highest ΔP in the forward movement occurred in numerical experiment 11. At the same time, the largest difference between the inlet and outlet temperatures in both reverse and forward flow directions was reported in numerical experiment 3.

Table 7. Data and results related to numerical method experiments.

Numerical Experiment Number	Numerical Simulation Data					
	ΔT_f (K)	ΔT_r (K)	ΔP_f (Pa)	ΔP_r (Pa)	TDR	PDR
1	6.31	7.31	254.28	284.32	1.158	1.118
2	3.83	6.08	2479.4	5790.9	1.587	2.336
3	6.94	8.96	240.18	269.04	1.291	1.120
4	5.46	7.13	128.58	149.04	1.306	1.159
5	3.46	5.6	7259.5	18,286	1.618	2.519
6	4.22	6.69	2995.2	7646.8	1.585	2.553
7	4.41	6.45	14,233	24,887	1.463	1.749
8	3.47	6.27	2260.1	7614.9	1.807	3.369
9	2.67	4.77	2767.3	12,578	1.787	4.545
10	3.50	5.59	1407.2	4622.9	1.597	3.285
11	4.16	4.65	15,159	18,580	1.118	1.226
12	5.75	8.14	120.32	209.9	1.416	1.745
13	3.55	5.71	6522.3	19,819	1.608	3.039
14	4.45	5.06	4414.8	4283.5	1.137	0.970
15	4.49	7.13	48.879	135.79	1.588	2.778

Figure 6 shows the pressure contours, and it can be seen that the inlet pressure of reverse flows is higher than the forward flows direction. Additionally, the fluid pressure in the reverse flow is high due to the longer fluid movement path, more friction, more vortices, and more fluid collisions. Figure 7 displays the velocity contours, and it is evident that in the reverse direction of fluid flow in the Tesla valve, the fluid passes through the bent channels with a higher mass flow rate and velocity than the forward flow. Using these two contours, it was concluded that much flow passes through the bent paths in the reverse direction movement. Therefore, the thermal performance of the device is expected to be higher in the direction of the reverse flow, which is confirmed by Table 6. Figure 8 presents the temperature contours for checking the thermal performance. As shown in this figure, in the Tesla valve with the forward flow, the heat transfer is weaker since less fluid enters the bent channels, and the temperature of the liquid in the bent channels is higher. Of course, in the reverse flow temperature contour of experiment 11, the fluid temperature in the second bent channel is also high due to the geometry of the valve, which causes not much fluid to enter the bend, and the fluid temperature in that section increases. The temperature distribution is generally better in the Tesla valve with reverse flow. The results of tests 9, 11, and 13 are presented in all contours. Experiments 9 and 11 were compared to observe the effect of changing variable L on Tesla valve performance, and experiments 9 and 13 were compared to observe the impact of changing variable S on valve performance. In the following, the ANN method and predictions of this method were used for a more detailed investigation of the effect of variables and optimization of geometry and responses.

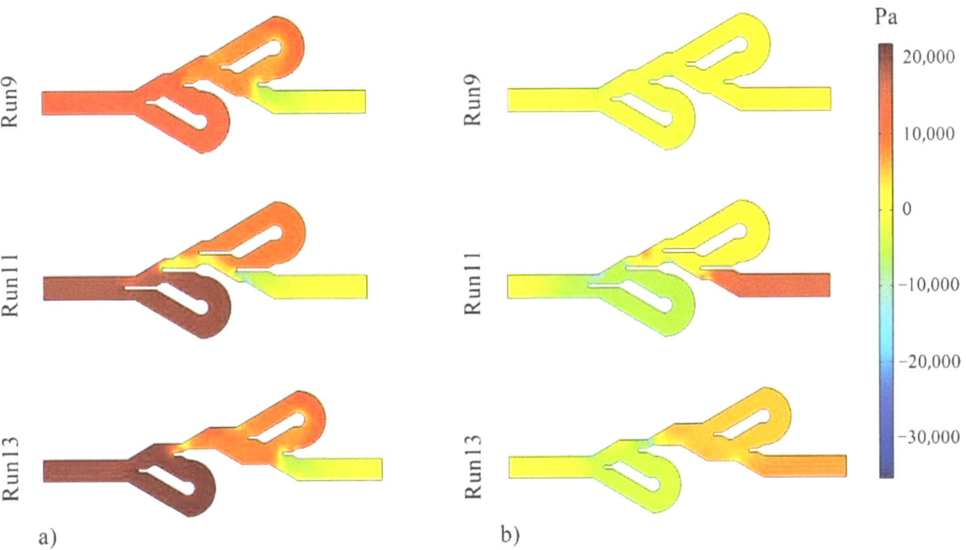

Figure 6. Pressure contours for (**a**) Tesla valves with fluid flow in the reverse direction and (**b**) Tesla valves with fluid flow in the forward direction.

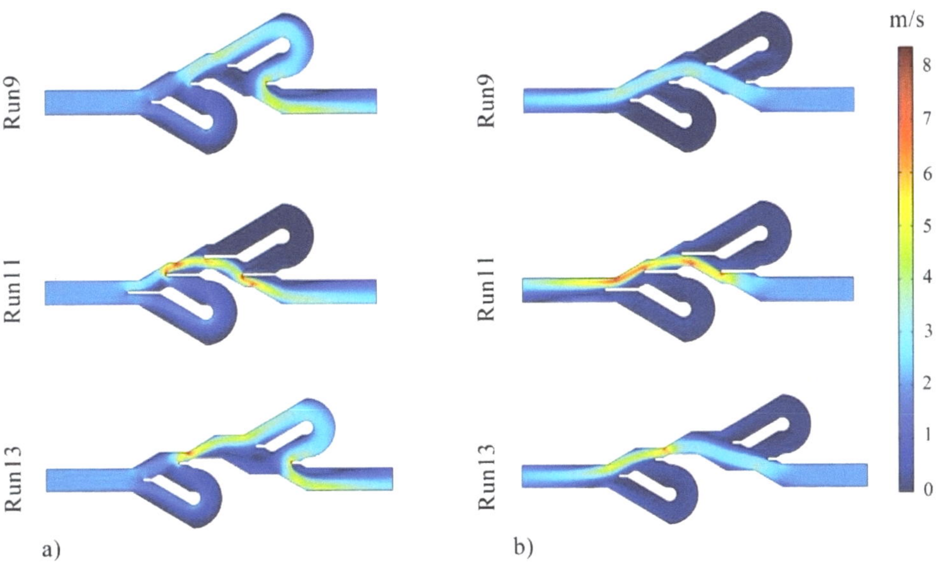

Figure 7. Velocity contours for (**a**) Tesla valves with fluid flow in the reverse direction and (**b**) Tesla valves with fluid flow in the forward direction.

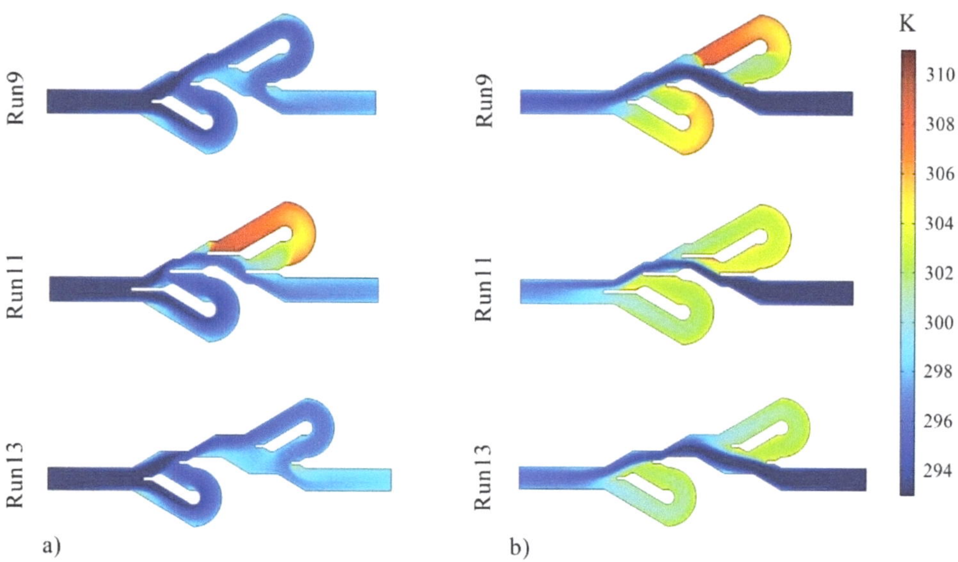

Figure 8. Temperature contours for (**a**) Tesla valves with fluid flow in the reverse direction and (**b**) Tesla valves with fluid flow in the forward direction.

3.2. The ANN Results

In this section, the results of the ANN method are discussed. The predicted results for TDR and PDR responses using the models obtained from the ANN method are compared with the numerical analysis results and presented in Figure 9a,b. It can be seen that the results obtained by both methods are similar. The models for the TDR and PDR responses developed through the ANN method are described in Equations (13) and (15), respectively. The optimized weights and biases of the models related to the responses were obtained using the ANN method and LM algorithm. The R^2 parameter was examined to assess the accuracy of the models (Table 8), and it was observed that the TDR model correctly predicts 99.1% of the responses. Also, the R^2 error function with a value of 0.992 for the PDR model indicates that this model can correctly predict 99.2% of responses and has only 0.8% error.

To show the effect of each input variable (L, S, and V) on the output responses, the plots in Figure 9c,d were presented. In these plots, one variable was changed from -1 to $+1$ levels while the other two remained constant at zero levels. Also, the reference point in these graphs is the point where all three variables are at zero level. As seen in Figure 9c, changes in the L variable up to a certain level do not affect TDR, and from that level onwards, it causes TDR to decrease. Changes in the S parameter initially cause a decrease in TDR, and the intensity of this reduction gradually decreases. With an increase in the value of V, TDR also increases, but the slope of this plot is higher in the initial part, indicating a greater intensity. In the following, the effects of the independent input variables on the response of the diodicity are discussed, shown in Figure 9d. According to the high slope of the L parameter graph, it seems that the effect of this variable on the diodicity is more significant than the rest of the variables. After the L parameter, the Tesla valve input velocity variable has the most impact on PDR, and this response rises with the increase in velocity.

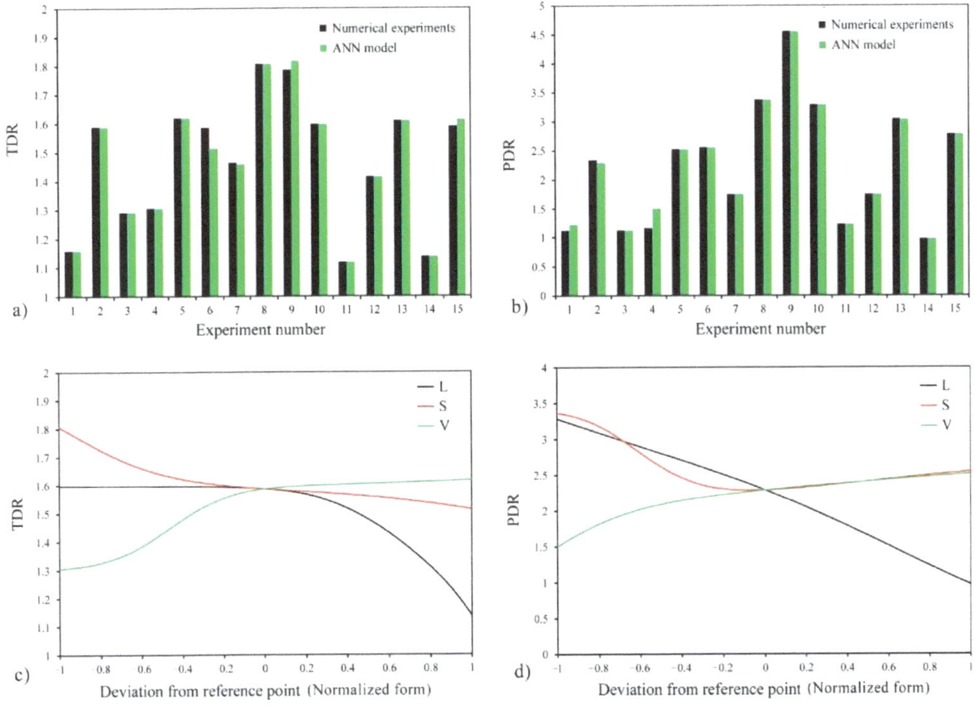

Figure 9. Comparing the results obtained by ANN and numerical methods for (**a**) TDR and (**b**) PDR. Perturbation plots of (**c**) TDR and (**d**) PDR.

$$TDR = \left([1.359 \quad 0.433 \quad -0.634 \quad -0.073] \times \begin{bmatrix} \frac{2}{1+exp(-2A_1)} - 1 \\ \frac{2}{1+exp(-2A_2)} - 1 \\ \frac{2}{1+exp(-2A_3)} - 1 \\ \frac{2}{1+exp(-2A_4)} - 1 \end{bmatrix} + [-0.703] + 1 \right) \times \frac{1.807 - 1.118}{2} + 1.118 \tag{13}$$

$$\begin{bmatrix} A_1 \\ A_2 \\ A_3 \\ A_4 \end{bmatrix} = \begin{bmatrix} -2.913 & 1.239 & -0.825 \\ -2.622 & -1.047 & 3.314 \\ -0.775 & 2.054 & -0.647 \\ 1.159 & 0.256 & -1.438 \end{bmatrix} \times \begin{bmatrix} 2 \times \frac{L-2}{8-2} - 1 \\ 2 \times \frac{S-2}{8-2} - 1 \\ 2 \times \frac{V-0.2}{2-0.2} - 1 \end{bmatrix} + \begin{bmatrix} 3.496 \\ 1.541 \\ 2.027 \\ 2.786 \end{bmatrix} \tag{14}$$

$$PDR = \left([0.074 \quad -0.981 \quad 0.445 \quad -0.472] \times \begin{bmatrix} \frac{2}{1+exp(-2B_1)} - 1 \\ \frac{2}{1+exp(-2B_2)} - 1 \\ \frac{2}{1+exp(-2B_3)} - 1 \\ \frac{2}{1+exp(-2B_4)} - 1 \end{bmatrix} + [-0.674] + 1 \right) \times \frac{4.545 - 0.970}{2} + 0.970 \tag{15}$$

$$\begin{bmatrix} B_1 \\ B_2 \\ B_3 \\ B_4 \end{bmatrix} = \begin{bmatrix} 1.356 & -0.686 & -1.462 \\ 0.807 & -0.265 & -0.205 \\ 1.391 & 0.550 & 2.249 \\ 1.421 & 2.936 & 0.309 \end{bmatrix} \times \begin{bmatrix} 2 \times \frac{L-2}{8-2} - 1 \\ 2 \times \frac{S-2}{8-2} - 1 \\ 2 \times \frac{V-0.2}{2-0.2} - 1 \end{bmatrix} + \begin{bmatrix} -2.375 \\ -0.555 \\ 2.542 \\ 1.825 \end{bmatrix} \tag{16}$$

Table 8. Nonlinear error functions for prediction models obtained by ANN method.

Error Function	Mathematical Form	ANN	
		TDR	**PDR**
R^2	$1 - \dfrac{\sum_{i=1}^{m}\left(r_{num}-r_{pred}\right)}{\sum_{i=1}^{m}\left(r_{num}-\overline{r}_{pred}\right)^2}$	0.991	0.992
MAE	$\dfrac{1}{m}\sum_{i=1}^{m}\left|r_{pred}-r_{num}\right|$	0.009	0.034
RMSE	$\sqrt{\dfrac{1}{m}\sum_{i=1}^{m}\left(r_{pred}-r_{num}\right)^2}$	0.021	0.095

In the following, the two-by-two effects of the variables on the responses were investigated according to Figure 10. Figure 10a shows that as L increases, the TDR value remains somewhat stable and then decreases, and the effect of this decrease is more significant at lower S. Figure 10b confirms this observation and shows that the effect of the L parameter on TDR is more pronounced at higher inlet velocities. Figure 10c reveals that the heat transfer capability of the Tesla valve increases with the increase of the inlet velocity (V), and this increase is more significant at lower S values. Furthermore, it can be seen that increasing S leads to a decrease in TDR. Turning to the three-dimensional plots for PDR response, Figure 10d shows that increasing L reduces PDR, and this effect is more pronounced at smaller S values. Figure 10e,f demonstrate that increasing the inlet velocity increases PDR, which is more prominent at smaller L and S values. It is worth mentioning that increasing the L parameter leads to a decrease in the diodicity of the Tesla valve.

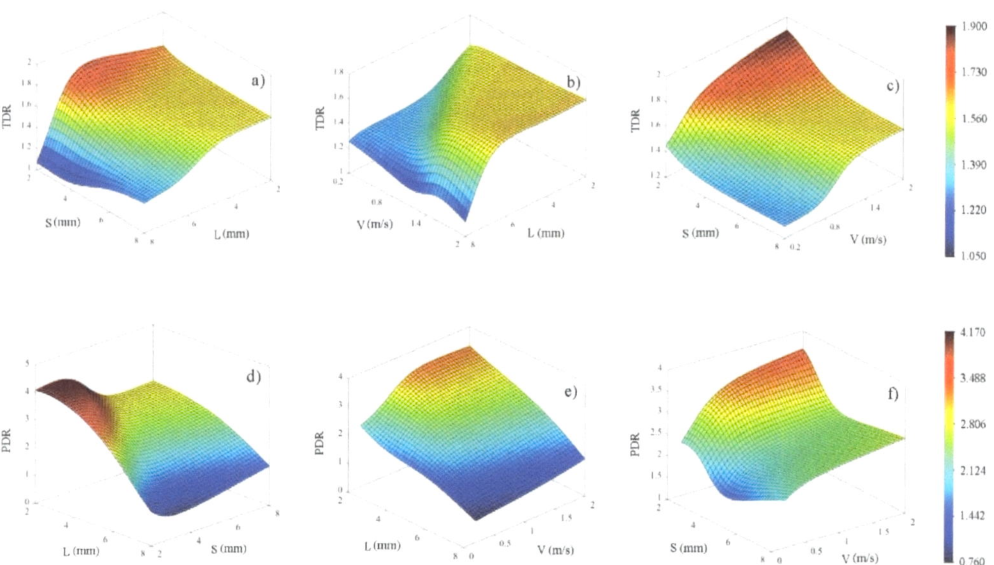

Figure 10. The relationship between (**a**) TDR, L, and S, (**b**) TDR, L, and V, (**c**) TDR, S, and V, (**d**) PDR, L, and S, (**e**) PDR, L, and V, (**f**) PDR, S, and V was illustrated in these 3D plots.

This section discusses the influence of variables on the performance of the Tesla valve and the analysis of these effects using streamlines (Figure 11). In this figure, the primary issue that can be noticed is that in the direction of forward movement, compression of the lines in the main channel is greater than bent channels, indicating a large fluid flow through this path. While in the reverse flow direction, the fluid flow is spread in all the channels of the Tesla valve. The effect of changing the divider baffle length on PDR can be observed based on the fluid flow streamlines of the 9th and 11th numerical experiments. According to the streamlines in the reverse direction, it is apparent that increasing the divider baffle length causes most of the fluid flow to pass without entering the second bent channel, and this function reduces the ΔP and diodicity. However, it should be noted that as L increases, the dimensions of the central path reduce, leading to increased ΔP in both directions of fluid movement. Comparing the streamlines of experiments 9th and 13th reveals the effect of changing the parameter S on the Tesla valve's performance. It is clearly seen that in the 9th test, more flow enters the bent channels in reverse flow than in the 13th test, and due to this matter, the ΔP ratio of reverse flow to the forward flow is higher in this design. Additionally, it should be mentioned that increasing the S variable leads to a larger Tesla valve, causing more friction and ΔP in both directions.

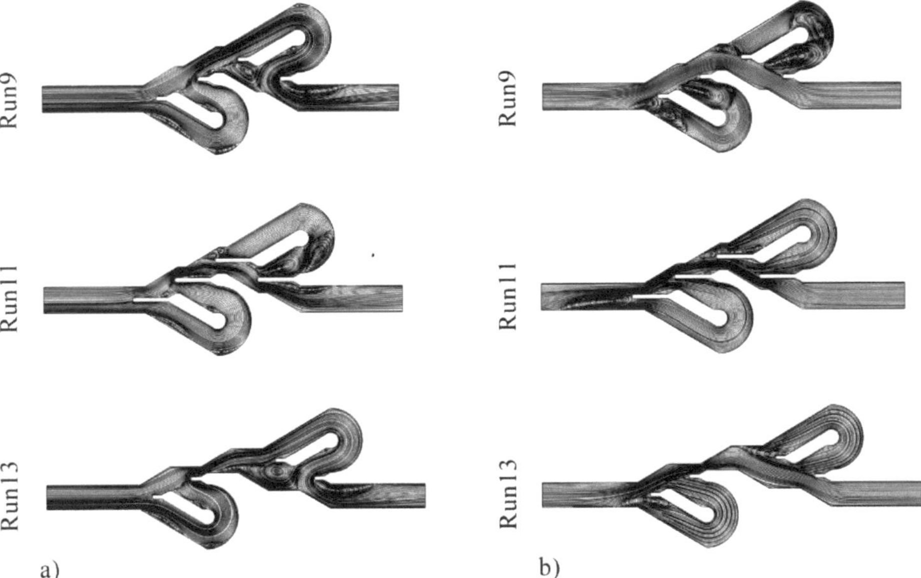

Figure 11. Streamlines of (**a**) reverse flow direction and (**b**) forward flow direction for different numerical experiments.

As mentioned in the previous sections, one of the most important goals of the current research is to provide optimal designs for the Tesla valve in different conditions. The goal was to maximize the values of PDR and TDR, and genetic algorithm and ANN models were used to predict the optimal design parameters. According to Table 9, the designs with maximum PDR and TDR values were predicted, and the predicted results for the values of the responses were also presented. Next, the given Tesla valves were designed and then numerically tested to ensure that the predictions were correct and that these designs had the most optimal response values. The data relating to these numerical analyzes were also reported in Table 9. As can be seen, the predicted results have good accuracy, and at the same time, these designs have the maximum value of diodicity and TDR in reality. The maximum value of TDR in the Tesla valve with the design of L = 4.502 mm, S = 2 mm, and V = 2 m/s was obtained and had a predicted value of 1.908. The numerical results reported the TDR value for this design as 1.862, and it can be said that the ANN model has a 2.5% error in predicting this number. Using this number, it can be argued that in this design, the heat transfer in the reverse direction is 86.2% more than in the forward movement. One of the essential parameters in Tesla valves is diodicity, and it was predicted that the maximum value of this response is achieved in the design of L = 2 mm, S = 2.048 mm, and V = 2 m/s. The predicted and calculated values for PDR were 4.546 and 4.581, respectively, with a low prediction error of 0.8%. The number of 4.581, obtained numerically for diodicity, indicates that the ΔP obtained in the reverse direction is 358.1% higher than the ΔP obtained in the forward direction. Overall, the approach of using a genetic algorithm and ANN models to predict the optimal design parameters for the Tesla valve proved successful, and the numerical analysis confirmed the accuracy of the predictions.

Table 9. Comparison of predicted and numerical data for the predicted optimal design of the Tesla valve.

Optimized Parameter	Optimal Design			ANN Predicted Data		Numerical Data	
	L (mm)	S (mm)	V (m/s)	TDR	PDR	TDR	PDR
TDR (maximization)	4.502	2	2	1.908	3.848	1.862	3.835
PDR (maximization)	2	2.048	2	1.810	4.546	1.776	4.581

Using the models provided by ANN, it is possible to predict designs with the best performance at different velocities. In this case, the desired velocity value is entered into the prediction model, and then the genetic algorithm is used to optimize the models. The optimal designs at different inlet velocities for TDR and PDR responses were presented in Tables 10 and 11, respectively. As observed in the tables, the predicted results agree with those obtained from numerical methods in all cases. Furthermore, upon careful examination of these results, it becomes evident that both responses can be enhanced by increasing the inlet velocity and selecting an appropriate geometric design for the Tesla valve. Next, the performance of the two-stage Tesla valve designed by Bao and Wang [10] was reported in Table 12 to compare with the results of the designs presented in this study. It is evident from the tables that the optimal designs exhibit significantly superior performance compared to the reference design, and this improvement is more visible in PDR.

Table 10. Optimal geometric design for TDR optimization at different velocities.

Inlet Velocity (m/s)	Optimal Geometric Design		ANN Predicted Data		Numerical Data	
	L (mm)	S (mm)	TDR	PDR	TDR	PDR
0.2	2.066	2	1.613	2.766	1.594	2.762
0.5	4.004	2	1.663	2.963	1.740	3.014
1	5.504	2	1.801	3.040	1.794	3.092
1.5	5.117	2	1.860	3.415	1.838	3.448
2	4.502	2	1.908	3.848	1.862	3.835

Table 11. Optimal geometric design for PDR optimization at different velocities.

Inlet Velocity (m/s)	Optimal Geometric Design		ANN Predicted Data		Numerical Data	
	L (mm)	S (mm)	TDR	PDR	TDR	PDR
0.2	2	3.107	1.586	2.837	1.664	2.787
0.5	2	3.284	1.599	3.108	1.678	3.360
1	2.270	2.978	1.629	4.034	1.745	3.976
1.5	2	2.453	1.691	4.449	1.768	4.398
2	2	2.048	1.810	4.546	1.776	4.581

Table 12. Performance of the two-stage Tesla valve presented in Bao and Wang's work at different velocities.

Inlet Velocity (m/s)	Designed by Bao and Wang [10]	
	TDR	**PDR**
0.2	1.323	1.310
0.5	1.354	1.611
1	1.358	1.787
1.5	1.375	1.845
2	1.386	1.874

4. Conclusions

This research aimed to optimize the design of a two-stage Tesla valve and study its fluid flow and heat transfer characteristics numerically. The input variables were selected as L, S, and V, and two responses, TDR and PDR, were used to evaluate thermal and hydraulic performances. Then, an ANN was trained for each response using data obtained from numerical experiments to predict the responses for different designs. By using these models, without doing experimental and numerical work and complex calculations, and in the shortest time, the responses related to the design of the desired Tesla valve can be predicted. In the following, the optimal designs of this device for different conditions were presented using models trained by the ANN method and genetic algorithm. According to the findings, the following was determined:

- It was shown that the models obtained using the ANN method could correctly predict the results for the thermal and hydraulic diodicities of the Tesla valve with a determination coefficient of 99.1% and 99.2%, respectively;
- It was found that increasing the length of the divider baffle decreases PDR. Also, increasing this variable from a specific limit reduces the value of the TDR response;
- The fluid inlet velocity parameter positively affects the responses, and generally, with its increase, PDR and TDR also increase;
- The highest value of diodicity was predicted for the Tesla valve with $L = 2$ mm, $S = 2.048$ mm, and $V = 2$ m/s parameters. This prediction was confirmed by performing numerical tests. The predicted diodicity value for this design is 4.546, and numerical tests reported this number to be 4.581. The prediction error of this response is very low and equal to 0.8%. This Tesla valve with a PDR of 4.581 indicates that the pressure drop in the reverse flow direction is 358.1% more than in the forward flow direction;
- The most optimal TDR response value was predicted to be 1.908, obtained in the condition of $L = 4.502$ mm, $S = 2$ mm, and $V = 2$ m/s. Numerical tests were performed on the designed Tesla valve, and the actual TDR was 1.862, a 2.5% difference from the predicted value. TDR with this value shows that the Tesla valve with the mentioned design has 86.2% more heat transfer in the reverse direction.

Author Contributions: Conceptualization, K.V.; methodology, M.V.; software, F.S.M.; validation, A.S.; formal analysis, K.V.; investigation, K.V.; resources, H.A.; data curation, H.J.; writing—original draft preparation, M.V. and F.S.M.; writing—review and editing, M.V. and F.S.M.; visualization, K.V. and H.J.; supervision, M.V., F.S.M. and H.N.; project administration, H.N.; funding acquisition, H.N. All authors have read and agreed to the published version of the manuscript.

Funding: This research received no external funding.

Data Availability Statement: The data presented in this study are available on logical request from the corresponding author.

Conflicts of Interest: The authors declare no conflict of interest.

Abbreviations

Nomenclature

b	bias
C_p	specific heat capacity (J/kg·K)
k	turbulent kinetic energy
L	divider baffle length (mm)
Ma	Mach number
m	number of tests
$\dot{m}$	mass flow rate (g/min)
ΔP	pressure drop (Pa)
Pr	Prandtl number
r	response
S	step length (mm)
T	temperature (K)
V	velocity (m/s)
V_τ	friction velocity (m/s)
W	weight
Z	height of fluid column (mm)

Symbols

ρ	density (kg/m^3)
μ	dynamic viscosity (Pa·s)
λ	thermal conductivity (W/m·K)
ε	energy dissipation rate

Subscripts

exp	experimental data
f	forward flow direction
in	inlet
num	numerical data
out	outlet
$pred$	predicted data
r	reverse flow direction
t	turbulence
w	wall

References

1. Porwal, P.R.; Thompson, S.M.; Walters, D.K.; Jamal, T. Heat Transfer and Fluid Flow Characteristics in Multistaged Tesla Valves. *Numer. Heat Transf. Part A Appl.* **2018**, *73*, 347–365. [CrossRef]
2. Liu, Z.; Shao, W.Q.; Sun, Y.; Sun, B.H. Scaling Law of the One-Direction Flow Characteristics of Symmetric Tesla Valve. *Eng. Appl. Comput. Fluid Mech.* **2022**, *16*, 441–452. [CrossRef]
3. Doddamani, H.; Samad, A. Dynamic Performance of a Fluidic Diode Subjected to Periodic Flow. *Ocean Eng.* **2023**, *268*, 113381. [CrossRef]
4. Hithaish, D.; Siddique, M.H.; Samad, A. A Pareto Optimal Front of Fluidic Diode for a Wave Energy Harnessing Device. *Ocean Eng.* **2022**, *260*, 111821. [CrossRef]
5. Qian, J.Y.; Chen, M.R.; Liu, X.L.; Jin, Z.J. A Numerical Investigation of the Flow of Nanofluids through a Micro Tesla Valve. *J. Zhejiang Univ. Sci. A* **2019**, *20*, 50–60. [CrossRef]
6. Nigro, A.; Algieri, A.; De Bartolo, C.; Bova, S. Fluid Dynamic Investigation of Innovative Intake Strategies for Multivalve Internal Combustion Engines. *Int. J. Mech. Sci.* **2017**, *123*, 297–310. [CrossRef]
7. Yang, X.; Liu, Z.; Liu, Z.; Feng, Z.; Simon, T. Turbine Platform Phantom Cooling from Airfoil Film Coolant, with Purge Flow. *Int. J. Heat Mass Transf.* **2019**, *140*, 25–40. [CrossRef]
8. Nowak, A.J. Selected papers presented during the Numerical Heat Transfer 2012 International Conference (NHT2012) held on 4–6 September 2012 in Wroclaw, Poland. *Int. J. Numer. Methods Heat Fluid Flow* **2014**, *24*, 949–968. [CrossRef]
9. Pakatchian, M.R.; Saeidi, H.; Ziamolki, A. CFD-Based Blade Shape Optimization of MGT-70(3)Axial Flow Compressor. *Int. J. Numer. Methods Heat Fluid Flow* **2020**, *30*, 3307–3321. [CrossRef]
10. Bao, Y.; Wang, H. Numerical Study on Flow and Heat Transfer Characteristics of a Novel Tesla Valve with Improved Evaluation Method. *Int. J. Heat Mass Transf.* **2022**, *187*, 122540. [CrossRef]
11. Liosis, C.; Sofiadis, G.; Karvelas, E.; Karakasidis, T.; Sarris, I. A Tesla Valve as a Micromixer for Fe3O4 Nanoparticles. *Processes* **2022**, *10*, 1648. [CrossRef]

12. Jin, Z.J.; Gao, Z.X.; Chen, M.R.; Qian, J.Y. Parametric Study on Tesla Valve with Reverse Flow for Hydrogen Decompression. *Int. J. Hydrogen Energy* **2018**, *43*, 8888–8896. [CrossRef]
13. Qian, J.Y.; Wu, J.Y.; Gao, Z.X.; Wu, A.; Jin, Z.J. Hydrogen Decompression Analysis by Multi-Stage Tesla Valves for Hydrogen Fuel Cell. *Int. J. Hydrogen Energy* **2019**, *44*, 13666–13674. [CrossRef]
14. Lu, Y.; Wang, J.; Liu, F.; Liu, Y.; Wang, F.; Yang, N.; Lu, D.; Jia, Y. Performance Optimisation of Tesla Valve-Type Channel for Cooling Lithium-Ion Batteries. *Appl. Therm. Eng.* **2022**, *212*, 118583. [CrossRef]
15. Monika, K.; Chakraborty, C.; Roy, S.; Sujith, R.; Datta, S.P. A Numerical Analysis on Multi-Stage Tesla Valve Based Cold Plate for Cooling of Pouch Type Li-Ion Batteries. *Int. J. Heat Mass Transf.* **2021**, *177*, 121560. [CrossRef]
16. de Vries, S.F.; Florea, D.; Homburg, F.G.A.; Frijns, A.J.H. Design and Operation of a Tesla-Type Valve for Pulsating Heat Pipes. *Int. J. Heat Mass Transf.* **2017**, *105*, 1–11. [CrossRef]
17. Qian, J.Y.; Chen, M.R.; Gao, Z.X.; Jin, Z.J. Number and Energy Loss Analysis inside Multi-Stage Tesla Valves for Hydrogen Decompression. *Energy* **2019**, *179*, 647–654. [CrossRef]
18. Yang, A.S.; Chuang, F.C.; Chen, C.K.; Lee, M.H.; Chen, S.W.; Su, T.L.; Yang, Y.C. A High-Performance Micromixer Using Three-Dimensional Tesla Structures for Bio-Applications. *Chem. Eng. J.* **2015**, *263*, 444–451. [CrossRef]
19. Sun, L.; Li, J.; Xu, H.; Ma, J.; Peng, H. Numerical Study on Heat Transfer and Flow Characteristics of Novel Microchannel Heat Sinks. *Int. J. Therm. Sci.* **2022**, *176*, 107535. [CrossRef]
20. Esen, H.; Inalli, M.; Sengur, A.; Esen, M. Performance Prediction of a Ground-Coupled Heat Pump System Using Artificial Neural Networks. *Expert Syst. Appl.* **2008**, *35*, 1940–1948. [CrossRef]
21. Sivaprasad, H.; Lekkala, M.R.; Latheef, M.; Seo, J.; Yoo, K.; Jin, C.; Kim, D.K. Fatigue Damage Prediction of Top Tensioned Riser Subjected to Vortex-Induced Vibrations Using Artificial Neural Networks. *Ocean Eng.* **2023**, *268*, 113393. [CrossRef]
22. Ahmad, I.; M'zoughi, F.; Aboutalebi, P.; Garrido, I.; Garrido, A.J. Fuzzy Logic Control of an Artificial Neural Network-Based Floating Offshore Wind Turbine Model Integrated with Four Oscillating Water Columns. *Ocean Eng.* **2023**, *269*, 113578. [CrossRef]
23. Olabi, A.G.; Abdelkareem, M.A.; Semeraro, C.; Al Radi, M.; Rezk, H.; Muhaisen, O.; Al-Isawi, O.A.; Sayed, E.T. Artificial Neural Networks Applications in Partially Shaded PV Systems. *Therm. Sci. Eng. Prog.* **2023**, *37*, 101612. [CrossRef]
24. Yulia, F.; Chairina, I.; Zulys, A.; Nasruddin. Multi-Objective Genetic Algorithm Optimization with an Artificial Neural Network for CO2/CH4 Adsorption Prediction in Metal–Organic Framework. *Therm. Sci. Eng. Prog.* **2021**, *25*, 100967. [CrossRef]
25. Gao, J.; Hu, Z.; Yang, Q.; Liang, X.; Wu, H. Fluid Flow and Heat Transfer in Microchannel Heat Sinks: Modelling Review and Recent Progress. *Therm. Sci. Eng. Prog.* **2022**, *29*, 101203. [CrossRef]
26. Mollalo, A.; Rivera, K.M.; Vahedi, B. Artificial Neural Network Modeling of Novel Coronavirus (COVID-19) Incidence Rates across the Continental United States. *Int. J. Environ. Res. Public Health* **2020**, *17*, 4204. [CrossRef]
27. Abiodun, O.I.; Kiru, M.U.; Jantan, A.; Omolara, A.E.; Dada, K.V.; Umar, A.M.; Linus, O.U.; Arshad, H.; Kazaure, A.A.; Gana, U. Comprehensive Review of Artificial Neural Network Applications to Pattern Recognition. *IEEE Access* **2019**, *7*, 158820–158846. [CrossRef]
28. Polat, M.E.; Cadirci, S. Artificial Neural Network Model and Multi-Objective Optimization of Microchannel Heat Sinks with Diamond-Shaped Pin Fins. *Int. J. Heat Mass Transf.* **2022**, *194*, 123015. [CrossRef]
29. Kanesan, J.; Arunasalam, P.; Seetharamu, K.N.; Azid, I.A. Artificial Neural Network Trained, Genetic Algorithms Optimized Thermal Energy Storage Heatsinks for Electronics Cooling. In Proceedings of the ASME 2005 Pacific Rim Technical Conference and Exhibition on Integration and Packaging of MEMS, NEMS, and Electronic Systems collocated with the ASME 2005 Heat Transfer Summer Conference, San Francisco, CA, USA, 17–22 July 2005; pp. 1389–1395. [CrossRef]
30. Mahmoudabadbozchelou, M.; Eghtesad, A.; Jamali, S.; Afshin, H. Entropy Analysis and Thermal Optimization of Nanofluid Impinging Jet Using Artificial Neural Network and Genetic Algorithm. *Int. Commun. Heat Mass Transf.* **2020**, *119*, 104978. [CrossRef]
31. Kuang, Y.; Han, F.; Sun, L.; Zhuan, R.; Wang, W. Saturated Hydrogen Nucleate Flow Boiling Heat Transfer Coefficients Study Based on Artificial Neural Network. *Int. J. Heat Mass Transf.* **2021**, *175*, 121406. [CrossRef]
32. Heng, S.Y.; Asako, Y.; Suwa, T.; Nagasaka, K. Transient Thermal Prediction Methodology for Parabolic Trough Solar Collector Tube Using Artificial Neural Network. *Renew. Energy* **2019**, *131*, 168–179. [CrossRef]
33. Ermis, K.; Erek, A.; Dincer, I. Heat Transfer Analysis of Phase Change Process in a Finned-Tube Thermal Energy Storage System Using Artificial Neural Network. *Int. J. Heat Mass Transf.* **2007**, *50*, 3163–3175. [CrossRef]
34. Xie, G.; Sunden, B.; Wang, Q.; Tang, L. Performance Predictions of Laminar and Turbulent Heat Transfer and Fluid Flow of Heat Exchangers Having Large Tube-Diameter and Large Tube-Row by Artificial Neural Networks. *Int. J. Heat Mass Transf.* **2009**, *52*, 2484–2497. [CrossRef]
35. Beigmoradi, S.; Hajabdollahi, H.; Ramezani, A. Multi-Objective Aero Acoustic Optimization of Rear End in a Simplified Car Model by Using Hybrid Robust Parameter Design, Artificial Neural Networks and Genetic Algorithm Methods. *Comput. Fluids* **2014**, *90*, 123–132. [CrossRef]
36. Li, B.; Lee, Y.; Yao, W.; Lu, Y.; Fan, X. Development and Application of ANN Model for Property Prediction of Supercritical Kerosene. *Comput. Fluids* **2020**, *209*, 104665. [CrossRef]
37. George, A.; Poguluri, S.K.; Kim, J.; Cho, I.H. Design Optimization of a Multi-Layer Porous Wave Absorber Using an Artificial Neural Network Model. *Ocean Eng.* **2022**, *265*, 112666. [CrossRef]

38. Zhu, K.; Shi, H.; Han, M.; Cao, F. Layout Study of Wave Energy Converter Arrays by an Artificial Neural Network and Adaptive Genetic Algorithm. *Ocean Eng.* **2022**, *260*, 112072. [CrossRef]
39. Vaferi, K.; Vajdi, M.; Nekahi, S.; Nekahi, S.; Sadegh Moghanlou, F.; Azizi, S.; Shahedi Asl, M. Numerical Simulation of Cooling Performance in Microchannel Heat Sinks Made of AlN Ceramics. *Microsyst. Technol.* **2023**, *29*, 141–156. [CrossRef]
40. Li, S.; Li, C.; Li, Z.; Xu, X.; Ye, C.; Zhang, W. Design Optimization and Experimental Performance Test of Dynamic Flow Balance Valve. *Eng. Appl. Comput. Fluid Mech.* **2020**, *14*, 700–712. [CrossRef]
41. Kuzmin, D.; Mierka, O.; Turek, S. On the Implementation of the Fr-Fturbulence Model in Incompressible Flow Solvers Based on a Finite Element Discretisation. *Int. J. Comput. Sci. Math.* **2007**, *1*, 193–206. [CrossRef]
42. Savicki, D.L.; Goulart, A.; Becker, G.Z. A Simplified K- ε Turbulence Model. *J. Brazilian Soc. Mech. Sci. Eng.* **2021**, *43*, 1–16. [CrossRef]
43. Micale, D.; Ferroni, C.; Uglietti, R.; Bracconi, M.; Maestri, M. Computational Fluid Dynamics of Reacting Flows at Surfaces: Methodologies and Applications. *Chemie Ing. Tech.* **2022**, *94*, 634–651. [CrossRef]
44. Demidov, D.; Mu, L.; Wang, B. Accelerating Linear Solvers for Stokes Problems with C++ Metaprogramming. *J. Comput. Sci.* **2021**, *49*, 101285. [CrossRef]
45. Schenk, O.; Gärtner, K.; Fichtner, W.; Stricker, A. PARDISO: A High-Performance Serial and Parallel Sparse Linear Solver in Semiconductor Device Simulation. *Futur. Gener. Comput. Syst.* **2001**, *18*, 69–78. [CrossRef]
46. Vaferi, K.; Vajdi, M.; Nekahi, S.; Nekahi, S.; Sadegh Moghanlou, F.; Shahedi Asl, M.; Mohammadi, M. Thermo-Mechanical Simulation of Ultrahigh Temperature Ceramic Composites as Alternative Materials for Gas Turbine Stator Blades. *Ceram. Int.* **2021**, *47*, 567–580. [CrossRef]
47. Salim, S.M.; Ariff, M.; Cheah, S.C. Wall Y+ Approach for Dealing with Turbulent Flows over a Wall Mounted Cube. *Prog. Comput. Fluid Dyn. Int. J.* **2010**, *10*, 341. [CrossRef]
48. Blocken, B.; Stathopoulos, T.; Carmeliet, J. CFD Simulation of the Atmospheric Boundary Layer: Wall Function Problems. *Atmos. Environ.* **2007**, *41*, 238–252. [CrossRef]
49. Klewicki, J.; Saric, W.; Marusic, I.; Eaton, J. Wall-Bounded Flows. In *Springer Handbook of Experimental Fluid Mechanics*; Springer: Berlin/Heidelberg, Germnay, 2007; pp. 871–907.
50. Hou, D.; Chen, D.; Wang, X.; Wu, D.; Ma, H.; Hu, X.; Zhang, Y.; Wang, P.; Yu, R. RSM-Based Modelling and Optimization of Magnesium Phosphate Cement-Based Rapid-Repair Materials. *Constr. Build. Mater.* **2020**, *263*, 567–580. [CrossRef]

Article

Design and Performance Evaluation of Integrating the Waste Heat Recovery System (WHRS) for a Silicon Arc Furnace with Plasma Gasification for Medical Waste

Yuehong Dong [1,†], Lai Wei [2,†], Sheng Wang [1], Peiyuan Pan [2,*] and Heng Chen [2]

1   State Key Laboratory of Clean and Efficient Coal-Fired Power Generation and Pollution Control, China Energy Science and Technology Research Institute Co., Ltd., Nanjing 210023, China
2   Beijing Key Laboratory of Emission Surveillance and Control for Thermal Power Generation, North China Electric Power University, Beijing 102206, China; l.wei@ncepu.edu.cn (L.W.); heng@ncepu.edu.cn (H.C.)
*   Correspondence: peiyuanpan@ncepu.edu.cn
†   These authors contributed equally to this work.

Abstract: A hybrid scheme integrating the current waste heat recovery system (WHRS) for a silicon arc furnace with plasma gasification for medical waste is proposed. Combustible syngas converted from medical waste is used to drive the gas turbine for power generation, and waste heat is recovered from the raw syngas and exhaust gas from the gas turbine for auxiliary heating of steam and feed water in the WHRS. Meanwhile, the plasma gasifier can also achieve a harmless disposal of the hazardous fine silica particles generated in polysilicon production. The performance of the proposed design is investigated by energy, exergy, and economic analysis. The results indicate that after the integration, medical waste gave rise to 4.17 MW net power at an efficiency of up to 33.99%. Meanwhile, 4320 t of the silica powder can be disposed conveniently by the plasma gasifier every year, as well as 23,040 t of medical waste. The proposed design of upgrading the current WHRS to the hybrid system requires an initial investment of 18,843.65 K$ and has a short dynamic payback period of 3.94 years. Therefore, the hybrid scheme is feasible and promising for commercial application.

Keywords: waste heat recovery; plasma gasification; silicon arc furnace; medical waste; system integration

Citation: Dong, Y.; Wei, L.; Wang, S.; Pan, P.; Chen, H. Design and Performance Evaluation of Integrating the Waste Heat Recovery System (WHRS) for a Silicon Arc Furnace with Plasma Gasification for Medical Waste. *Entropy* **2023**, *25*, 595. https://doi.org/10.3390/e25040595

Academic Editors: Daniel Flórez-Orrego, Meire Ellen Ribeiro Domingos and Rafael Nogueira Nakashima

Received: 19 February 2023
Revised: 25 March 2023
Accepted: 28 March 2023
Published: 31 March 2023

1. Introduction

Solar energy is a widely distributed renewable energy and becoming increasingly popular for power generation. Photovoltaics (PV) is at present the most used and cost-effective technology of solar energy, which converts sunlight into electricity, directly based on the photovoltaic effect. In the last decade, PV production witnessed great growth. In 2021, solar PV capacity increased by 17% globally, accounting for ~60% of the total renewable power expansion [1]. In order to meet the fast-growing solar PV demand, global production of PV-related products is expected to more than double by 2030 [1]. Today, China dominates the global solar PV supply chains and contributes to an 80% decline in the price of solar panels, making solar PV an affordable electricity generation technology [2,3].

Crystalline silicon modules have dominated the current solar PV market at more than 95% of the installed capacity in the last five years [4]. Solar PV manufacturing is energy-intensive and mostly powered by fossil fuels. Polysilicon production is the largest energy-consuming segment of the solar PV supply chain, accounting for up to ~40% of the total energy consumption [4]. The first stage of polysilicon production is to extract metallurgical-grade silicon by melting quartz ore and reducing silica in a large electric arc furnace, which requires a great deal of heat at a high temperature (~2000 °C) and a lengthy time [5,6]. Metallurgical-grade silicon of 98% silicon purity is the fundamental material of subsequent silicon products [2,7]. Generally, 10 to 13 MWh of electricity is needed to produce one ton of metallurgical-grade silicon, but only ~30% of the total energy input is contained in the silicon product, whereas the rest of the energy is taken away as thermal energy by the off gas and the cooling water [6,8]. The temperature of the exhaust gas leaving the arc furnace mainly depends on the furnace load and air excess, at 500 to 700 °C. Therefore, there is great potential to utilize the waste heat in the exhaust gas, and a heat recovery steam generator (HRSG) with a steam turbine is a suitable and cost-effective method [9]. Currently, many large-scale silicon arc furnaces are equipped with some sort of waste heat recovery system (WHRS), and the generated electricity can offset some of their power consumption. Metallurgical-grade silicon is subsequently purified into solar-grade polysilicon of 6–13 N purity [7,10].

Solar PV systems have obvious superiority to traditional power generation methods due to near-zero emission of atmospheric pollutants and greenhouse gases during operation, but the vast majority of their environmental burdens caused by solar PV are released during their manufacturing processes [11,12]. For instance, a lot of silica particles are formed in the silicon arc furnace and entrained by the exhaust gas, which needs to be gathered and removed before going into the atmosphere [13,14]. These particles are quite small in size, and their high specific surface area and volume result in easy contamination and difficult transportation. Emission or leakage of these fine particles aggravates atmospheric pollution and is hazardous to human health.

Due to good stability, silica cannot be easily tackled by common chemical methods, although melting followed by consolidation seems to be an effective way. Nowadays, plasma gasification is regarded as a superior and promising waste to energy (WTE) technique of solid waste [15–18].

Direct current (DC) plasma is widely considered due to its better stability and load adaptability than other types of plasma. In a DC plasma gasifier, strong DC electric arc is created by electrodes, resulting in thermal plasma and a high-temperature environment, which can destroy nearly all chemical bonds in the substances, releasing free electrons, ions, radicals, and molecules, allowing many reactions that cannot proceed in normal conditions to occur [19,20]. Consequently, the input organics are quickly decomposed into their component elements, which subsequently react to form a synthetic gas mostly consisting of H_2, CO, CH_4, and some other light hydrocarbons, and meanwhile, the inorganics are completely melted and transformed into inert and nontoxic glassy slags [21–23]. Plasma gasification has two remarkable advantages: (1) producing combustible syngas of high calorific value, as a clean and valuable fuel for power generation; (2) disposing of a broad variety of solid wastes safely and harmlessly, especially some hazardous wastes, while consolidating solid residues [18,24,25].

Plasma gasification has been widely considered in the treatment of municipal solid wastes (MSW) as an alternative to traditional incineration and landfill, due to its outstanding environmental benefits and flexibility [16,18,26–28]. The disposal fee of wastes and electricity selling are the major sources of income for the power plant based on plasma gasification. However, at present, plasma gasification plants are quite scarce, owing to their high investments and operating costs, and therefore, energy conversion efficiency and economic viability are emphasized.

As far as net electrical efficiency is concerned, a plasma gasification power plant containing only one single-stage steam turbine or gas turbine does not obviously perform better than conventional incineration plants. Using MSW as the feedstock, the net electrical efficiency of a plasma gasification power plant based on an individual steam cycle (Rankine cycle) is only about 14 to 21%, and utilization of a single gas turbine has the efficiency of 13 to 24%, compared to 20 to 30% of direct incineration [24,27,29–33]. Therefore, an integrated plasma gasification combined cycle (IPGCC) is essential for efficiency improvement. In an IPGCC plant, the direct conversion of chemical energy of syngas into electricity is conducted by a gas turbine, and meanwhile, a steam cycle is established for heat recovery from the exhaust gas of the gas turbine, thereby promoting overall power generation and economic competitiveness. The net electrical efficiency of an IPGCC plant can be close to or even exceed 30% [34,35]. Minutillo et al. [24] developed a thermochemical model of an

IPGCC plant and pointed out that the system efficiency of power generation could be up to 31%. Montiel-Bohórquez et al. [20] assessed the technical and economic performance of an IPGCC plant fueled with MSW, finding out its highest efficiency was 32.49%, and one-third of the total gross power output was from the steam turbine.

Recently, the development of healthcare facilities and the breaking COVID-19 pandemic made medical waste management a great environmental issue [36–38]. Medical waste is mainly composed of plastics, paper, textiles, glass, and some organics, this means that they have similar characteristics to typical MSW [39–41]. But due to the relatively low moisture content, high volatile content, and high lower calorific value (LHV) compared to MSW, medical waste is more worth exploiting in view of WTE [37,38]. Furthermore, because medical waste contains some infectious, pathological, chemical, pharmaceutical, or cytotoxic matters, there are potential risks of environmental pollution and infection, which lead to high disposal costs [30,42,43].

In this study, medical waste rather than common MSW is taken as the major feedstock in the plasma gasifier [36,38]. The plasma gasifier is designed to be built close to the silicon arc furnace so that simultaneous disposal of medical waste and silica powder can be performed.

Over the past several decades, a WHRS for power generation based on the steam cycle has been extensively investigated and applied in many industrial processes. The concept of incorporating a conventional WHRS into the gasification system also has been proposed, because one of the HRSGs could be saved by sharing the equipment, thereby lowering the overall costs and land occupation. Chen et al. [44] designed a novel medical WTE system based on plasma gasification integrated with an MSW incineration plant, using exhaust gas from the gas turbine to heat the live steam and feed water in the incineration plant, thereby increasing WTE efficiency up to 37.83%. Yang et al. [45] proposed and techno-economically assessed a WTE process based on combined heat and power plant and intermediate pyrolysis technology, finding that the levelized cost of electricity was £0.063/kWh. In the authors' previous work [46], plasma waste gasification was integrated with a coal-fired power plant, promoting WTE efficiency by feeding the syngas directly into the coal-fired boiler. However, in the previous literature, the gasification system is usually integrated with large-scale power plants, so that the thermal energy contained by the syngas can be utilized at a high temperature, benefiting its efficiency. A WHRS for small-scale industrial boilers generating live steam of relatively low parameters is seldom considered. On the other hand, there is hardly any research on the treatment of the silica powder generated in polysilicon production by plasma gasification, because the melting of SiO_2 brings down the gasification efficiency. Therefore, the economic performance of integrating the WHRS of a silicon arc furnace with plasma gasification is still questionable.

In this work, in view of the expanding polysilicon production and increasing demand for medical waste treatment, a design that integrates the current WHRS of a silicon arc furnace with plasma gasification for medical waste is proposed. The advantage of this integration mainly includes: (1) The heat recovered in the plasma gasification system is exploited for extra heating of the steam and feed water in WHRS, thereby promoting WTE efficiency of the medical waste without affecting the power output by the exhaust gas from the silicon arc furnace. (2) The silica powder collected from the flue gas leaving the silicon furnace can be fed into the plasma gasifier and consolidated into vitrified slags, thereby avoiding pollution caused by the leakage of fine particles. (3) An HRSG system in a typical IPGCC scheme can be substituted by the existing equipment in a WHRS, so that investment, operational costs, and land occupation of the plasma gasification system can be significantly lowered.

2. System Description

A 33 MVA submerged arc furnace producing metallurgical-grade silicon with its current WHRS was selected as the reference plant, which is now operational at a large-scale manufacturing base of silicon PV in northwestern China. The manufacturing base has

32 silicon arc furnaces and dozens of polysilicon purification and monocrystalline silicon production lines. At present, each arc furnace has already been equipped with WHRS for power generation. The furnaces and WHRS have both been operating reliably for the past five years, and the operating parameters collected online coincided well with design values. In this study, the design diagrams and data provided by the manufacturers were used for analysis. The existing WHRS and the proposed WHRS integrated with plasma gasification for medical waste are respectively described in Sections 2.1 and 2.2.

2.1. Current WHRS for the Silicon Arc Furnace

Figure 1 shows the current WHRS for the silicon arc furnace. Quartz ore that mainly consists of SiO_2 is intermittently fed into the submerged arc furnace and reduced by carbon reductants (coke) at ~2000 °C to produce polysilicon. The major chemical reactions are as follows [6]:

$$2C(s) + SiO_2(s) \rightarrow Si(l) + SiO(g) + CO(g), \tag{1}$$

$$SiO(g) + 2C(s) \rightarrow SiC(s) + CO(g), \tag{2}$$

$$SiC(s) + SiO_2(s) \rightarrow Si(l) + SiO(g) + CO(g), \tag{3}$$

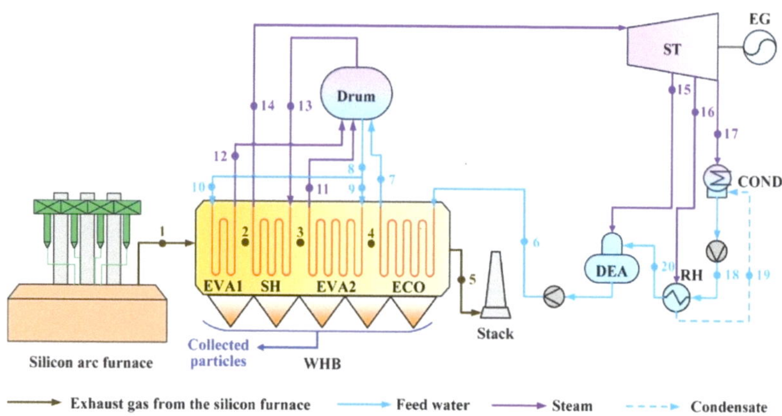

Figure 1. Diagram of the current WHRS for the silicon arc furnace.

These reactions occur in the packed bed and the required heat is provided by the arc created by three AC electrodes that are half submerged in the raw material. Molten silicon is taken out through the tapping at the bottom, while the hot exhaust gas continuously flows upward and leaves the furnace through the top hole. Table 1 lists the main gaseous components of the exhaust gas. The average temperature of the exhaust gas is ~650 °C, making its heat worth recovering. Meanwhile, the exhaust gas has 4 to 10 g/m^3 of fly ash, and the main components of the fly ash are listed in Table 2. Fine amorphous silica particles dominate in the fly ash, and have an average particle size of less than 1 μm.

Table 1. Main gaseous components of the exhaust gas leaving the silicon arc furnace.

Components	N_2	O_2	H_2O	CO_2
Concentration (volume fraction)	78.48%	18.53%	1.47%	1.52%

Table 2. Main components of the fly ash in the exhaust gas leaving the silicon arc furnace.

Components	SiO_2	SiC	FeO	Al_2O_3	CaO	MgO
Content (mass fraction)	93.30%	4.80%	0.04%	0.18%	0.40%	1.00%

As depicted in Figure 1, current WHRS is composed of a waste heat boiler (WHB) for waste heat recovery and steam generation, a steam turbine (ST) driven by the live steam, an electricity generator (EG) for power generation, and heat regeneration equipment. Their basic parameters are referred by the values in Tables A1 and A2 in Appendix A. The exhaust gas leaving the silicon arc furnace flows through WHB and is cooled from 650 °C to 187.3 °C, and meanwhile, the feed water from the outlet of the deaerator (DEA) at 104.8 °C/4.02 MPa is heated to a superheated steam at 450 °C/3.82 MPa. Due to the high content and small particle size of the fly ash in the gas flow, efficient and uninterrupted ash removal from heat transfer surfaces is necessary in WHB. Mechanical striking using steel balls is applied and the falling ash is collected by the ash hoppers at the bottom.

The feed water delivered to WHB is first heated in the economizer (ECO) to approximate saturated water. Water/steam separation is performed in the drum. Some of the saturated water from the drum flows into the evaporator 2 (EVA2) between the superheater (SH) and ECO, and the rest of the saturated water is sent to EVA1 at the inlet of WHB. The purpose of EVA1 is to cool the high-temperature gas rapidly in view of slagging prevention. The saturated steam separated by the drum is sent into SH for final heating and then flows into ST.

2.2. Proposed WHRS Integrated with Plasma Gasification

The current WHRS can be integrated with a plasma gasification system fueled with medical waste, exploiting the heat recovered from syngas treating processes to further raise its power output. The combustible syngas generated by gasification is used for power generation through the gas turbine (GT). In the meantime, the collected silica powder is fed into the gasifier together with medical waste and ends up in vitrified slags that are harmless and transportable.

This proposed hybrid scheme is illustrated in Figure 2. From the perspective of power generation based on plasma gasification, a combination with WHRS is an alternative to conventional IPGCC, making full use of the syngas and combustion gas and requiring lower capital investment.

The plasma gasification system can be roughly divided into two parts, the plasma gasifier subsystem, and the gas turbine subsystem.

The plasma gasifier subsystem includes the DC plasma gasifier and syngas conditioning equipment [20]. The medical waste and the collected silica powders are fed from the top of the gasifier, and the organic components quickly decompose when being heated, generating volatiles. O_2 separated from the air is injected into the bottom of the gasifier as the oxidizing agent of the gasification process, and the extremely high temperature environment (~4000 °C) created by plasma torches transforms the residual carbon, hydrogen, and other combustible elements into micro-molecular gases. The remaining inorganic solids, including the fine silica particles in the feedstock, are completely melted, and the effluent slags are cooled and solidified.

The syngas formed in the gasifier contains various high calorific value components, making it a good fuel. However, the formation of contaminants, such as particulates, condensable hydrocarbons, sulfur compounds, nitrogen compounds, halides, and trace heavy metals, is inevitable in gasification, and thus the raw syngas needs to be cleaned to meet stringent emission regulations and protect the downstream equipment from fouling, corrosion, and erosion [47,48]. There is a multitude of technologies for syngas purification. Conventional syngas cleaning equipment includes the cyclone separator (CS), wet scrubber, carbonyl sulfide (COS) hydrolysis, acid gas removal (AGR), and filters [47,49]. Syngas needs to be properly cooled and heated in order to meet the temperature requirements of different cleaning processes [48].

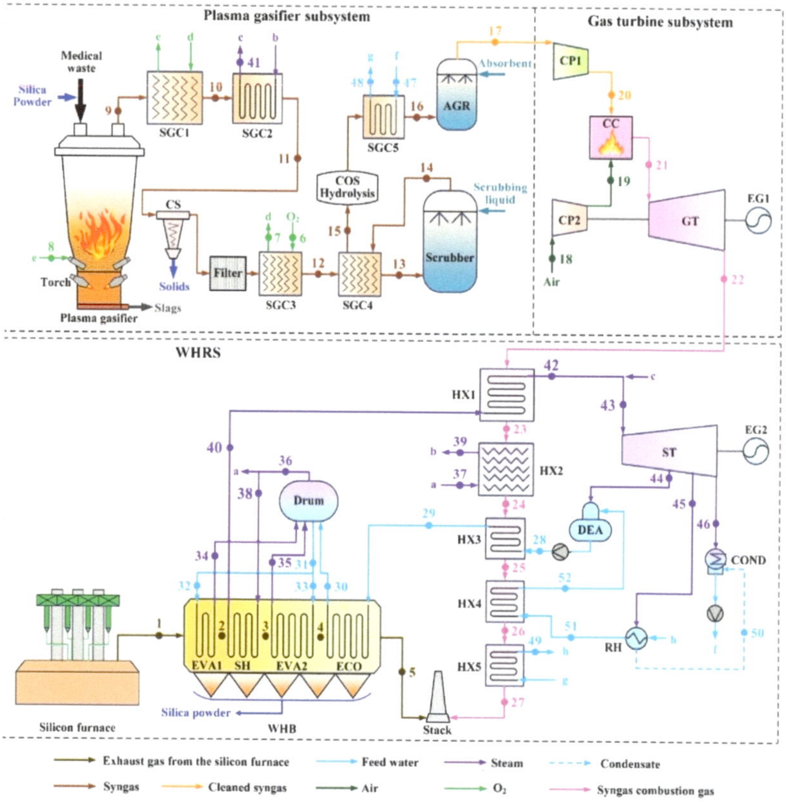

Figure 2. Diagram of the proposed WHRS integrated with plasma gasification for medical waste.

Heat recovery in syngas conditioning is essential for attaining system efficiency, and thus heat exchangers are required. As shown in Figure 2, the raw syngas is first cooled in a gas-gas heat exchanger, named syngas cooler 1 (SGC1). O_2 entering the plasma gasifier needs to be well-preheated to facilitate gasification. In this scheme, O_2 has two-stage preheating, and SGC1 is used for the second stage. SGC2 heats a steam flow from WHRS and supplies superheated steam to ST directly. Syngas from SGC2 flows through CS and a filter to remove bulk particles. CS is a widely used inertial separation device and is able to operate in a wide temperature range while requiring low energy [48,50]. In view of syngas, a temperature of 300 to 500 °C benefits the operation of CS and filters because particulate matters can stay in the solid state, avoiding problems caused by melting or moisture absorption.

The filtered syngas is then used for the first-stage preheating of O_2 in SGC3. SGC4 is placed downstream of SGC3, cooling the syngas further to ~130 °C, while reheating the syngas at the outlet of the wet scrubber to 250.0 °C. Water scrubbing is an easily-operated and effective method for syngas decontamination, removing NH_3 compounds, halides, fine particles, some H_2S, and other trace contaminants simultaneously [47,49]. However, a large flow of scrubbing water is required so that the syngas is rapidly cooled to its dew point temperature, ensuring the finest particles can be removed by acting as the nuclei for condensation [51].

Sulfur compounds are the main residual contaminants in the scrubbed syngas. Although most sulfur in the feedstock is converted to H_2S in gasification, about 3 to 10% of the sulfur is converted to carbonyl sulfide, which is the main organic sulfur component and cannot be efficiently removed by the downstream AGR due to its low solubility in most solvents. Therefore, a catalytic hydrolysis reactor is set up to convert over 99% of COS in the syngas to H_2S, according to the flowing reaction [47,52].

$$COS(g) + H_2O(g) \leftrightarrow H_2S(g) + CO_2(g), \tag{4}$$

The scrubbed syngas needs to be reheated to ~250 °C in view of the efficiency of alumina-based catalysts used in COS hydrolysis [53,54]. The conversion of COS to H_2S is an exothermic reaction, technically, but the passing of the syngas through COS hydrolysis can be regarded as an isothermal process in heat calculation due to the low concentration of COS.

AGR is the endmost syngas cleaning process, using regenerative solvents in an absorber column to remove various sulfur-bearing gases, including H_2S and SO_2 surviving from the wet scrubbing, H_2S converted from COS by hydrolysis, and some organic sulfur compounds [55,56]. Most of the common chemical solvents, such as piperazine-activated methyl diethanolamine (MDEA) and aqueous alkaline salt solutions, are effective over a wide range of acid gas concentrations at near-room temperatures, and thus AGR is typically designed to operate at slightly above the ambient temperature [56]. In order to minimize the heat loss due to AGR, the syngas from COS hydrolysis is used to heat the low-temperature condensate from condenser (COND) of WHRS in SGC5.

In the gas turbine subsystem, the cleaned syngas is sent to compressor 1 (CP1) for compression, and meanwhile, the feeding air is compressed in CP2. Syngas is burned in the combustion chamber (CC), and the formed high-temperature and -pressure combustion gas enters GT, generating power by electric generator 1 (EG1).

The exhaust gas of GT still has a high temperature and is delivered to the WHRS of the silicon arc furnace for heat exploitation, so as to save HRSG equipment in the conventional IPGCC scheme. The combustion gas leaving GT flows through gas coolers HX1 to HX5 successively.

The live steam supplied by WHB is further heated in HX1 before entering ST. HX2 is to heat a saturated steam flow from the drum, which is subsequently sent to SGC2 for final heating. HX4 and HX5 are used for additional heating of the feed water in the heat regeneration subsystem so as to save the extracted steam from ST. The condensate from COND flows through SGC5, HX5, RH, and HX4 successively before entering DEA. At last, the temperature of the feed water into WHB is promoted by HX3.

Via integration with plasma gasification, the steam cycle in WHRS is largely assisted by the heat recovered from the syngas and combustion gas, and thus its power generation capacity is improved. Meanwhile, the combination with WHRS makes it possible for the plasma gasification power generation system to save some high-cost components compared to conventional IPGCC systems. Furthermore, with a view to waste management, the troublesome silica fines generated by polysilicon production can be handled by the plasma gasifier harmlessly.

3. Methodology

The models and analysis methods used in this work are described in this chapter. The current WHRS, a plasma gasifier, and a gas turbine system are individually simulated, and the obtained parameters are compared with the operational data or data from references to validate the reliability of these models.

3.1. Analysis Methods

3.1.1. Energy Analysis

The power generation efficiency ($\eta_{\mathrm{p,wh}}$, %) of WHRS is defined as follows.

$$\eta_{\mathrm{p,wh}} = \frac{P_{\mathrm{net,wh}}}{Q_{\mathrm{rec,wh}}}, \tag{5}$$

where $P_{\mathrm{net,wh}}$ is the net power generation by the exhaust gas from the silicon arc furnace, MW. $Q_{\mathrm{rec,wh}}$ is the thermal energy recovered from the exhaust gas, MW. They are respectively calculated as follows.

$$P_{\mathrm{net,wh}} = P_{\mathrm{gr,wh}} - P_{\mathrm{ax,wh}}, \tag{6}$$

$$Q_{\mathrm{rec,wh}} = m_{\mathrm{eg}} \times \left(h_{\mathrm{fg,in}} - h_{\mathrm{fg,out}} \right), \tag{7}$$

where $P_{\mathrm{gr,wh}}$ is the gross power output by ST, MW. $P_{\mathrm{ax,wh}}$ is the estimated power consumption by the auxiliaries in WHRS, MW. m_{eg} is the flow rate of the exhaust gas from the silicon arc furnace, kg/s. $h_{\mathrm{fg,in}}$ and $h_{\mathrm{fg,out}}$ are specific enthalpies of the inlet and outlet flue gas of WHB, respectively, kJ/kg.

In the plasma gasifier, solid waste is used as the feedstock, and air and steam are used to assist gasification. Plasma torches are the main heat source and power consumer in the gasifier, and the torch thermal efficiency is estimated to be 86%. The plasma gasification efficiency (η_{pg}, %) is calculated as follows [57].

$$\eta_{\mathrm{pg}} = \frac{m_{\mathrm{syn}} \times \mathrm{LHV}_{\mathrm{syn}}}{m_{\mathrm{waste}} \times \mathrm{LHV}_{\mathrm{waste}} + \frac{P_{\mathrm{tor}}}{\eta_{\mathrm{tor}} \times \eta_{\mathrm{e}}}}, \tag{8}$$

where m_{syn} is the flow rate of the syngas generated by plasma gasification, kg/s. $\mathrm{LHV}_{\mathrm{syn}}$ is the lower heating value of the syngas, MJ/kg. m_{waste} is the feed rate of waste into the plasma gasifier, kg/s. $\mathrm{LHV}_{\mathrm{waste}}$ is the lower heating value of the waste, MJ/kg. P_{tor} is the power consumption by the plasma torches, MW. η_{tor} is torch thermal efficiency, %. η_{e} is the overall power generation efficiency of the plasma gasification plant, %, considered as 35% according to the average level of IPGCC plants [57].

In the hybrid system, assuming the power generation by the exhaust gas of the arc furnace ($P_{\mathrm{net,wh}}$, MW) is constant, the net power generation by the plasma gasification of medical waste ($P_{\mathrm{net,pg}}$, MW) and its efficiency ($\eta_{\mathrm{p,pg}}$, %) are calculated as follows.

$$P_{\mathrm{net,pg}} = P_{\mathrm{net,tot}} - P_{\mathrm{net,wh}}, \tag{9}$$

$$\eta_{\mathrm{p,pg}} = \frac{P_{\mathrm{net,pg}}}{m_{\mathrm{mw}} \times \mathrm{LHV}_{\mathrm{mw}}}, \tag{10}$$

where $P_{\mathrm{net,tot}}$ is the total net output by ST and GT, MW.

The overall power generation efficiency of the hybrid system ($\eta_{\mathrm{p,tot}}$, %) is calculated as follows.

$$\eta_{\mathrm{p,tot}} = \frac{P_{\mathrm{net,tot}}}{Q_{\mathrm{rec,wh}} + m_{\mathrm{mw}} \times \mathrm{LHV}_{\mathrm{mw}}}, \tag{11}$$

Power consumption by auxiliaries in WHRS is estimated to be 15% of the current gross power output by ST. Power consumption by the syngas conditioning equipment is estimated to be 5% of the gross power output of GT. In the plasma gasifier, power consumption by O_2 separation from the air is estimated to be 0.261 kWh/kg$_{\mathrm{pure\,O2}}$ [24]. Torch thermal efficiency when O_2 is used for gasification assistance is estimated to be 90%.

3.1.2. Exergy Analysis

Exergy is an indicator of both the quantity and quality of the energy, which can be used to assess the utilization potential of an energy source and the performance of a system [58,59]. The exergy input by the exhaust gas of the silicon arc furnace (EX_{fg}, MW) and by the medical waste (EX_{mw}, MW) are respectively calculated as follows [44].

$$EX_{fg} = m_{fg} \times \left[h_{fg,in} - h_{fg,0} - T_0 \times \left(s_{fg,in} - s_{fg,0} \right) \right], \tag{12}$$

$$EX_{mw} = m_{mw} \times LHV_{mw} \times \left(1.0064 + 0.1519 \times \frac{\omega_H}{\omega_C} + 0.0616 \times \frac{\omega_O}{\omega_C} + 0.0429 \times \frac{\omega_N}{\omega_C} \right), \tag{13}$$

where T_0 is the environmental temperature, assigned as 293.15 K. $h_{fg,0}$ is specific enthalpy of the flue gas at T_0, kJ/kg. $s_{fg,in}$ and $s_{fg,0}$ are specific entropies of the flue gas at the inlet state and at T_0, respectively, kJ/(kg·K). ω_C, ω_H, ω_O, and ω_N are the mass contents of elements C, H, O, and N in the medical waste.

There is an exergy balance in energy processes, which can be applied to the components, subsystem, and entire system.

$$EX_{in} + W_{in} = EX_{out} + W_{out} + EX_{des}, \tag{14}$$

where EX_{in} and EX_{out} are the exergy input and output, MW. W_{in} and W_{out} are the work input and output, MW. EX_{des} is the exergy destruction, MW.

The exergy efficiency of power generation by plasma gasification of the medical waste ($\eta_{ex,pg}$, %) and the overall exergy efficiency of the hybrid system ($\eta_{ex,tot}$, %) are calculated as follows.

$$\eta_{ex,pg} = \frac{P_{net,pg}}{EX_{mw}}, \tag{15}$$

$$\eta_{ex,tot} = \frac{P_{net,tot}}{EX_{fg} + EX_{mw}}, \tag{16}$$

3.1.3. Economic Analysis

The main economic income of the integrated system includes disposal fees for medical waste, electricity selling, and slag selling. Economic analysis is conducted based on the assumptions given in Table 3. The construction period includes 0.5 years for the reconstruction of the current WHRS, assuming in the second year, which causes a decrease in electricity generation of that year, and thus compensation for the income decrease is considered.

Table 3. Assumptions used for the economic analysis of the system [22,44,60].

Item	Unit	Value
Construction period	year	2
Economic period	year	23
Annual operating time	hour	8000
Operating cost	-	10% of the total investment
Discount rate	%	12
Price of electricity	$/MWh	96.51
Price of slags	$/t	53.78
Tipping fee for medical waste	$/t	463.86

Upgrade of the current WHRS to the proposed hybrid system requires new equipment, and the cost of these components is estimated based on the methods in Tables 4 and 5. The scaling-up methods in Table 5 are conducted as follows [61].

$$C = C_0 \times \left(\frac{S}{S_0} \right)^f \tag{17}$$

where C_0 is the basic cost of the reference equipment, k\$; C is the capital cost of the target equipment, k\$; S_0 is the basic scale of the reference equipment; S is the scale of the target equipment; f is the scale factor.

Table 4. Investment estimation methods for upgrading the current WHRS to the proposed hybrid system using cost function methods.

Component	Estimate Function (\$)	Reference
SGC1 SGC3 SGC4	$C = 130\left(\frac{A}{0.093}\right)^{0.78}$	[62]
CP1 CP2	$C = \frac{71.1 m_{wf}}{0.9 - \eta_c} r_p \ln(r_p)$	[62]
CC	$C = \frac{25.65 m_{air}}{0.995 - \frac{P_{out}}{P_{in}}}\left[e^{(0.018 T_{out} - 26.4}} + 1 \right]$	[44]
EVA1 EVA2	$C = 1010(A)^{0.78}$	[62]
EG1	$C = 60 E_p{}^{0.95}$	[63]

Table 5. Investment estimation methods for upgrading the current WHRS to the proposed hybrid system using scaling-up methods.

Component	Basic Cost (k\$)	Basic Scale	Scaling Factor	Scale Unit	Reference
Plasma gasifier	78,000.00	39.20	0.67	kg/s	[44]
Syngas cleaning section	33,650.00	4232.70	0.65	kmol/s	[64]
SGC2 SH HX1 HX2	45.84	500.00	0.74	m^2	[65]
SGC5 ECO HX3 HX4 HX5	44.91	500.00	0.68	m^2	[65]
GT	1100.00	1.00	1.00	MW	[44]

The dynamic payback period (DPP) of the integration projected is an important and widely used economic performance indicator, representing the least necessary time to recover the initial investment. Meanwhile, the net present value (NPV, k\$), referring to the variation of the cash inflow over the lifetime of the project, is usually employed in DPP calculation, in order to consider the discounting risk of cash. NPV and DPP are estimated as follows [44,66].

$$\text{NPV} = \sum_{y=1}^{k} \frac{C_{\text{inflow}} - C_{\text{outflow}}}{(1 + r_{\text{dis}})^y}, \tag{18}$$

$$\sum_{y=1}^{\text{DPP}} \frac{C_{\text{inflow}} - C_{\text{outflow}}}{(1 + r_{\text{dis}})^y} = 0, \tag{19}$$

where k is the lifetime of this project, assigned as 25 years. y is the year number in the lifetime. r_{dis} is the discount rate, estimated as 12%. C_{inflow} and C_{outflow} are the cash inflow and cash outflow in year y, k\$.

A shorter DPP and a higher NPV are favored for a project.

3.2. Model Development and Validation

WHRS and the plasma gasification system are mainly modeled and simulated using the EBSILON Professional platform. The behavior of each modeled equipment is described by thermodynamic laws, and the thermodynamic cycle is tackled through a group of linear equations, which are solved iteratively. Accordingly, parameters of the system with low uncertainties can be derived. Moreover, the plasma gasification process is modeled by Aspen Plus, due to its specialty in chemical simulation. Figure 3 illustrates the established models of the proposed hybrid system.

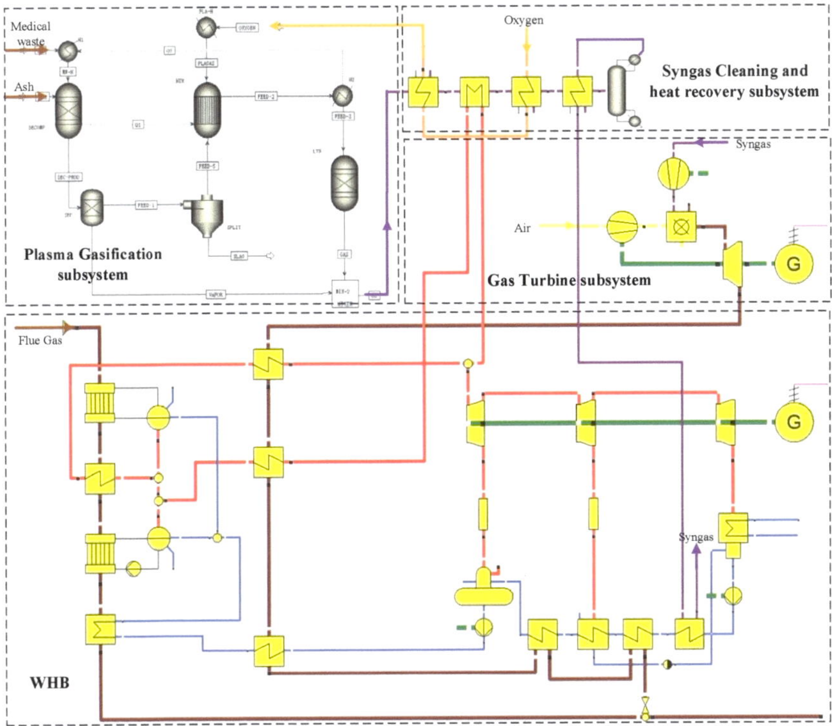

Figure 3. Models for simulation of the proposed WHRS integrated with plasma gasification based on EBSILON Professional and Aspen Plus.

The composition of medical waste varies. One type of medical waste is selected for modeling and its properties are shown in Table 6.

Table 6. Properties of the medical waste fed into the plasma gasifier used for simulation (as received basis) [67].

Item		Unit	Value
	C	wt%	45.71
	H	wt%	5.96
Elements	O	wt%	37.18
	N	wt%	0.16
	S	wt%	0.12
Moisture		wt%	7.01
Ash		wt%	3.85
LHV		MJ/kg	15.35

The current WHRS for the 33 MVA submerged arc furnace illustrated in Figure 1 is modeled to validate the reliability of models, and the boundary conditions are from its design parameters. Simulation results obtained by iteration are compared with the design or operating parameters in Table A1 in Appendix A. Detailed parameters of stream flows in Figure 1 are listed in Table A2.

In order to validate the reliability of models for the plasma gasifier and equipment in GT section, a plasma gasifier fueled with solid waste is simulated by Aspen Plus, and a GT system containing CP, CC, and GT is simulated by EBSILON Professional. The simulation results are listed in Tables A3 and A4, compared with the reference values [68,69]. Waste power output by GT has excluded the power consumed by CP to compress air because they are coaxial.

The comparison reveals good reliability of the established models so that precise parameters of the proposed system can also be obtained based on these models.

4. Results and Discussion

4.1. Parameters of the Proposed Hybrid System

4.1.1. Plasma Gasifier Subsystem

Table 7 presents the main parameters of the plasma gasifier in the hybrid system. The silica powder is fed at 0.15 kg/s, so that most of the silica fines collected in WHB can be treated harmlessly. However, power consumption by the torches increases because extra energy is required to melt SiO_2. The raw syngas leaving the gasifier has an initial temperature of 818.4 °C and a lower heating value of 10.57 MJ/kg.

Table 7. Parameters of the plasma gasifier in the proposed hybrid system.

Item		Unit	Value
Feed rate of medical waste		kg/s	0.80
Feed rate of silica powder		kg/s	0.15
Feed rate of O_2		kg/s	0.34
Raw syngas composition	H_2	vol%	35.35
	CO		41.18
	CH_4		0.61
	CO_2		12.03
	N_2		0.08
	H_2O		10.70
	H_2S		0.05
Raw syngas properties	Temperature	°C	818.4
	Flow rate	kg/s	1.11
	Higher heating value	MJ/kg	11.66
	Lower heating value	MJ/kg	10.57
Output rate of slags		kg/s	0.18
Torch thermal efficiency		%	90
Torch power consumption		MW	1.43
Gasification efficiency		%	71.57

Along the conditioning processes, the syngas has significant temperature changes only in SGCs, wet scrubber, and AGR, and other procedures can be regarded as nearly isothermal. Composition change of the syngas is inconsiderable and neglectable from the perspective of energy calculation, due to relatively low concentrations of the pollutants. The outlet temperature of the wet scrubber is 47.1 °C, approximately equal to the syngas dewpoint. The outlet temperature of AGR is treated as 40 °C, a little higher than typical environmental temperature, which means the syngas witnesses moisture condensation, and the condensates are left in AGR. The main parameters of SGCs, wet scrubber, and AGR obtained by simulation are listed in Table 8.

Table 8. Parameters of SGCs, wet scrubber, and AGR in the plasma gasifier subsystem.

	Item		Unit	Value
SGC1	Syngas	Inlet temperature	°C	818.4
		Outlet temperature	°C	746.0
		Flow rate	kg/s	1.11
	Oxygen	Inlet temperature	°C	370.0
		Outlet temperature	°C	780.0
		Flow rate	kg/s	0.34
	Log mean temperature difference		°C	148.0
	Heat capacity		MW	0.15
SGC2	Syngas	Inlet temperature	°C	746.0
		Outlet temperature	°C	394.4
	Steam	Inlet temperature	°C	390.0
		Inlet pressure	MPa	3.87
		Outlet temperature	°C	480.0
		Outlet pressure	MPa	3.82
		Flow rate	kg/s	3.30
	Log mean temperature difference		°C	63.6
	Heat capacity		MW	0.69
SGC3	Syngas	Inlet temperature	°C	394.4
		Outlet temperature	°C	334.0
	Oxygen	Inlet temperature	°C	25.0
		Outlet temperature	°C	370.0
		Flow rate	kg/s	0.34
	Log mean temperature difference		°C	112.1
	Heat capacity		MW	0.11
SGC4	Syngas (cooled)	Inlet temperature	°C	334.0
		Outlet temperature	°C	133.0
	Syngas (heated)	Inlet temperature	°C	47.1
		Outlet temperature	°C	250.0
	Log mean temperature difference		°C	85.0
	Heat capacity		MW	0.37
Wet scrubber	Syngas	Inlet temperature	°C	133.0
		Outlet temperature	°C	47.1
SGC5	Syngas	Inlet temperature	°C	250.0
		Outlet temperature	°C	50.0
	Water	Inlet temperature	°C	36.2
		Outlet temperature	°C	44.8
		Flow rate	kg/s	9.91
	Log mean temperature difference		°C	71.0
	Heat capacity		MW	0.36
AGR	Syngas	Inlet temperature	°C	50.0
		Outlet temperature	°C	40.0

4.1.2. Gas Turbine Subsystem

Table 9 presents the main parameters of the gas turbine subsystem. Before combustion, the clean syngas is compressed in CP1, and simultaneously, O_2 is fed at a ratio of 9.50 kg/s and compressed in CP2. The compressed gases of 1.42 MPa are mixed and burned in CC to form hot combustion gas at 1303.3 °C, which subsequently drives GT for power generation and has a net power output of 4.70 MW. The exhaust gas leaving GT is still at a high temperature of 666.6 °C.

Table 9. Parameters of the gas turbine subsystem.

	Item	Unit	Value
	Inlet temperature	°C	40.0
	Inlet pressure	MPa	0.10
	Outlet temperature	°C	395.3
CP1 (syngas)	Outlet pressure	MPa	1.42
	Flow rate	kg/s	1.06
	Isentropic efficiency	%	88.0
	Power consumption	MW	0.62
	Inlet temperature	°C	25.0
	Inlet pressure	MPa	0.10
	Outlet temperature	°C	393.9
CP2 (O_2)	Outlet pressure	MPa	1.42
	Flow rate	kg/s	9.50
	Isentropic efficiency	%	88.0
	Power consumption	MW	3.67
	Outlet temperature	°C	1303.3
CC	Outlet pressure	MPa	1.41
	Flow rate	kg/s	10.56
	Exhaust gas temperature	°C	666.6
Gas turbine	Exhaust gas pressure	MPa	0.10
	Isentropic efficiency	%	90.0
	Power output	MW	4.70

4.1.3. WHRS for the Silicon Arc Furnace

In the proposed WHRS integrated with plasma gasification, extra heat provided by the exhaust gas of GT and the syngas is used to assist in the temperature promotion of the steam entering ST and the water entering WHB. Table 10 lists the main parameters of the new WHRS, and Table 11 lists the parameters of the heat exchangers in WHB.

Table 10. Main parameters of the proposed WHRS for the silicon arc furnace integrated with plasma gasification.

	Item	Unit	Value
	Inlet temperature	°C	650.0
Flue gas flowing through WHB	Outlet temperature	°C	187.3
	Flow rate	kg/s	48.29
	Temperature	°C	167.0
Feed water into WHB	Pressure	MPa	4.02
	Flow rate	kg/s	9.95
	Temperature	°C	480.0
Superheated steam into ST	Pressure	MPa	3.82
	Flow rate	kg/s	9.95
	Temperature	°C	36.2
Exhaust steam out of ST	Pressure	MPa	0.01
	Flow rate	kg/s	9.90
Energy recovered from flue gas		MW	24.36
Power output by ST		MW	10.21

Table 11. Parameters of the heat exchangers in WHB in the proposed system.

Item			Unit	Value
Flow rate of exhaust gas from the silicon arc furnace			kg/s	48.29
ECO	Flue gas	Inlet temperature	°C	258.3
		Outlet temperature	°C	187.3
	Water	Inlet temperature	°C	167.0
		Inlet pressure	MPa	4.02
		Outlet temperature	°C	246.2
		Outlet pressure	MPa	3.92
		Flow rate	kg/s	9.95
	Log mean temperature difference		°C	15.9
	Heat capacity		MW	3.57
EVA1	Flue gas	Inlet temperature	°C	650.0
		Outlet temperature	°C	524.6
	Water/steam	Inlet temperature (water)	°C	246.2
		Inlet pressure	MPa	3.92
		Outlet temperature (steam)	°C	249.2
		Outlet pressure	MPa	3.92
		Flow rate	kg/s	3.95
	Log mean temperature difference		°C	334.2
	Heat capacity		MW	6.85
EVA2	Flue gas	Inlet temperature	°C	458.5
		Outlet temperature	°C	258.3
	Water/steam	Inlet temperature (water)	°C	246.2
		Inlet pressure	MPa	3.92
		Outlet temperature (steam)	°C	249.2
		Outlet pressure	MPa	3.92
		Flow rate	kg/s	6.00
	Log mean temperature difference		°C	63.9
	Heat capacity		MW	10.39
SH	Flue gas	Inlet temperature	°C	524.6
		Outlet temperature	°C	458.5
	Steam	Inlet temperature	°C	249.2
		Inlet pressure	MPa	3.92
		Outlet temperature	°C	450.0
		Outlet pressure	MPa	3.87
		Flow rate	kg/s	6.65
	Log mean temperature difference		°C	130.6
	Heat capacity		MW	3.53

As shown in Figure 2, in the hybrid scheme, the saturated steam leaving the drum is divided into two flows. One steam flow of 3.30 kg/s is heated in HX2 and SGC2 successively, from 249.2 °C to 480.0 °C. The other steam flow of 6.65 kg/s is sent to SH in WHB as before, but subsequently further heated in HX1 to 480.0 °C. The two superheated steam flows merge before entering ST. Compared with the current design, the power generation capacity of ST in the integrated WHRS is greatly promoted by the increase in temperature and flow rate of the inlet steam, from 8.08 MW to 10.21 MW. The steam temperature in WHB is controlled no higher than 450 °C, in view of the anti-fouling requirements of the heat transfer surfaces.

Tables 12 and 13 list the parameters of HX1 to 5 and the heat regeneration equipment. Because of auxiliary heating by HX3 to 5, the amount of steam extracted from ST and fed into RH and DEA is reduced from 1.00 kg/s to 0.05 kg/s, benefiting power generation by ST, and the temperature of the feed water entering WHB increases from 104.8 °C to 167.0 °C. Detailed parameters of the stream flows in the proposed system are listed in Table A5.

Table 12. Parameters of HXs in the proposed WHRS.

	Item		Unit	Value
	Flow rate of combustion gas leaving GT		kg/s	10.56
HX1	Combustion gas	Inlet temperature	°C	666.6
		Outlet temperature	°C	630.0
	Steam	Inlet temperature	°C	450.0
		Inlet pressure	MPa	3.87
		Outlet temperature	°C	480.0
		Outlet pressure	MPa	3.82
		Flow rate	kg/s	6.65
	Log mean temperature difference		°C	183.3
	Heat capacity		MW	0.46
HX2	Combustion gas	Inlet temperature	°C	630.0
		Outlet temperature	°C	525.8
	Steam	Inlet temperature	°C	249.2
		Inlet pressure	MPa	3.92
		Outlet temperature	°C	390.0
		Outlet pressure	MPa	3.87
		Flow rate	kg/s	3.30
	Log mean temperature difference		°C	257.9
	Heat capacity		MW	1.29
HX3	Combustion gas	Inlet temperature	°C	525.8
		Outlet temperature	°C	304.4
	Water	Inlet temperature	°C	104.8
		Inlet pressure	MPa	4.12
		Outlet temperature	°C	167.0
		Outlet pressure	MPa	4.02
		Flow rate	kg/s	9.95
	Log mean temperature difference		°C	271.5
	Heat capacity		MW	2.64
HX4	Combustion gas	Inlet temperature	°C	304.4
		Outlet temperature	°C	155.0
	Water	Inlet temperature	°C	61.4
		Inlet pressure	MPa	0.12
		Outlet temperature	°C	102.2
		Outlet pressure	MPa	0.12
		Flow rate	kg/s	9.91
	Log mean temperature difference		°C	141.0
	Heat capacity		MW	1.70
HX5	Combustion gas	Inlet temperature	°C	155.0
		Outlet temperature	°C	95.0
	Water	Inlet temperature	°C	44.8
		Inlet pressure	MPa	0.13
		Outlet temperature	°C	60.9
		Outlet pressure	MPa	0.13
		Flow rate	kg/s	9.91
	Log mean temperature difference		°C	69.9
	Heat capacity		MW	0.67

Table 13. Parameters of the equipment in the heat regeneration subsystem.

	Item		Unit	Value
COND	Inlet steam	Temperature	°C	36.2
		Pressure	MPa	0.01
		Flow rate	kg/s	9.90
	Outlet condensed water	Temperature	°C	36.2
		Pressure	MPa	0.01
		Flow rate	kg/s	9.91
RH	Inlet feed water	Temperature	°C	60.9
		Pressure	MPa	0.13
		Flow rate	kg/s	9.91
	Extracted steam from ST	Temperature	°C	67.5
		Pressure	MPa	0.03
		Flow rate	kg/s	0.01
	Outlet feed water	Temperature	°C	61.4
		Pressure	MPa	0.12
		Flow rate	kg/s	9.91
	Drain water	Temperature	°C	36.2
		Pressure	MPa	0.01
		Flow rate	kg/s	0.01
DEA	Inlet feed water	Temperature	°C	102.2
		Pressure	MPa	0.12
		Flow rate	kg/s	9.91
	Extracted steam from ST	Temperature	°C	133.1
		Pressure	MPa	0.14
		Flow rate	kg/s	0.04
	Outlet feed water	Temperature	°C	104.3
		Pressure	MPa	4.12
		Flow rate	kg/s	9.95

4.2. Energy Performance

The energy performance of the integrated system and current WHRS is examined and compared in Table 14. After the integration, the heat recovered from the exhaust gas leaving the silicon arc furnace is invariable. Medical waste is fed into the plasma gasifier and supplies extra energy by its conversion into combustible syngas. From this, 4.70 MW of power could be generated by GT, and meanwhile, ST has an increase of 2.13 MW in its gross power output. After the integration, auxiliary power consumption increases from 1.21 MW to 3.82 MW, due to the plasma torches, O_2 separation, syngas conditioning equipment, and CP1. Plasma torches are the biggest power consumer in the proposed WHRS because a great deal of heat is required to maintain a high-temperature environment in the gasifier, especially when high-melting SiO_2 exists, which is inevitable when consolidating silica fines using thermal methods. To sum up, 4.17 MW power can be attributed to plasma gasification for medical waste in this case, and the net power generation efficiency of medical waste exploitation is up to 33.99%, close to the power generation efficiency of conventional IPGCC plants, despite the extra heat caused by SiO_2 melting in gasification. Overall power generation efficiency of WHRS increases from 28.19% to 30.13%.

Figure 4 illustrates detailed energy flows in the current WHRS and the proposed integrated system. The exhaust gas from WHB and exhaust steam from ST are major causes of energy loss both in the current WHRS and the proposed system. In the proposed system, 1.05 MW and 6.76 MW of waste heat are recovered from the syngas conditioning processes and the exhaust gas from GT, respectively, to assist power generation by ST. The power generation efficiency through the steam cycle can also be promoted because the steam temperature entering ST increases from 450 °C to 480 °C.

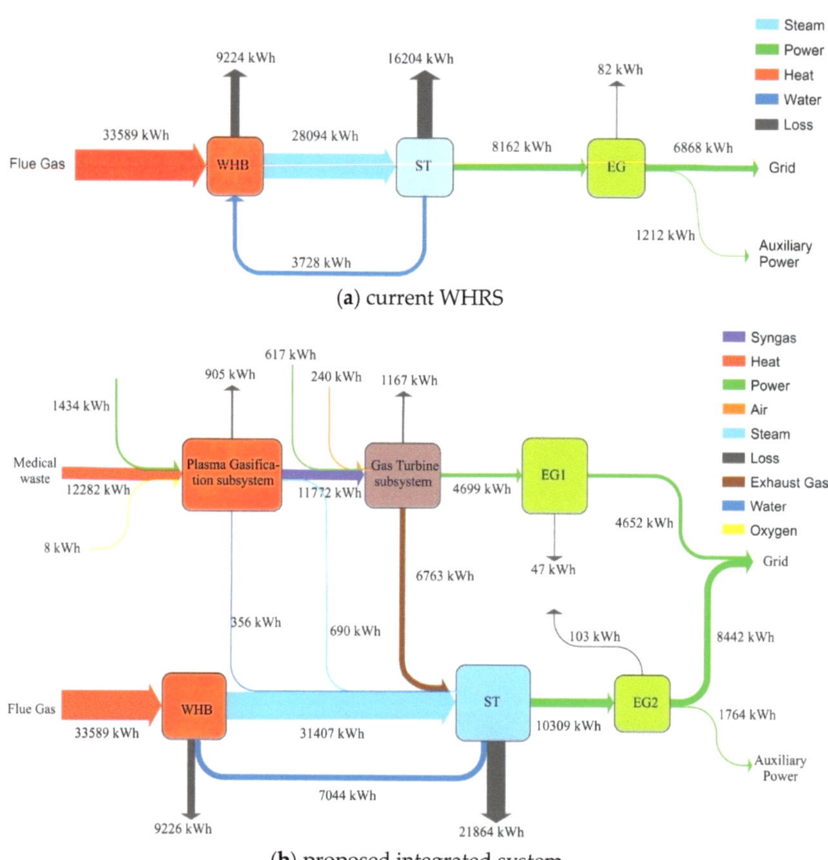

Figure 4. Energy flows in the current WHRS and the proposed integrated system.

Table 14. Energy performance of the proposed integrated system, compared with the current WHRS.

Item	Unit	Proposed Integrated System	Current WHRS
Energy input by exhaust gas from the silicon arc furnace	MW	33.59	33.59
Energy input by the medical waste	MW	12.28	/
Power output by ST	MW	10.21	8.08
Power output by GT	MW	4.70	/
Power consumption by auxiliaries in WHRS	MW	1.21	1.21
Power consumption by torches	MW	1.43	/
Power consumption by syngas conditioning equipment	MW	0.23	/
Power consumption by O_2 separation	MW	0.32	
Power consumption by CP1	MW	0.62	/
Total auxiliary power consumption	MW	3.82	1.21
Total net power output	MW	11.04	6.87
Net power generation by exhaust gas from the silicon arc furnace	MW	6.87	6.87
Net power generation by medical waste	MW	4.17	/
Net power generation efficiency by medical waste	%	33.99	/
Overall power generation efficiency	%	30.13	28.19

4.3. Exergy Performance

Table 15 lists the exergy analysis results of the proposed integrated system and the current WHRS. Exergy input by the exhaust gas from the arc furnace is unaltered and regarded as 100%, and in the hybrid scheme, gasification of medical waste offers an increase of 87.75% in exergy input. Table 16 displays detailed exergy destruction in the components of the system.

Table 15. Exergy performance of the proposed WHRS, compared with the current WHRS.

Item	Unit	Proposed Integrated System	Current WHRS
Exergy input by the flue gas from the silicon arc furnace	MW	15.07	15.07
	%	100.00	100.00
Exergy input by medical waste	MW	13.22	/
	%	87.75	/
Total exergy input	MW	28.29	15.07
	%	187.75	100.00
Exergy output by exhaust gas from silicon arc furnace (electricity)	MW	6.87	6.87
	%	45.58	45.58
Exergy output by medical waste (electricity)	MW	4.17	/
	%	27.71	/
Total exergy output (electricity)	MW	11.03	6.87
	%	73.29	45.58
Exergy efficiency of power generation by medical waste	%	31.58	/
Overall exergy efficiency	%	39.04	45.58

Table 16. Exergy destruction in the proposed system, compared with the current WHRS.

Item		Unit	Proposed Integrated System	Current WHRS
Plasma gasifier subsystem	Gasifier	MW	2.559	/
	SGC1	MW	0.011	/
	SGC2	MW	0.047	/
	SGC3	MW	0.022	/
	SGC4	MW	0.046	/
	Wet scrubber	MW	0.050	/
	SGC5	MW	0.060	
	AGR	MW	0.038	/
Gas turbine system	CP1	MW	0.040	/
	CP2	MW	0.238	/
	CC	MW	2.674	/
	GT	MW	0.395	/
	EG1	MW	0.047	/
WHB		MW	4.517	4.78
ST		MW	1.835	1.45
EG2		MW	0.103	0.08
Heat regeneration subsystem	COND	MW	0.789	0.59
	RH	MW	0.001	0.04
	DEA	MW	0.001	0.05
HXs	HX1	MW	0.049	/
	HX2	MW	0.212	/
	HX3	MW	0.733	/
	HX4	MW	0.416	/
	HX5	MW	0.596	/
Auxiliaries in WHRS		MW	1.764	1.21
Sum		MW	17.25	8.20

Exergy destruction in the current system mainly occurred in WHB and ST. In the proposed system, exergy destruction in WHB slightly decreases because of the higher average temperature for heat absorption. The inevitable increase in exergy destruction in ST and EG2 is caused by the promotion of power output. Among the newly added heat exchangers, HX2-5 working at low temperatures have relatively high exergy destruction. Besides, the gasifier and CC also lead to great exergy destruction. Overall, after integration, the total exergy destruction more than doubled, mainly due to the addition of necessary equipment for plasma gasification. The gasification and combustion reactions are responsible for the relatively low exergy efficiency of power generation by medical waste. So that the overall exergy efficiency of the integrated system is lower than the current system.

4.4. Economic Performance

Economic benefits are important advantages of the proposed integrated WHRS. When the plasma gasification system operates alone, in order to exploit heat from the exhaust gas of GT, combined cycles need to be adopted and the total investment is quite high. In the proposed hybrid scheme, HRSG and ST of WHRS for the silicon arc furnace can be used for waste heat recovery from the exhaust gas of GT, so that the equipment for the steam cycle becomes unnecessary in the plasma gasification system, saving at least 5000 k\$ of investment. However, the heat exchangers in WHB need to be replaced to meet the new requirements for steam generation.

The estimated investments for adding or replacing equipment in the upgrade project are listed in Table 17. The total investment for equipment is 16191.56 k\$. The plasma gasifier and the gas turbine subsystem account for most of the investment.

Table 17. Investment estimation of equipment for upgrading the current WHRS to the proposed hybrid system.

Item		Cost(k\$)
	Gasifier	6451.60
	SGC1	17.53
	SGC2	36.05
Plasma gasifier subsystem	SGC3	17.66
	SGC4	54.71
	SGC5	17.45
	Syngas cleaning section	22.85
	CP1	139.37
	CP2	1247.79
Gas turbine system	CC	1948.52
	GT	5169.35
	EG1	183.00
	EVA1	76.39
WHB	EVA2	384.33
	SH	71.00
	ECO	235.15
	HX1	12.22
	HX2	20.34
HXs	HX3	27.43
	HX4	31.77
	HX5	27.05
Total investment of equipment for upgrade		16,191.56

Table 18 shows the economic analysis results of the upgrade project of the current WHRS to the hybrid system. The upgrade costs 18,843.65 k\$ for equipment purchase in the first 2 years and 1542.47 k\$/year for the operation of the plasma gasifier subsystem, gas turbine system, and HXs in the following years, leading to an increase in the gross annual income of 14,189.31 k\$/ye. The tipping fee for medical waste contributes to about 75% of the total income increase. Meanwhile, in the integration system, 4320 t of silica powder collected in WHB can be disposed harmlessly by the plasma gasifier every year. Although the saving from silica particle treatment is not considered, DDP is just 3.94 years since the investment, revealing good economic efficiency of this upgrade. NPV of the upgrade project in its 25-year lifetime is 61,246.12 k$. Even if the tipping fee of the waste fed into the gasifier decreases sharply to 50 $/t, DDP of the upgrade project is 16.84 years, still delivering positive returns in view of its lifetime.

Table 18. Economic performance of upgrading the current WHRS to the hybrid system.

Item	Unit	Value
Investment for equipment	k$	16,191.56
Compensation for income loss of current WHRS	k$	2652.09
Annual operating cost increase	k$	1542.47
Medical waste treatment capacity	t/year	23,040.00
Income from medical waste treatment	k$/year	10,687.33
Power generation increase	MWh/year	33,397.38
Income increase from electricity selling	k$/year	3223.18
Slag production	t/year	5184.00
Income from slag selling	k$/year	278.80
Gross annual income increase	k$/year	14,189.31
Net annual income increase	k$/year	12,646.84
Dynamic payback period (DDP)	year	3.94
Net present value (NPV)	k$	61,246.12

The economic performance of the proposed system can be further improved by integrating two or more WHBs with one plasma gasification subsystem and one gas turbine subsystem, because the average investment for 1 MW capacity decreases as the scale of the gasifier, CC, and GT increases. The industrial base where the reference WHRS is located has 32 silicon arc furnaces and WHRS of the same design parameters, making it feasible to construct plasma gasifiers of a larger scale.

5. Conclusions

A novel design that integrates the current WHRS for a silicon arc furnace with plasma gasification for medical waste is proposed. In the hybrid scheme, heating of the steam and feed water in WHRS is promoted by the heat recovered from the high-temperature syngas converted from the medical waste and the exhaust gas from GT in the plasma gasification system. In view of WTE, the syngas not only drives GT for power generation but also increases power output by ST in WHRS. Meanwhile, the silica fines generated in polysilicon production can be treated harmlessly by being fed into the plasma gasifier with the medical waste. The system is modeled and simulated on EBSILON Professional and Aspen Plus platforms. Energy, exergy, and economic analyses are conducted to examine the feasibility of upgrading the current system to the proposed system.

The results indicate that in the integrated WHRS, assuming the net power output resulting from the exhaust gas leaving the arc furnace is constant, the net power generation attributed to medical waste is 4.17 MW, and the efficiency from medical waste to electricity is up to 33.99%, close to the efficiency of conventional IPGCC plants. The project of upgrading the current WHRS to the proposed hybrid system requires an initial investment of 18,843.65 k$ and attains a net annual income increase of 12,646.84 k$. DPP of the upgrade project is only 3.94 years. Meanwhile, the plasma gasifier can also dispose 4320 t of silica powder generated in the silicon arc furnace, which benefits the environmental friendliness of polysilicon production. These findings reveal good superiority and industrial prospects of the hybrid system.

Author Contributions: Conceptualization, P.P. and Y.D.; methodology, P.P.; software, L.W.; validation, Y.D., L.W. and S.W.; formal analysis, Y.D.; investigation, L.W.; resources, S.W.; data curation, H.C.; writing—original draft preparation, Y.D.; writing—review and editing, P.P. and L.W.; visualization, L.W.; supervision, H.C.; project administration, S.W.; funding acquisition, P.P. All authors have read and agreed to the published version of the manuscript.

Funding: This research was funded by the National Nature Science Fund of China, grant number 52106008, and Open Project Program of State Key Laboratory of Clean and Efficient Coal-Fired Power Generation and Pollution Control, grant number D2021Y001.

Data Availability Statement: Not applicable.

Conflicts of Interest: The authors declare no conflict of interest.

Nomenclature

C	Cash
EX	Exergy
P	Power
Q	Thermal energy
T	Temperature
W	Work
h	Specific enthalpy
m	Mass flow
r	Rate
ω	Content
η	Efficiency

Superscripts and Subscripts

0	environmental state
dis	Discount
des	Destruction
eg	Energy
ex	Exergy
fg	Flue gas
gr	gross
in	inlet
mw	Medical waste
out	outlet
p	Power
pg	Plasma gasification
rec	Recovery
syn	Syngas
tor	Torch
tot	Total
wh	Waste heat recovery system

Abbreviations

AGR	Acid gas removal
CC	Combustion chamber
COND	Condenser
COS	Carbonyl sulfide
CP	Compressor
CS	Cyclone separator
DC	Direct current
DDP	Dynamic payback period
DEA	Deaerator
ECO	Economizer
EG	Electric generator
EVA	Evaporator
GT	Gas turbine
HRSG	Heat recovery steam generator

HX	Heat exchanger
IPGCC	Integrated plasma gasification combined cycle
LHV	Lower calorific value
MDEA	Methyl diethanolamine
MSW	Municipal solid wastes
MW	Medical waste
NPV	Net present value
PV	Photovoltaics
RH	Regenerative heater
SGC	Syngas cooler
SH	Superheater
ST	Steam turbine
WHB	Waste heat boiler
WHRS	Waste heat recovery system
WTE	Waste to energy

Appendix A

Table A1. Simulation results of the current WHRS for the silicon arc furnace modeled by EBSILON Professional, compared with actual parameters.

Item		Unit	Reference Value	Simulation Value
	Inlet temperature	°C	650.0	650.0
Flue gas through WHB	Outlet temperature	°C	185.0	187.3
	Flow rate	kg/s	48.29	48.29
	Temperature	°C	104.0	104.8
Feed water into WHB	Pressure	MPa	4.02	4.02
	Flow rate	kg/s	8.33	8.33
	Temperature	°C	450.0	450.0
Superheated steam into ST	Pressure	MPa	3.82	3.82
	Flow rate	kg/s	8.43	8.43
	Temperature	°C	36.2	36.2
Exhaust steam of ST	Pressure	MPa	0.01	0.01
	Flow rate	kg/s	7.43	7.43
Energy recovered from flue gas		MW	24.46	24.36
Gross power output		MW	8.12	8.08
Auxiliary power consumption		MW	1.21	1.21
Net power output		MW	6.91	6.87
Power generation efficiency		%	28.35	28.19

Table A2. Properties of the main streams in the current WHRS in Figure 1.

Stream No.	Substance	Temperature (°C)	Pressure (MPa)	Flow Rate (kg/s)
1	Flue gas	650.0	0.10	48.29
2	Flue gas	597.0	0.10	48.29
3	Flue gas	514.4	0.10	48.29
4	Flue gas	291.5	0.10	48.29
5	Flue gas	187.3	0.10	48.29
6	Water	104.8	4.02	8.43
7	Water	246.2	3.92	8.43
8	Water	246.2	3.92	8.43
9	Water	246.2	3.92	6.75
10	Water	246.2	3.92	1.68
11	Steam	249.2	3.92	6.75
12	Steam	249.2	3.92	1.68
13	Steam	249.2	3.92	8.43

Table A2. *Cont.*

Stream No.	Substance	Temperature (°C)	Pressure (MPa)	Flow Rate (kg/s)
14	Steam	450.0	3.82	8.43
15	Steam	113.4	0.14	0.62
16	Steam	67.5	0.03	0.37
17	Steam	36.2	0.01	7.43
18	Water	36.2	0.12	7.81
19	Water	36.2	0.01	0.37
20	Water	61.4	0.12	7.81

Table A3. Simulation results of a plasma gasifier fueled with solid waste modeled by Aspen Plus, compared with reference parameters [68].

Item		Unit	Reference Value	Simulation Value
Feed rate of waste		kg/s	1.00	1.00
Lower heating value of waste		MJ/kg	23.67	23.67
Feed rate of air		kg/s	0.16	0.16
Feed rate of steam		kg/s	0.20	0.20
Raw syngas composition	H_2	vol%	43.50	43.16
	CO		34.50	33.37
	CH_4		0.01	0.56
	CO_2		0.03	0.90
	N_2		5.63	5.73
	H_2O		16.22	16.18
	H_2S		0.09	0.08
Raw syngas properties	Temperature	°C	1267.0	1204.1
	Flow rate	kg/s	1.21	1.21
	Higher heating value	MJ/kg	14.71	14.64
	Lower heating value	MJ/kg	13.44	13.36
Slag output		kg/s	0.15	0.15
Torch power consumption		MW	4.06	4.08
Torch thermal efficiency		%	86.0	86.0
Plasma gasification efficiency		%	43.30	44.11

Table A4. Simulation results of CPs, CC, and GT modeled by EBSILON Professional, compared with reference parameters [69].

Item		Unit	Reference Value	Simulation Value
CP (air)	Inlet temperature	°C	25.0	25.0
	Inlet pressure	MPa	0.10	0.10
	Compression ratio	-	14.87	14.87
	Flow rate	kg/s	73.23	73.23
	Isentropic efficiency [70]	%	88.0	88.0
	Power consumption	MW	29.12	29.34
CC	Outlet temperature	°C	1247.0	1242.2
	Outlet pressure	MPa	14.27	14.27
	Flow rate	kg/s	74.90	74.83
Gas turbine	Exhaust gas temperature	°C	631.0	627.9
	Exhaust gas pressure	MPa	0.11	0.11
	Exhaust gas flow rate	kg/s	74.90	74.83
	Isentropic efficiency	%	90.0	90.0
	Power output	MW	58.61	58.29

Table A5. Properties of the main streams in the hybrid scheme in Figure 2.

Stream No.	Substance	Temperature (°C)	Pressure (MPa)	Flow Rate (kg/s)
1	Flue gas	650.0	0.10	48.29
2	Flue gas	524.6	0.10	48.29
3	Flue gas	458.5	0.10	48.29
4	Flue gas	258.3	0.10	48.29
5	Flue gas	187.3	0.10	48.29
6	Oxygen	25.0	0.10	0.34
7	Oxygen	370.0	0.10	0.34
8	Oxygen	780.0	0.10	0.34
9	Syngas	818.4	0.10	1.11
10	Syngas	746.0	0.10	1.11
11	Syngas	394.4	0.10	1.11
12	Syngas	334.0	0.10	1.11
13	Syngas	133.0	0.10	1.11
14	Syngas	47.1	0.10	1.11
15	Syngas	250.0	0.10	1.11
16	Syngas	50.0	0.10	1.11
17	Syngas	40.0	0.10	1.06
18	Air	25.0	0.10	9.50
19	Air	393.9	1.42	9.50
20	Syngas	395.3	1.42	1.06
21	Combustion gas	1303.3	1.41	10.56
22	Combustion gas	666.6	0.10	10.56
23	Combustion gas	630.0	0.10	10.56
24	Combustion gas	525.8	0.10	10.56
25	Combustion gas	304.4	0.10	10.56
26	Combustion gas	155.0	0.10	10.56
27	Combustion gas	95.0	0.10	10.56
28	Water	104.8	4.12	9.95
29	Water	167.0	4.02	9.95
30	Water	246.2	3.92	9.95
31	Water	249.2	3.92	9.95
32	Water	249.2	3.92	3.95
33	Water	249.2	3.92	6.00
34	Steam	249.2	3.92	3.95
35	Steam	249.2	3.92	6.00
36	Steam	249.2	3.92	9.95
37	Steam	249.2	3.92	3.30
38	Steam	249.2	3.92	6.65
39	Steam	390.0	3.87	3.30
40	Steam	450.0	3.87	6.65
41	Steam	480.0	3.82	3.30
42	Steam	480.0	3.82	6.65
43	Steam	480.0	3.82	9.95
44	Steam	133.1	0.14	0.04
45	Steam	67.5	0.03	0.01
46	Steam	36.2	0.01	9.90
47	Water	36.2	0.14	9.91
48	Water	44.8	0.13	9.91
49	Water	60.9	0.13	9.91
50	Water	36.2	0.01	0.01
51	Water	61.4	0.12	9.91
52	Water	102.2	0.12	9.91

References

1. IEA. *Renewables*; IEA: Paris, France, 2021.
2. Shen, Z.; Ma, L.; Yang, Y.; Fan, M.; Ma, W.; Fu, L.; Li, M. A life cycle assessment of hydropower-silicon-photovoltaic industrial chain in china. *J. Clean Prod.* **2022**, *362*, 132411. [CrossRef]

3. Xin-Gang, Z.; You, Z. Technological progress and industrial performance: A case study of solar photovoltaic industry. *Renew. Sustain. Energy Rev.* **2018**, *81*, 929–936. [CrossRef]

4. IEA. *Solar PV Global Supply Chains*; IEA: Paris, France, 2022.

5. Eriksson, G.; Hack, K. Iv.7—Production of metallurgical-grade silicon in an electric arc furnace. In *The SGTE Casebook*, 2nd ed.; Hack, K., Ed.; Woodhead Publishing: Oxford, UK, 2008; pp. 415–424.

6. Satpathy, R.; Pamuru, V. Chapter 1—Manufacturing of polysilicon. In *Solar PV Power*; Satpathy, R., Pamuru, V., Eds.; Academic Press: Cambridge, MA, USA, 2021; pp. 1–29.

7. Chigondo, F. From metallurgical-grade to solar-grade silicon: An overview. *Silicon* **2018**, *10*, 789–798. [CrossRef]

8. Chen, Z.; Ma, W.; Wu, J.; Wei, K.; Lei, Y.; Lv, G. A study of the performance of submerged arc furnace smelting of industrial silicon. *Silicon* **2018**, *10*, 1121–1127. [CrossRef]

9. Takla, M.; Kamfjord, N.E.; Tveit, H.; Kjelstrup, S. Energy and exergy analysis of the silicon production process. *Energy* **2013**, *58*, 138–146. [CrossRef]

10. Filtvedt, W.O.; Javidi, M.; Holt, A.; Melaaen, M.C.; Marstein, E.; Tathgar, H.; Ramachandran, P.A. Development of fluidized bed reactors for silicon production. *Sol. Energy Mater. Sol. Cells* **2010**, *94*, 1980–1995. [CrossRef]

11. Hong, J.; Chen, W.; Qi, C.; Ye, L.; Xu, C. Life cycle assessment of multicrystalline silicon photovoltaic cell production in china. *Sol. Energy* **2016**, *133*, 283–293. [CrossRef]

12. Liu, F.; van den Bergh, J.C.J.M. Differences in CO_2 emissions of solar PV production among technologies and regions: Application to China, EU and USA. *Energy Policy* **2020**, *138*, 111234. [CrossRef]

13. Zhang, P.; Duan, J.; Chen, G.; Li, J.; Wang, W. Production of polycrystalline silicon from silane pyrolysis: A review of fines formation. *Sol. Energy* **2018**, *175*, 44–53. [CrossRef]

14. Nemchinova, N.V.; Mineev, G.G.; Tyutrin, A.A.; Yakovleva, A.A. Utilization of dust from silicon production. *Steel Transl.* **2017**, *47*, 763–767. [CrossRef]

15. Tavares, R.; Ramos, A.; Rouboa, A. A theoretical study on municipal solid waste plasma gasification. *Waste Manag.* **2019**, *90*, 37–45. [CrossRef] [PubMed]

16. Li, J.; Liu, K.; Yan, S.; Li, Y.; Han, D. Application of thermal plasma technology for the treatment of solid wastes in China: An overview. *Waste Manag.* **2016**, *58*, 260–269. [CrossRef] [PubMed]

17. Gomez, E.; Rani, D.A.; Cheeseman, C.R.; Deegan, D.; Wise, M.; Boccaccini, A.R. Thermal plasma technology for the treatment of wastes: A critical review. *J. Hazard. Mater.* **2009**, *161*, 614–626. [CrossRef] [PubMed]

18. Ramos, A.; Rouboa, A. Life cycle thinking of plasma gasification as a waste-to-energy tool: Review on environmental, economic and social aspects. *Renew. Sustain. Energy Rev.* **2022**, *153*, 111762. [CrossRef]

19. Zhang, Q.; Wu, Y.; Dor, L.; Yang, W.; Blasiak, W. A thermodynamic analysis of solid waste gasification in the plasma gasification melting process. *Appl. Energy* **2013**, *112*, 405–413. [CrossRef]

20. Montiel-Bohórquez, N.D.; Saldarriaga-Loaiza, J.D.; Pérez, J.F. Analysis of investment incentives for power generation based on an integrated plasma gasification combined cycle power plant using municipal solid waste. *Case Stud. Therm. Eng.* **2022**, *30*, 101748. [CrossRef]

21. Mountouris, A.; Voutsas, E.; Tassios, D. Solid waste plasma gasification: Equilibrium model development and exergy analysis. *Energy Conv. Manag.* **2006**, *47*, 1723–1737. [CrossRef]

22. Danthurebandara, M.; Van Passel, S.; Machiels, L.; Van Acker, K. Valorization of thermal treatment residues in enhanced landfill mining: Environmental and economic evaluation. *J. Clean Prod.* **2015**, *99*, 275–285. [CrossRef]

23. Sanito, R.C.; Bernuy-Zumaeta, M.; You, S.; Wang, Y. A review on vitrification technologies of hazardous waste. *J. Environ. Manag.* **2022**, *316*, 115243. [CrossRef]

24. Minutillo, M.; Perna, A.; Di Bona, D. Modelling and performance analysis of an integrated plasma gasification combined cycle (IPGCC) power plant. *Energy Conv. Manag.* **2009**, *50*, 2837–2842. [CrossRef]

25. Fabry, F.; Rehmet, C.; Rohani, V.; Fulcheri, L. Waste gasification by thermal plasma: A review. *Waste Biomass Valoriz.* **2013**, *4*, 421–439. [CrossRef]

26. Munir, M.T.; Mardon, I.; Al-Zuhair, S.; Shawabkeh, A.; Saqib, N.U. Plasma gasification of municipal solid waste for waste-to-value processing. *Renew. Sustain. Energy Rev.* **2019**, *116*, 109461. [CrossRef]

27. Subramanyam, V.; Gorodetsky, A. 5–municipal wastes and other potential fuels for use in igcc systems. In *Integrated Gasification Combined Cycle (IGCC) Technologies*; Wang, T., Stiegel, G., Eds.; Woodhead Publishing: Oxford, UK, 2017; pp. 181–219.

28. Awasthi, S.K.; Sarsaiya, S.; Kumar, V.; Chaturvedi, P.; Sindhu, R.; Binod, P.; Zhang, Z.; Pandey, A.; Awasthi, M.K. Processing of municipal solid waste resources for a circular economy in China: An overview. *Fuel* **2022**, *317*, 123478. [CrossRef]

29. Valmundsson, A.S.; Janajreh, I. Plasma gasification process modeling and energy recovery from solid waste. In Proceedings of the ASME 5th International Conference on Energy Sustainability, Washington, DC, USA, 7 August 2011.

30. Cudjoe, D.; Wang, H. Plasma gasification versus incineration of plastic waste: Energy, economic and environmental analysis. *Fuel Process. Technol.* **2022**, *237*, 107470. [CrossRef]

31. James, I.R.E.; Keairns, D.; Turner, M.; Woods, M.; Kuehn, N.; Zoelle, A. *Cost and Performance Baseline for Fossil Energy Plants Volume 1: Bituminous Coal and Natural Gas to Electricity*; National Energy Technology Laboratory (NETL): Pittsburgh, PA, USA, 2019.

32. Robil Zwart, S.V.D.H. *Tar Removal from Low-Temperature Gasifiers*; Energy Research Centre of the Netherlands (ECN): Petten, The Netherlands, 2010.

33. Pérez-Fortes, M.; Bojarski, A.D.; Velo, E.; Nougués, J.M.; Puigjaner, L. Conceptual model and evaluation of generated power and emissions in an IGCC plant. *Energy* **2009**, *34*, 1721–1732. [CrossRef]
34. Montiel-Bohórquez, N.D.; Agudelo, A.F.; Pérez, J.F. Effect of origin and production rate of msw on the exergoeconomic performance of an integrated plasma gasification combined cycle power plant. *Energy Conv. Manag.* **2021**, *238*, 114138. [CrossRef]
35. Mazzoni, L.; Janajreh, I. Plasma gasification of municipal solid waste with variable content of plastic solid waste for enhanced energy recovery. *Int. J. Hydrogen Energy* **2017**, *42*, 19446–19457. [CrossRef]
36. Erdogan, A.A.; Yilmazoglu, M.Z. Plasma gasification of the medical waste. *Int. J. Hydrogen Energy* **2021**, *46*, 29108–29125. [CrossRef]
37. Sikarwar, V.S.; Hrabovský, M.; Van Oost, G.; Pohořelý, M.; Jeremiáš, M. Progress in waste utilization via thermal plasma. *Prog. Energy Combust. Sci.* **2020**, *81*, 100873. [CrossRef]
38. Su, G.; Ong, H.C.; Ibrahim, S.; Fattah, I.M.R.; Mofijur, M.; Chong, C.T. Valorisation of medical waste through pyrolysis for a cleaner environment: Progress and challenges. *Environ. Pollut.* **2021**, *279*, 116934. [CrossRef]
39. Deng, N.; Zhang, Y.; Wang, Y. Thermogravimetric analysis and kinetic study on pyrolysis of representative medical waste composition. *Waste Manag.* **2008**, *28*, 1572–1580. [CrossRef]
40. Du, Y.; Ju, T.; Meng, Y.; Lan, T.; Han, S.; Jiang, J. A review on municipal solid waste pyrolysis of different composition for gas production. *Fuel Process. Technol.* **2021**, *224*, 107026. [CrossRef]
41. Ding, Z.; Chen, H.; Liu, J.; Cai, H.; Evrendilek, F.; Buyukada, M. Pyrolysis dynamics of two medical plastic wastes: Drivers, behaviors, evolved gases, reaction mechanisms, and pathways. *J. Hazard. Mater.* **2021**, *402*, 123472. [CrossRef] [PubMed]
42. Saxena, P.; Pradhan, I.P.; Kumar, D. Redefining bio medical waste management during COVID-19 in india: A way forward. *Mat. Today Proc.* **2022**, *60*, 849–858. [CrossRef]
43. Windfeld, E.S.; Brooks, M.S. Medical waste management—A review. *J. Environ. Manag.* **2015**, *163*, 98–108. [CrossRef]
44. Chen, H.; Li, J.; Li, T.; Xu, G.; Jin, X.; Wang, M.; Liu, T. Performance assessment of a novel medical-waste-to-energy design based on plasma gasification and integrated with a municipal solid waste incineration plant. *Energy* **2022**, *245*, 123156. [CrossRef]
45. Yang, Y.; Wang, J.; Chong, K.; Bridgwater, A.V. A techno-economic analysis of energy recovery from organic fraction of municipal solid waste (MSW) by an integrated intermediate pyrolysis and combined heat and power (CHP) plant. *Energy Conv. Manag.* **2018**, *174*, 406–416. [CrossRef]
46. Pan, P.; Peng, W.; Li, J.; Chen, H.; Xu, G.; Liu, T. Design and evaluation of a conceptual waste-to-energy approach integrating plasma waste gasification with coal-fired power generation. *Energy* **2022**, *238*, 121947. [CrossRef]
47. Courson, C.; Gallucci, K. 8—Gas cleaning for waste applications (syngas cleaning for catalytic synthetic natural gas synthesis). In *Substitute Natural Gas from Waste*; Materazzi, M., Foscolo, P.U., Eds.; Academic Press: Cambridge, MA, USA, 2019; pp. 161–220.
48. Woolcock, P.J.; Brown, R.C. A review of cleaning technologies for biomass-derived syngas. *Biomass Bioenerg.* **2013**, *52*, 54–84. [CrossRef]
49. Kosstrin, H.M. 10—wet scrubbing and gas filtration of syngas in igcc systems. In *Integrated Gasification Combined Cycle (IGCC) Technologies*; Wang, T., Stiegel, G., Eds.; Woodhead Publishing: Oxford, UK, 2017; pp. 375–383.
50. Sharma, S.D.; Dolan, M.; Park, D.; Morpeth, L.; Ilyushechkin, A.; Mclennan, K.; Harris, D.J.; Thambimuthu, K.V. A critical review of syngas cleaning technologies—Fundamental limitations and practical problems. *Powder Technol.* **2008**, *180*, 115–121. [CrossRef]
51. Giuffrida, A.; Romano, M.C. On the effects of syngas clean-up temperature in IGCCs. In Proceedings of the ASME Turbo Exp 2010, Glasgow, Scotland, UK, 14 June 2010.
52. Frilund, C.; Tuomi, S.; Kurkela, E.; Simell, P. Small- to medium-scale deep syngas purification: Biomass-to-liquids multi-contaminant removal demonstration. *Biomass Bioenerg.* **2021**, *148*, 106031. [CrossRef]
53. Huang, H.; Young, N.; Williams, B.P.; Taylor, S.H.; Hutchings, G. High temperature cos hydrolysis catalysed by gamma-Al_2O_3. *Catal. Lett.* **2006**, *110*, 243–246. [CrossRef]
54. Shangguan, J.; Liu, Y.; Wang, Z.; Xu, Y. The synthesis of magnesium-aluminium spinel for catalytic hydrolysis of carbonyl sulphur at the middle temperature. *IOP Conf. Ser. Mat. Sci. Eng.* **2019**, *585*, 12039. [CrossRef]
55. Al Ani, Z.; Thafseer, M.; Gujarathi, A.M.; Vakili-Nezhaad, G.R. Towards process, energy and safety based criteria for multi-objective optimization of industrial acid gas removal process. *Process Saf. Environ. Protect.* **2020**, *140*, 86–99. [CrossRef]
56. Bhattacharyya, D.; Turton, R.; Zitney, S.E. 11—acid gas removal from syngas in IGCC plants. In *Integrated Gasification Combined Cycle (IGCC) Technologies*; Wang, T., Stiegel, G., Eds.; Woodhead Publishing: Oxford, UK, 2017; pp. 385–418.
57. Indrawan, N.; Mohammad, S.; Kumar, A.; Huhnke, R.L. Modeling low temperature plasma gasification of municipal solid waste. *Environ. Technol. Innov.* **2019**, *15*, 100412. [CrossRef]
58. Chaiyat, N. Energy, exergy, economic, and environmental analysis of an organic rankine cycle integrating with infectious medical waste incinerator. *Therm. Sci. Eng. Prog.* **2021**, *22*, 100810. [CrossRef]
59. Mahian, O.; Mirzaie, M.R.; Kasaeian, A.; Mousavi, S.H. Exergy analysis in combined heat and power systems: A review. *Energy Conv. Manag.* **2020**, *226*, 113467. [CrossRef]
60. Danthurebandara, M.; Van Passel, S.; Vanderreydt, I.; Van Acker, K. Environmental and economic performance of plasma gasification in enhanced landfill mining. *Waste Manag.* **2015**, *45*, 458–467. [CrossRef]
61. Zang, G.; Jia, J.; Tejasvi, S.; Ratner, A.; Lora, E.S. Techno-economic comparative analysis of Biomass Integrated Gasification Combined Cycles with and without CO_2 capture. *Int. J. Greenh. Gas Control* **2018**, *78*, 73–84. [CrossRef]
62. Ogorure, O.J.; Oko, C.O.C.; Diemuodeke, E.O.; Owebor, K. Energy, exergy, environmental and economic analysis of an agricultural waste-to-energy integrated multigeneration thermal power plant. *Energy Conv. Manag.* **2018**, *171*, 222–240. [CrossRef]
63. Lian, Z.T.; Chua, K.J.; Chou, S.K. A thermoeconomic analysis of biomass energy for trigeneration. *Appl. Energy* **2010**, *87*, 84–95. [CrossRef]

64. Peng, W.; Chen, H.; Liu, J.; Zhao, X.; Xu, G. Techno-economic assessment of a conceptual waste-to-energy CHP system combining plasma gasification, SOFC, gas turbine and supercritical CO_2 cycle. *Energy Conv. Manag.* **2021**, *245*, 114622. [CrossRef]
65. Elsido, C.; Martelli, E.; Kreutz, T. Heat integration and heat recovery steam cycle optimization for a low-carbon lignite/biomass-to-jet fuel demonstration project. *Appl. Energy* **2019**, *239*, 1322–1342. [CrossRef]
66. Zhao, X.; Jiang, G.; Li, A.; Wang, L. Economic analysis of waste-to-energy industry in China. *Waste Manag.* **2016**, *48*, 604–618. [CrossRef]
67. Zhu, H.M.; Yan, J.H.; Jiang, X.G.; Lai, Y.E.; Cen, K.F. Study on pyrolysis of typical medical waste materials by using TG-FTIR analysis. *J. Hazard. Mater.* **2008**, *153*, 670–676. [CrossRef] [PubMed]
68. Janajreh, I.; Raza, S.S.; Valmundsson, A.S. Plasma gasification process: Modeling, simulation and comparison with conventional air gasification. *Energy Conv. Manag.* **2013**, *65*, 801–809. [CrossRef]
69. Hou, S.; Zhou, Y.; Yu, L.; Zhang, F.; Cao, S.; Wu, Y. Optimization of a novel cogeneration system including a gas turbine, a supercritical CO_2 recompression cycle, a steam power cycle and an organic Rankine cycle. *Energy Conv. Manag.* **2018**, *172*, 457–471. [CrossRef]
70. Roy, D.; Samanta, S.; Ghosh, S. Performance assessment of a biomass-fuelled distributed hybrid energy system integrating molten carbonate fuel cell, externally fired gas turbine and supercritical carbon dioxide cycle. *Energy Conv. Manag.* **2020**, *211*, 112740. [CrossRef]

MDPI AG
Grosspeteranlage 5
4052 Basel
Switzerland
Tel.: +41 61 683 77 34

Entropy Editorial Office
E-mail: entropy@mdpi.com
www.mdpi.com/journal/entropy